STRENGTH OF MATERIALS

Volume 2: Advanced Theory and Applications

J. M. ALEXANDER DSc, PhD, FCGI, FRSA, FICE, FIMechE, FIProdE, FIM, MASME, FEng
formerly Professor of Applied Mechanics, Imperial College of Science and
Technology, University of London, *and* Stocker Visiting Professor of Engineering
and Technology, Ohio University, Athens, USA

J. S. GUNASEKERA PhD, MSc, DIC, MASME, MAIIE, MIMechE, MIProdE, MIE(Aust), MIE(SC)
Professor of Mechanical Engineering, Ohio University, Athens, USA

ELLIS HORWOOD
NEW YORK LONDON TORONTO SYDNEY TOKYO SINGAPORE

First published in 1991 by
ELLIS HORWOOD LIMITED
Market Cross House, Cooper Street,
Chichester, West Sussex, PO19 1EB, England

A division of
Simon & Schuster International Group
A Paramount Communications Company

Printed and bound in Great Britain
by Redwood Press Limited, Melksham, Wiltshire

British Library Cataloguing in Publication Data

J. M. Alexander
Strength of materials: Vol 2: Advanced theory and applications. —
(Ellis Horwood series in mechanical engineering)
I. Title. II. Gunasekera, G. S. III. Series.
620.1001

ISBN 0–13–853714–3

Library of Congress Cataloging-in-Publication Data

(Revised for Vol. 2)
Alexander, John Malcolm.
Strength of materials.
(Ellis Horwood series in engineering science)
Includes bibliographical references and index.
ISBN 0–13–853714–3
Contents: v. 1. Fundamentals — v. 2. Advanced theory and applications.
1. Strength of materials. I. Title. II. Series.
TA405.A514   620.1'12   80–42009

STRENGTH OF MATERIALS
Volume 2: Advanced Theory and Applications

Books are to be returned on or before
the last date below.

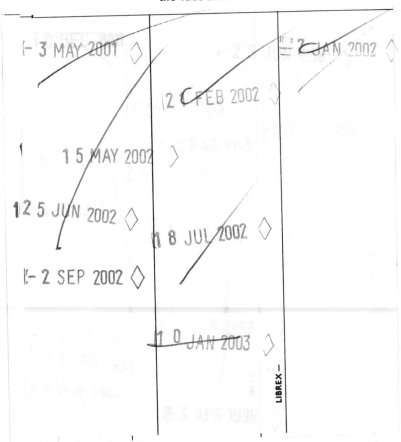

- 3 MAY 2001

2 JAN 2002

2 FEB 2002

1 5 MAY 2002

2 5 JUN 2002

1 8 JUL 2002

- 2 SEP 2002

1 0 JAN 2003

LIBREX —

Contents

This is the second of two volumes, appearing under the titles:

STRENGTH OF MATERIALS

Volume 1: Fundamentals,
Volume 2: Advanced Theory and Applications.

In the Preface to Volume 1, I mentioned the motivation for writing these two volumes, which are intended to cover the whole of a three or four year university undergraduate course and also provide useful background reading for postgraduate courses.

These books are complementary to one another and it is desirable to have both. In view of the high cost of publishing books these days, my publisher and I decided that it would be best to split the material into these two volumes. This was to enable us to include more up–to–date material than can be found in other books on the subject without the necessity for students to pay a large sum of money in the one amount required to buy a single very large volume. However, it has inevitably delayed the publication of the second volume.

One modern development is the use of simple computer programs to solve some of the more complex problems, and we have tried to include a reasonable number of these in both volumes. We consider that these programs will provide models for the materials engineer, to help him predict and better understand the precise behaviour of materials under the often arduous conditions of service required of them, and also to help in the safe design of the manifold machines and structures used by engineers of all persuasions, and in determining the required specifications of the equipment necessary for it to carry out its particular functions. They have been written in a format which will be suitable for any small personal computer capable of handling FORTRAN or BASIC algorithms.

Regarding this second volume, I am indebted to Professor Jay Sarath Gunasekera, without whose expertise it would never have been completed. Dr. Gunasekera is at present the Moss Professor of Engineering in the Department of Mechanical Engineering at the University of Ohio. He was one of the best of the many excellent Ph.D students I advised at Imperial College, London, and also at University College of Swansea, and I am very pleased that he agreed to collaborate in the writing of this second volume. We have written alternate chapters, each concentrating on those topics which are of most interest to us.

Although modern finite difference techniques and more recently, with the advent of large computers, finite element methods have enabled solutions to be obtained to all such problems, there is, in the opinion of the authors, still a place for the more approximate solutions described in this work. In the field, large computers are often not available, so the simple solutions which can be obtained from small desk–top computers or even programmable pocket calculators are useful to the engineering designer.

In this book a few worked examples are included, solutions being given generally to four significant figures. These are deemed to be sufficiently accurate in view of the general approximations made within the basic assumptions originally proposed in such well–known and accepted theories as those for bending and torsion established by great mathematicians like the Bernoullis and Euler.

Both Professor Gunasekera and I hope that readers of this second volume will find it helpful and to have been worth the unduly long delay in publication.

J. M. Alexander,
Emeritus Professor, University of Wales,
Rowan Cottage, Furze Hill Road,
Headley Down, BORDON, Hampshire GU35 8NP.
England, U.K.

also:

Adjunct Professor of Engineering,
College of Engineering and Technology,
Ohio University,
ATHENS, Ohio 45701–2979, U.S.A.
22nd November 1990.

CHAPTER 1

The bending of beams

1.1 BASIC IDEAS

In Volume 1 of this book the main fundamental principles and concepts underlying the subject of strength of materials were developed, starting from the simplest ideas of elasticity and finishing with the more complex aspects of the mechanics of solid continua. The principles of equilibrium and compatibility, both at a point and between neighbouring points in a solid homogeneous and isotropic continuum, were established. The concepts of statically determinate and statically indeterminate structures were briefly introduced. The ideas of effective or equivalent stress and strain during plastic and elastic–plastic deformation and their application to the small–strain incremental deformation of solids using stress–strain relationships (or constitutive equations), generally derived from uniaxial tests, were described.

In this and the following Chapters the way in which these fundamental concepts can be applied to real problems will be discussed, again moving from the simple to the complex. Most practising professional engineers and materials scientists approach problems in this way, firstly by obtaining a rough idea of how the materials available to them will behave in a given situation and then making more difficult analyses and estimations of the behaviour of the material according to the requirements. It should always be remembered that there is little point in obtaining precise solutions to any problem if the available material data is inadequate or limited, as is often the case. For many situations a statistical analysis of the data and of the imposed duty will be necessary if any scientific analysis of material behaviour is to be made. This is most evident in problems involving the use of materials in engineering comp-onents and structures which are to be subjected to cycles of loading or straining which may involve fracture, fatigue or creep. Such aspects of material behaviour are very specialized and ideas about the way materials of different atomic and molecular structure will respond to such imposed conditions are continually changing and form a very specialized body of knowledge, generally to be found only in the learned journals. It is possible in a book of reasonable size only to discuss the more important aspects of such problems.

Perhaps the most important aspect of such problems is the fact that they are at present foremost in research studies. The generic term **constitutive laws** covers the attempts of many research workers to develop equations generally of the form used for creep, viz:

$$\bar{\epsilon} = f(\bar{\sigma}, t, T) \qquad (1.1)$$

with $\bar{\epsilon}$ = **effective strain**, $\bar{\sigma}$ = **effective stress**, t = **time** and T = **absolute temperature**, using the notation of Volume 1. Considerable experimental and theoretical work has been carried out on these effects in the past, probably starting with extensive investigations of the **creep strength** of metals. With the development of modern technology, particularly in aero-space applications, problems of **fatigue and fracture combined with creep** have required even more research to determine the strength of materials, particularly metals. These aspects will be discussed in later chapters in more detail – for the present introduction it may be worth pointing out that simple equations of the general form of equation (1.1) have often been used for creep, e.g:

$$\dot{\bar{\epsilon}} = A\bar{\sigma}^n \, \exp-(Q/RT). \qquad (1.2)$$

In equation (1.2) n lies between 3 and 8 and the **homologous temperature** T_H (= T/T_M where T_M is the **absolute temperature of melting**) is greater than about 0.35 for metals and 0.45 for ceramics. In practice, "one to one" functional relationships of the type given in equation (1.2) are very convenient to use in the **mathematical modelling** of material behaviour but certainly do not give a true picture of the real behaviour of materials, particularly polymers which are very "path-sensitive".

Put in another way, materials are sensitive to the history of their deformation and, if a particular over-all large deformation is reached by different histories of straining, stressing, temperature variation or strain rate, the final effective yield stress or flow stress of the material may differ significantly as between one history and another. These effects are due to metallurgical or other physical or chemical effects such as phase changes in metals or the equivalent molecular interactions (e.g. "cross-linking") in polymers or ceramics. Thus, **constitutive**

equations should be formulated to take the history of deformation into account, but this is rarely done. Whenever deformation beyond the elastic region occurs, constitutive equations are useful for specifying the flow stress as a function of strain, strain–rate and temperature and will be used in certain of the topics to be discussed in this and subsequent chapters.

Most engineering components or structures have, within them, members which are subjected to loading which will cause them to bend and deflect. Typical examples are the crankshaft of an engine or a girder in a building. Many components are made up of long prismatic bars which are easy to manufacture in that form and the bending of such a bar is the simplest example of the beam problem and will now be considered in some detail.

A long prismatic bar can obviously be subjected to different types of loading. The simplest to visualize was discussed in the companion volume (see References) in relation to the tensile test. A bar subjected to tension along its axis is often referred to as a **tie,** whilst one subjected to compression along its axis is termed a **strut.** If the strut is long in relation to its cross–sectional dimensions it will tend to **bend** or **buckle** and **bending stresses** will be introduced. If the bar is firmly clamped at one end and subjected to lateral forces it will obviously bend and become a **beam.** A beam clamped at one end (or both ends) is said to be **fixed** or **encastré** (from the French word meaning 'encased') at those ends. In practice it is physically impossible to achieve 'fixed' ends because of the (generally elastic) deformation of the clamping devices at those ends. A beam clamped at one end only and laterally loaded is called a **cantilever.**

A beam which is supported (but not clamped) at both ends and is free to move along its length (e.g. by being supported on rollers), is said to be **simply supported.** Depending on the direction in which the beam is loaded it may bend in more than one plane. In such cases the bending of the beam can be resolved into components in the planes of the loads considered and determined by the **principal of superposition** if the beam remains elastic. Initially it will be assumed that all loading is in the same plane, including the **support reactions.** Loads may be either **point loads** or **distributed loads.** Distributed loads may be either **uniform** or **non–uniform** and are described in terms of the **load per unit length** or **intensity of loading** which will be constant if the load is uniformly distributed.

Cantilevers and simply supported beams are clearly statically determinate systems, so that the **reactions** (which are a moment and a force in the case of a cantilever and two forces in the case of a simply supported beam) may be determined by applying the principles of statical equilibrium. On the other hand, a beam encastré at both ends or a beam with three simple supports is statically indeterminate because one of the supports could be removed without the beam collapsing under the applied load. Mathematically, this becomes evident as soon as the simple principles of equilibrium are applied because it is found that there are more unknowns than there are equations from which to determine the support reactions. In such cases it is necessary to specify additional boundary conditions which can only involve the deflections or slopes at the supports. Thus a theoretical analysis is required which will link the deflection and slope at any point in a beam with the applied loading system and this will be discussed later. First of all, however, it is necessary to establish the concept of the **shear force** and **bending moment** at any section of a beam subjected to applied loading.

1.2 SHEAR FORCE AND BENDING MOMENT AT ANY SECTION

Consider the simply supported beam of length l illustrated in Fig. 1.1, carrying a single point load W distant a from the left hand end, from which x is measured. The reactions at $x = 0$ and $x = l$ are unknown at the outset and will be designated R_0 and R_l, as shown.

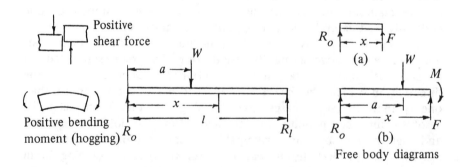

Fig. 1.1 Simply supported beam

As already mentioned, this is a statically determinate problem so the forces can be found by the principles of statics, thus:

Vertical equilibrium of forces (often written as $\Sigma V = 0$)

$$R_0 + R_l = W \qquad\qquad (1.3)$$

Moment equilibrium ($\Sigma M = 0$)

$$Wa = R_l\, l. \tag{1.4}$$

Hence:

From equation (1.3),
$$R_l = Wa/l \tag{1.5}$$
$$R_0 = W - R_l = W(l - a/l). \tag{1.6}$$

An example of a uniformly distributed load w/unit length is shown in Fig. 1.2. Equilibrium of forces ($\Sigma V = 0$), gives:

$$R_0 + R_l = wb. \tag{1.7}$$

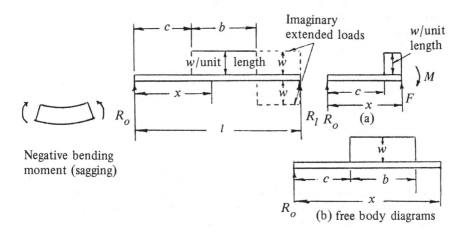

Negative bending moment (sagging)

Fig. 1.2 Distributed loading

Equilibrium of moments ($\Sigma M = 0$), gives (noting that the total load wb may be regarded as acting at its centre of gravity which is at a distance of $c + b/2$ from the left-hand end):

$$wb(c + b/2) = R_l l. \tag{1.8}$$

Hence:
$$R_l = (wb/l)(c + b/2) \tag{1.9}$$

and:
$$R_0 = wb\{1 - (1/l)(c + b/2)\}. \tag{1.10}$$

The **shear force** and **bending moment** at any section (distant x from the left-hand end) may be determined by considering the statical equilibrium of the free body either to the left or to the right of the section concerned. The usual convention adopted is such as to give deflection positive downwards (as shown later) and involves adopting the shear force F and bending moment M as positive when acting in the senses shown in Figs. 1.1 and 1.2. F and M are the transverse force (shear force) and bending moment which have to be applied to the right-hand end of the free body illustrated, to produce equilibrium. Alternatively, positive shear force is that which tends to shear the *right hand side* of any section of the beam *upwards* and positive bending moment is that which tends to cause **hogging** of the beam as illustrated in Fig. 1.1. Many texts adopt the opposite conventions. With this system, negative bending moment is that which tends to cause **sagging** of the beam as illustrated in Fig. 1.2. An alternative definition of the shear force (or bending moment) at any section is the sum of the shear forces (or bending moments) to one side of that section.

Considering Fig. 1.1, R_0 obviously would tend to move the left-hand side of any section distant x from the left-hand end <u>upwards</u> and is therefore negative for the stated convention. Thus, for $x \leqslant a$:

$$F = -R_l, \tag{1.11}$$

as can also be seen by considering the equilibrium of the free body illustrated in Fig. 1.1(a).

Likewise, if $x \geqslant a$, the shear force at section x is:

$$F = -R_l + W. \tag{1.12}$$

In considering the slope and deflection of beams it is convenient to have only one equation to represent F and M for any value of x and it is therefore necessary to introduce a discontinuous unit function, denoted usually by square brackets. Thus it is possible to write:

$$F = -R_l + W[x-a]^0 \tag{1.13}$$

where it is understood that $[x-a]$ is zero if $x \leqslant a$ and $[x-a]^0 = 1$ when $x \geqslant a$.

Considering moment equilibrium, for Fig. 1.1(b), again the reaction R_0 will clearly tend to cause sagging of the beam so that the moment of R_0 about the section distant x from the left-hand end is negative, i.e. $-R_0 x$. Hence:

$$M = -R_0 x + W[x-a] \tag{1.14}$$

and, since $[x-a] = 0$ when $x < a$, $M = -R_0 x$ when $x < a$, as required. The same result is obtained by determining the moment equilibrium of the free body diagram shown in Fig. 1.1(b).

In the case of Fig. 1.2(a), the expressions are:

$$F = -R_0 + w[x-c] \quad \text{and} \tag{1.15}$$

$$M = -R_0 x + w[x-c]^2/2 \tag{1.16}$$

since the centre of gravity of the load distributed over the length $(x-c)$ is at a distance of $(x-c)/2$ from the section x and $[x-c] =$

0 when $x<c$, as required.

When $x>b+c$, as shown in Fig. 1.2(b) however, a problem occurs because:

$$F = -R_0+wb \quad \text{and} \tag{1.17}$$

$$M = -R_0x+wb\{x-(c+b/2)\} \tag{1.18}$$

and these equations are very different from equations (1.15) and (1.16).

The problem of establishing an equation which will apply to the whole length of the beam in this case is solved by introducing an additional upward uniformly distributed load to compensate for an extended uniformly distributed downward load, as illustrated by the broken lines in Fig. 1.2. Then, the term $w[x-c]^2/2$ will apply for the bending moment of the whole of the extended downward u.d.l. (uniformly distributed load) and an upward u.d.l. represented by the term $-w[x-(b+c)]^2/2$ can be introduced to counterbalance the moment due to the extended downward u.d.l. Thus the required expressions for the whole beam will be:

$$F = -R_0+w[x-c]-w[x-(b+c)] \tag{1.19}$$

and $$M = -R_0x+w[x-c]^2/2-w[x-(b+c)]^2/2 \tag{1.20}$$

where $[x-(b+c)] = 0$, if $x < (b+c)$.

To check this result, when $x > (b+c)$,

$$F = -R_0+w(x-c-x+b+c) = -R_0+wb, \text{ as before} \tag{1.17}$$

and $M = -R_0x+(w/2)\{x^2-2xc+c^2-x^2+2x(b+c)-(b^2+2bc+c^2)\}$

$$= -R_0x+(w/2)(2xb-b^2-2bc) = -R_0x+wb\{x-(c+b/2)\}, \tag{1.18}$$

as required.

An equation representing the distribution of load intensity w along the beam is easily seen to be:

$$w = w[x-c]^0-w[x-(b+c)]^0 \tag{1.21}$$

since,   when $x > b+c$, $w = w-w = 0$,
             when $c < x < b+c$, $w = w$ and,
             when $x < c$, $w = 0$, as required.

1.3 INTER–RELATIONSHIP OF LOAD, SHEAR FORCE AND BENDING MOMENT

It may be observed from equations (1.19), (1.20) and (1.21) that $F = dM/dx$ and $w = dF/dx$. These relationships apply because of the change of origin introduced into each of the discontinuous unit functions within the square brackets, which was the reason for introducing them and for raising them to the power zero when the variables involved are independent of x, as in equations (1.13) and (1.21). These relationships are *generally* true, as will now be proved.

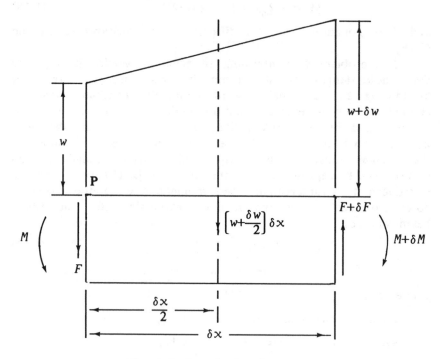

Fig. 1.3 Free body element

     Consider an element of any beam which is carrying a distributed load of varying intensity w, as illustrated in Fig. 1.3, which shows a free body element subjected to shear forces and bending moments at each end, changing from F to $F+\delta F$ and from M to $M+\delta M$ over the elemental length δx, whilst the intensity of loading also changes from w to $w+\delta w$ over δx. Since δx is small the distribution of w over δx may be assumed to be linear so that the total downward force from the distributed load may be taken as $(w+\delta w/2)/\delta x$, acting at a distance of $\delta x/2$ from the left–hand end of the element (the point P).

     Equilibrium of the vertical forces gives the equation:

$$F+(w+\delta w/2)\delta x = F+\delta F \ ,$$

hence $w\delta x+\delta w\,\delta x/2 = \delta F$

and $w+\delta w/2 = \delta F/\delta x.$ As δx and $\delta w \to 0$,

$$w = dF/dx, \quad \text{Q.E.F.} \tag{1.22}$$

Moment equilibrium about the point P requires:

$$(w+\delta w/2)\delta x.\delta x/2-(F+\delta F)\delta x-M+(M+\delta M) = 0,$$

hence: $w\delta x/2+\delta w\delta x/4-F-\delta F+\delta M/\delta x = 0.$

As δx, δw and $\delta F \to 0$, $F = dM/dx$, Q.E.F.   (1.23)

Conversely, $F = \int wdx+\text{Constant}_1,$ (1.24)

and $M = \int Fdx+\text{Constant}_2.$ (1.25)

In the analysis of beams it is useful to sketch, or even to draw accurately to scale, shear force and bending moment diagrams and these relationships [defined by the four equations (1.22) to (1.25)] are very helpful for that purpose. In that respect, it is probably best to try always to work with equations (1.22) and (1.23) only, otherwise it is very easy to make the mistake of omitting the constants of integration in equations (1.24) and (1.25). In any case, it is easier not to determine constants of integration. Thus, in practice, the shear force diagram is drawn first of all, having determined the support reactions and/or support moments, and the shear force diagram can then be viewed as representing the slope of the bending moment diagram at every point along the beam.

This procedure is illustrated in Fig. 1.4. which relates to the beam and loading shown in Fig. 1.1. Sufficient space is left for the construction of the three diagrams, the top one being a sketch of the actual beam and its load distribution. For **concentrated** or **point loads** such as those shown in Fig. 1.1. at the support reactions and at the load W the **load intensity** w tends to infinity. Of course, in practice it is not possible to have infinite load intensity or a point load and the actual load distributed at these points would appear somewhat as sketched in Fig. 1.4(a) with the broken lines. If each of the point loads is imagined to be spread out over a very short distance δx then the load intensity at each of the point loads will be $w = W/\delta x$, $-R_0/\delta x$ and $-R_1/\delta x$ as shown. The actual load intensity distribution of each load over the length δx will be determined mainly by the interaction between the (presumably elastic) surface deformations of the applied loads and supporting beam.

Regarding the load intensity w diagram (actually, the **physical plane**) as a diagram showing the slope of the shear force F−diagram [since $w = dF/dx$, equation (1.22)], it is possible to derive the F−diagram from it by systematically proceeding along the beam from the left−hand end, where $x = 0$. Initially at the point A, w has a large (infinite for the hypothetical point load) negative value equal to $-R_0/\delta x$, which tends to minus infinity as δx tends to zero. Thus the slope of the F-diagram is infinite and negative as indicated in Fig. 1.4(b) for the hypothetical point load but, in fact, equal to $-R_0/\delta x$ for the actual distributed load. At the section $x = \delta x$, therefore, the total load on the beam to the left of the section x (i.e. the shear force at section x) will be $-(R_0/\delta x).\delta x = -R_0$, as shown.

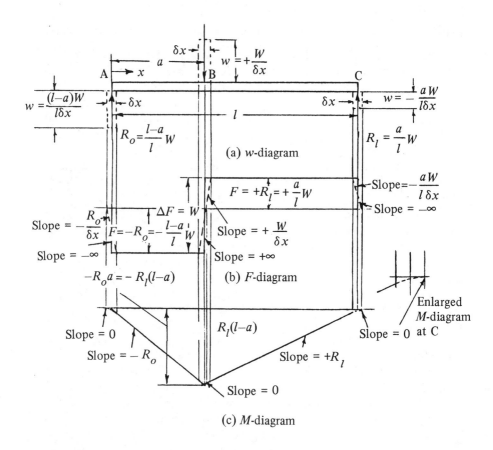

Fig. 1.4 Typical beam, illustrating relationships
between w, F and M for a point load

At the point B the load intensity is positive and tends to infinity, thus at B the slope of the F-diagram is positive and infinite. Actually, the slope is equal to $+W/\delta x$ so that in the distance δx the shear force will have changed by an amount $+(W/\delta x).\delta x = +W$, from the value $-(l-a)W/l$ to the value $-(l-a)W/l+W = aW/l$, as shown on Fig. 1.4(b). At the point C the slope is actually $-(aW)/(l\delta x)$, hypothetically $-\infty$, as shown.

This procedure has been described at some length because it

emphasizes the elegant simplicity of the simple beam theory which has been developed over the years by a select group of distinguished mathematicians and engineers. The process of integration would have given the same result and the F-diagram represents the area of the w-diagram up to any value of x, as can easily be verified by the reader. Integrating the F-diagram leads to the M-diagram (since $F = dM/dx$) and, as will be shown later, the M-diagram can be integrated, to give the slope i of the beam at any point and the slope diagram (i-diagram) can be integrated to give the deflection y at any point (since $i = dy/dx$).

Proceeding in the same way, to integrate the F-diagram, the slope of the M-diagram (i.e. F) between points A and B is constant and negative, equal to $-R_0$. Between B and C it is constant and positive, equal to $+R_l$. Thus the M-diagram must be as indicated in Fig. 1.4(c) and the maximum bending moment is seen to be equal to slope × distance $= -R_0 \times a = -R_l \times (l-a)$, (or the change of area of the F-diagram up to the point considered, which gives the same result). The effect of 'spreading out' each point load over the small distance δx, say, is again illustrated – it 'rounds off' the sharp corners of the bending moment diagram.

It may be seen from this simple example that the maximum shear force is less than the applied load W (equal, in fact, to R_0, the larger of the support reactions) and it exists over a large length of the beam. On the other hand, the bending moment is a maximum at only one point of the beam, namely under the load, of magnitude $-R_0 a$. When the load is at the centre of the beam, the magnitude of the shear force is $W/2$ and the maximum bending moment is $Wl/4$, values which it is useful to remember. It will also be observed that the sign conventions originally chosen for the shear force and the bending moment are quite consistent with the integration procedure which has been used, starting from a load diagram in which downward acting loads such as W are represented by a <u>positive</u> (infinite for a point load) load intensity w.

The shear force and bending moment diagrams for the simply supported beam shown in Fig. 1.2 are illustrated in Fig. 1.5. As drawn, the total load $W = wb$ is nearer to the right-hand end of the beam so the reaction R_l will be bigger than R_0, as indicated. The locations and values of the maximum shear forces and bending moments are easily seen from the diagrams. The maximum (negative or sagging) bending moment occurs at the point of zero slope on the M-diagram and its numerical magnitude can readily be determined from the area of the F–diagram up to that point, as:

$$R_0 c + \frac{1}{2} \frac{bR_0{}^2}{R_0 + R_l}.$$

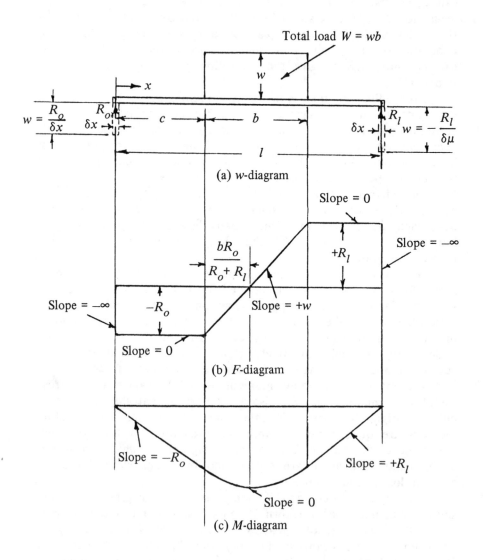

Fig. 1.5 Typical beam, illustrating relationships
between w, F and M for a distributed load

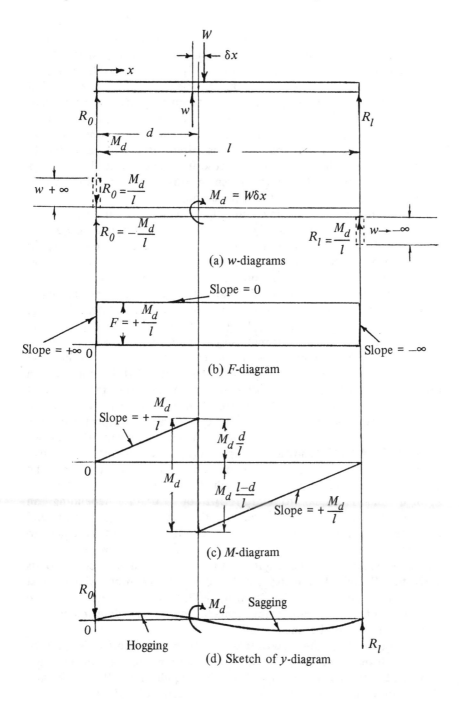

(a) w-diagrams

(b) F-diagram

(c) M-diagram

(d) Sketch of y-diagram

Fig. 1.6 Typical beam, illustrating relationships
between w, F and M for an applied torque

It can be seen from these two simple examples that graphical representation of the distributions of load, shear force and bending moment across a beam is very desirable since it improves the understanding of the problem, facilitates evaluation of the position and magnitude of maximum values and reduces the possibility of errors. Of course, the representation of distributions of w, F and M by mathematical expressions such as equations (1.19), (1.20) and (1.21) is often necessary in order to obtain exact solutions, as will be described later, but it is always advisable to sketch the diagrams since this often indicates errors which have been made in the calculations.

An interesting loading situation which often occurs in practice is that of the application of a concentrated torque or moment. Conceptually this is difficult to visualize – one way in which such a moment could be applied is illustrated in Fig. 1.6(a), δx being very small and the loads W very large.

Thus the applied loads W may effectively be replaced by a concentrated clockwise bending moment M_d as shown in Fig. 1.6(a), where $M_d = W\delta x$. Taking moments about any point and equilibrating the vertical forces shows that $R_0 = -R_l$ and $R_l = M_d/l$. Hence R_0 actually acts as indicated by the broken lines in Fig. 1.6(a) and the load diagram is also indicated by the broken lines in Fig. 1.6(a). The F–diagram is found from the fact that its slope is equal to w everywhere. Since w is zero everywhere except for values of $-\infty$ and $+\infty$ at $x = 0$ and l respectively, the F–diagram is the simple rectangle shown. Likewise the rather unusual M–diagram shown as Fig. 1.6(c) has a constant positive slope equal to M_d/l and the maximum bending moment on the beam occurs at the point at which the concentrated moment is applied and is equal to $-M_d(l-d)/l$ or $+M_d d/l$, depending on whether the moment is nearer to the left– or right–hand end respectively. This sudden change of bending moment is equal to the applied moment M_d and the fact that it is positive to the left and negative to the right of the 'point moment' corresponds with the 'hogging' and 'sagging' modes of bending which must occur, as shown in Fig. 1.6(d). This is a sketch of the displacement– or y–diagram which can be determined by calculation as will be shown later.

Another **statically determinate** problem of interest is that of the **cantilever**. The w–, F– and M–diagrams for a concentrated end load are shown in Fig. 1.7. For equilibrium $R_0 = W$ and these point loads may be imagined as loads of intensity $-\infty$ and $+\infty$ respectively. Thus the F–diagram may be constructed immediately, as shown, and the M–diagram is found, by observing that its slope must be negative and equal to $-W$. The maximum value of the bending moment is obviously Wl, at $x = 0$.

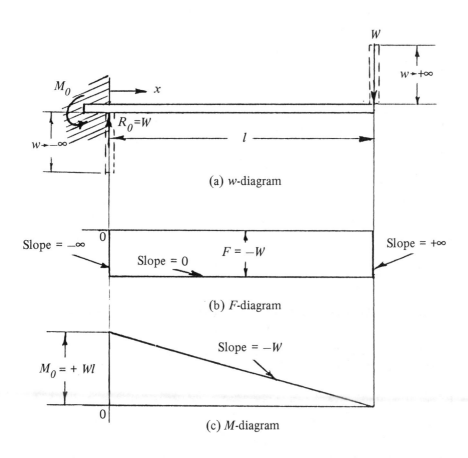

(a) w-diagram

(b) F-diagram

(c) M-diagram

Fig. 1.7 Typical cantilever, illustrating relationships
between w, F and M for a concentrated end load

The case of a cantilever with a uniformly distributed load is
shown in Fig. 1.8. If x is measured from the left–hand end,
which is encastré, $F = -R_0 + wx = -wl + wx$ and $M = -R_0 x + M_0 + wx^2/2 = -wlx + (l^2 + x^2)w/2$ which are the equations describing the
diagrams shown.

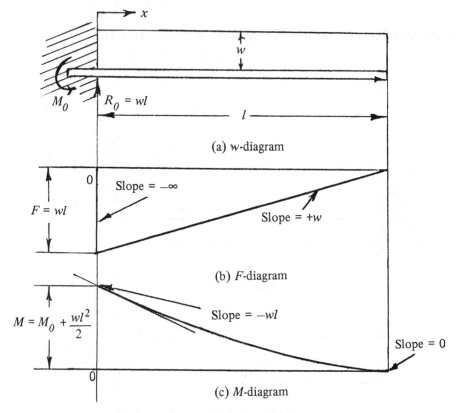

Fig. 1.8. Typical cantilever, illustrating relationships between w, F and M for a distributed load.

1.4. PURE BENDING, ASSUMPTIONS, RADIUS OF CURVATURE

If equal couples are applied to each end of a straight beam, as illustrated in Figs. 1.9 or 1.10, the beam is being subjected to **pure bending**.

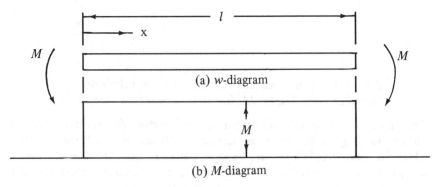

Fig. 1.9 Pure bending

The system of loading shown in Fig. 1.10 is known as **four-point bending** and is often used for calibrating **strain gauges**. This is because four-point bending provides, between the inner support reactions, the maximum possible length of beam along which the bending moment and hence the bending stresses are constant. As will now be shown the bending stresses are aligned along the beam and have their maximum values at the surface of the beam, to which the strain gauges (which may be electrical resistance, mechanical etc.) under test can be conveniently attached. Four-point bending also provides a very favourable constant known stress field in rotating bar fatigue machines. Clearly, the portion of the beam over which the bending moment is constant has zero shear force across its section, since the slope of the M-diagram is zero there. The positive bending moments shown will cause hogging and the beam will bend uniformly along the portion subjected to constant bending moment so that it will, if of uniform cross-section, bend into a circular arc with a large radius of curvature R, say. Since a beam of any strength must have depth, hogging will induce longitudinal tensile stresses in the top layers and longitudinal compressive stresses in the bottom layers of the beams shown in Figs. 1.9 and 1.10.

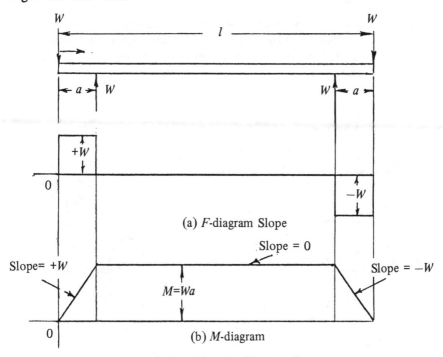

Fig. 1.10. Four-point bending

The assumptions which are made in developing a simple theory suitable for predicting stresses, strains, curvatures and deflections of beams are briefly as follows:

(i)   The beam is initially straight and the radius of curvature at any section is very much greater than its cross–sectional dimensions.

(ii)   The material of the beam is homogeneous and obeys Hooke's law whilst elastic.

(iii)   Young's Modulus is the same in tension and compression.

(iv)   In pure bending the stresses vary only through the depth of the section.

(v)   The cross–section is symmetrical about the plane of bending.

(vi)   Deflections, slopes and curvature of the beam are small.

(vii)   Plane transverse sections of the beam remain plane during bending (only precisely true for zero shear force).

(viii) Each longitudinal fibre can be regarded as a tiny uniform section under tension or compression and is free to extend or compress as though it were a simple rod.

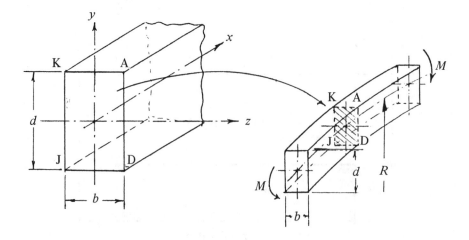

Fig. 1.11 Simple bending

As a simple example, consider a prismatic beam of rectang-ular cross–section KADJ of breadth b and depth d as illustrated in Fig. 1.11. The mode of bending is shown in the inset diagram, under the action of the applied (hogging) bending moment M, which is constant along the length of the beam, resulting in a uniform change of **curvature** from zero (for the initially straight beam) to $1/R$, where R is the **radius of curvature** of the beam which may conveniently be measured from the centre of curvature to the fibre whose length remains unchanged during the bending, as will now be discussed.

1.5 NEUTRAL SURFACE OF BENDING, SECOND MOMENT OF AREA

Using the rectangular cartesian coordinate system shown, consider an elemental length δx of the beam, as illustrated in Fig. 1.12, before and after bending. Since there is no longitudinal load on the beam there will be a fibre whose length δx will remain δx after bending and this fibre is designated E'F' in the figure, with curvature R. All such fibres across the beam form what is called the **neutral surface** of bending. Above the neutral surface, fibres such as AB are stretched or extended to become A'B' and below the neutral surface the fibres are compressed, as illustrated by the extreme fibres CD which become C'D'. The original positions of the transverse planes AD and BC are illustrated by the broken lines in Fig. 1.12(b).

Any generic fibre such as GH, distant y from the neutral surface will extend to a length G'H' $= (R+y)\delta x/R$, since $\delta x/R$ must be the angular change of circular arc in traversing the length δx of the beam. Thus the strain in any fibre such as GH = (final length–original length)/original length, i.e.

$$\epsilon = \frac{(R+y)\dfrac{\delta x}{R} - \delta x}{\delta x} = \frac{y}{R} \qquad (1.26)$$

and the stress in the fibre GH will be given by:

$$\sigma = E\epsilon = Ey/R, \text{ where } E = \text{Young's Modulus.} \qquad (1.27)$$

If the origin for y is coincident with the neutral surface, then positive y will designate fibres in tension, for hogging moments, and lead to positive stress, as required. Below the neutral surface, y is negative giving negative (compressive) stresses corresponding with compression in the case of hogging under discussion. This is another justification for regarding 'hogging' bending moments as positive, as will now be discussed.

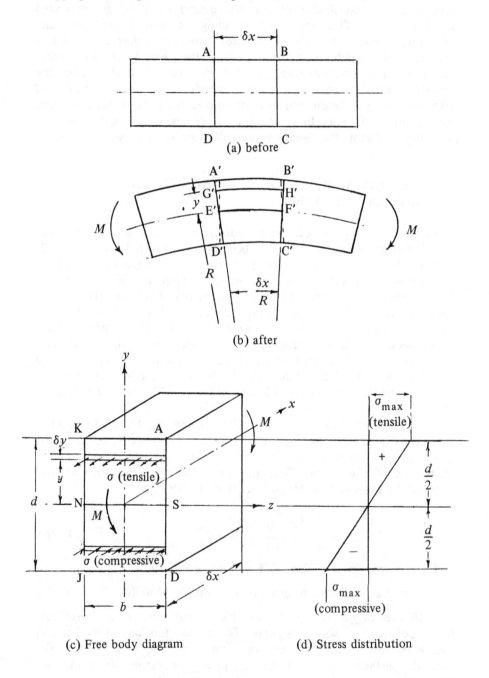

(a) before

(b) after

(c) Free body diagram

(d) Stress distribution

Fig. 1.12 Distribution of stresses due to bending

Considering the portion ABCD of the beam of length δx and width b (in the case of the rectangular section) to be a free body which is in statical equilibrium under the action of the forces applied to it, Fig. 1.12(c) illustrates the only stresses which can be acting on transverse plane sections of this beam, subjected to **pure bending** only. There are no shear forces and therefore no shear stresses. The bending moment M remains constant along the length of the beam so there can be no change in the longitudinal direct stresses (σ) along the beam. It is also clear that there can be no change in these stresses across the beam since they depend only on y.

Equilibrium of horizontal forces on the transverse rectangular cross-section of this beam requires that the total load on the end of the beam, P, say, should be zero. Thus:

$$P = \Sigma \sigma b \delta y = 0. \qquad (1.28)$$

From equation (1.27), therefore, in the limit:

$$P = \int_d \frac{Eyb}{R} \, dy = 0. \qquad (1.29)$$

For any section, E and R are constant for all values of y, thus equation (1.29) reduces to:

$$\int_d by\,dy = 0. \qquad (1.30)$$

The left-hand side of equation (1.30) can be seen to be simply the **first moment of area** about the axis $y = 0$. Since this has to be equated to zero, the axis of $y = 0$ must be coincident with the centroid of the cross-section which, for a rectangular cross-section, is · precisely at the geometrical centre of the section.

Considering now the equilibrium of moments applied to the section it is clear that the only stresses available for providing the moment M are again the longitudinal stresses. Referring to Fig. 1.12(c) it is necessary to determine the resultant moment of these stresses over the section and it is convenient to take moments in Fig. 1.12(d) about the neutral surface designated NS in Fig. 1.12(c). Since NS in that figure is a line rather than a surface it is often referred to as the **neutral axis** and could therefore be designated NA, or ZZ (since it coincides with the z-axis) or CC (for centroid) or GG (for centre of gravity of the area).

Thus, if A is the area of the cross-section:

$$M = \Sigma \sigma y \delta A = \int_{-d/2}^{+d/2} \sigma b y \, dy \qquad (1.31)$$

Substituting again from equation (1.27) gives:

$$M = \int_{-d/2}^{+d/2} \frac{Ey}{R} \, by \, dy = \frac{E}{R} \int_{-d/2}^{+d/2} by^2 \, dy = \frac{EI}{R} , \qquad (1.32)$$

where:

$$I = \int_{-d/2}^{+d/2} by^2 \, dy$$

is the **second moment of area** about the centroid.

Equations 1.27 and 1.32 are generally telescoped together to form one of the most important expressions for all engineers to remember, namely:

$$\frac{M}{I} = \frac{\sigma}{y} = \frac{E}{R} , \qquad (1.33)$$

which is really <u>three</u> equations, of course.

The product EI is very important also, particularly in civil and structural engineering, and is termed the **bending stiffness** of the prismatic beam section.

The quantity $M/\sigma_{max} = I/y_{max}$ is also important. I/y_{max} is often referred to as the **section modulus** and denoted by the quantity Z, where:

$$Z = I/y_{max} \qquad (1.34)$$

The discussion so far has been restricted to pure bending and a simple rectangular cross-section. When bending moment varies along the length of the beam, shear forces will be introduced as already discussed, and hence shear stresses over transverse sections. Complementary shear stresses will exist on longitudinal sections, and this will lead to departure from the assumed condition that plane sections will remain plane. **Warping** of sections will occur. However, in what follows, it is assumed for the simple theory of bending that the distribution of longitudinal stresses over the section is unaffected by any warping and remains linear as depicted in Fig. 1.12(d). Also, it will be obvious that equations (1.26) to (1.33) will then apply to <u>any</u> cross-sectional shape of prismatic beam with the exception that the limits of integration will be from y_1 (negative) to y_2

(positive) rather than from $-d/2$ to $+d/2$. In such a beam there will be two section moduli, $Z_1 = I/y_1$ and $Z_2 = I/y_2$, since $y_2 \neq y_1$ as was the case in the equation (1.34) for Z.

1.6 THEOREMS OF PARALLEL AND PERPENDICULAR AXES, PRODUCT MOMENT OF AREA, EXAMPLES

It is often necessary in beam problems to determine the position of the centroid and the second moment of area of an awkward cross–sectional shape. The problem is illustrated in Fig. 1.13 for the general case of a cross–sectional area A which is normal to the x–direction shown in Fig. 1.12.

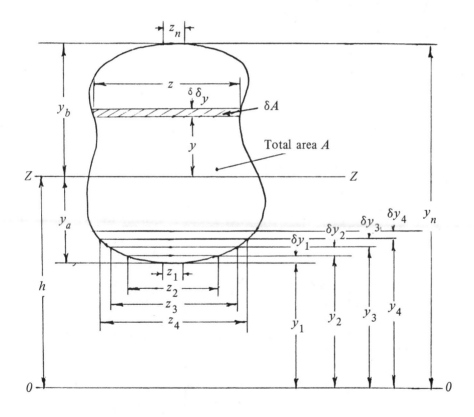

Fig. 1.13 Elemental strips in a lamina

If a reference axis ZZ, parallel to the z-axis, is chosen, the elemental area δA of interest in the bending analysis is the strip of width z and thickness δy, distant y from the axis ZZ. Then, the second moment of area I_{ZZ} of the whole area about the axis ZZ is given by the summation:

$$I_{ZZ} = \sum_A y^2 \delta A = \sum_{y_a}^{y_b} z y^2 \delta y. \qquad (1.35)$$

To determine the second moment of area I_{OO} about some other parallel axis OO distant h from ZZ as shown in Fig. 1.13, then, if Σ represents summation over the whole area A:

$$I_{OO} = \Sigma(y+h)^2 \delta A = \Sigma y^2 \delta A + \Sigma 2 y h \delta A + \Sigma h^2 \delta A =$$

$$\Sigma y^2 \delta A + 2h\Sigma y \delta A + h^2 \Sigma \delta A \qquad (1.36)$$

since h is a constant for all the strips.

Now if ZZ is the **centroidal axis**, designated GG, say, then $\Sigma y \delta A$ becomes the first moment of the area A about its own centroid, which is zero, $\Sigma \delta A = A$ and $\Sigma y^2 \delta A = I_{GG}$.

Hence: $I_{OO} = I_{GG} + Ah^2.$ $\qquad\qquad\qquad$ (1.37)

This is known as the **theorem of parallel axes.**

The second moment of any area of irregular shape about any axis is conveniently determined numerically by dividing the whole area into the large number of parallel thin strips indicated in Fig. 1.13. The method can be demonstrated in tabular form as shown in Table 1.1.

TABLE 1.1

y_1	z_1	δy_1	$z_1 \delta y_1$	$z_1 y_1 \delta y_1$	$z_1 y_1^2 \delta y_1$
y_2	z_2	δy_2	$z_2 \delta y_2$	$z_2 y_2 \delta y_2$	$z_2 y_2^2 \delta y_2$
--	--	---	-----	-------	--------
--	--	---	-----	-------	--------
--	--	---	-----	-------	--------
y_n	z_n	δy_n	$z_n \delta y_n$	$z_n y_n \delta y_n$	$z_n y_n^2 \delta y_n$

The position of the centroidal axis GG parallel to and distant h from the arbitrary axis OO is thus determined from the quotient of the summations of the fourth and fifth columns, $[\Sigma(\text{fifth})/\Sigma(\text{fourth})]$, since $h = Ah/A$. Then the second moment of the area A about the parallel centroidal axis is given from equation (1.37) as:

$$I_{GG} = I_{OO} - Ah^2. \qquad (1.37a)$$

It is worth mentioning here that the present discussion relates solely to the properties of a laminar plane of infinitesimal thickness in the context of stress analysis. In the theory of dynamics the inertial properties of bodies having mass involve **moments of inertia** which are effectively second moments of body force. It is incorrect to refer to second moments of area as moments of inertia, although that is often done. A concept often used in dynamics is that of the **radius of gyration**. Radii of gyration of the lamina under discussion may be denoted by k (not to be confused with shear yield stress) where, for example,

$$I_{OO} = Ak^2_{OO}, \text{ and } I_{GG} = Ak^2_{GG}. \qquad (1.38)$$

Thus, from equation (1.37), the theorem of parallel axes may be stated as follows:

$$Ak^2_{OO} = Ak^2_{GG} + Ak^2, \text{ or } k^2_{OO} = k^2_{GG} + k^2. \qquad (1.37b)$$

The radius of gyration may be thought of as the distance from the reference axis of the point at which the whole area may be considered to be concentrated, to give the required second moment.

The numerical analysis depicted in tabular form in Table 1.1 can be carried out to any desired degree of precision by reducing the thickness of the strips and thus the calculations are best carried out by a computer. It is most convenient to split the lamina area up into a number of strips of equal thickness δy and to arrange the strips so as to give an odd number of values (or 'ordinates') so that a more accurate numerical integration procedure can be adopted such as Simpson's rule. A subroutine written in the Fortran language for Simpson's rule is included in Program 1.1 shown later and it is a simple matter to write a main program by which the data (i.e. values of the interval δy and the 'ordinates' $z\delta y$, $zy\delta y$ and $zy^2\delta y$) can be successively fed into the subroutine to give the required summations.

Of importance in dynamics and also in the torsion problem of stress analysis is the second moment of area of the lamina under discussion about an axis <u>perpendicular</u> to the plane of the lamina. This situation is illustrated in Fig. 1.14 in which a lamina of area A is shown. An arbitrary system of rectangular cartesian coordinates x,y,z is shown, with the y and z axes in the plane of the lamina and the x–axis perpendicular to the lamina, to correspond with the axes already chosen for the beam problems being discussed in this Chapter. An elemental area δA is also shown, at a distance z' from the origin O of the original system of axes and forming one of the axes of a <u>new</u> system of rectangular cartesian coordinates, viz. x',y',z'. The second moment of area of the lamina about the origin O, in a rotational mode about the chosen x–axis, viz. XX, as shown in Fig. 1.14, is:

$$I_{XX} = \Sigma z'^2 \delta A. \tag{1.39}$$

The second moments of area of the lamina about the axes YY and ZZ shown in the figure are:

$$I_{YY} = \Sigma z^2 dA \tag{1.40}$$

and: $\qquad\qquad I_{ZZ} = \Sigma y^2 \delta A. \tag{1.35}$

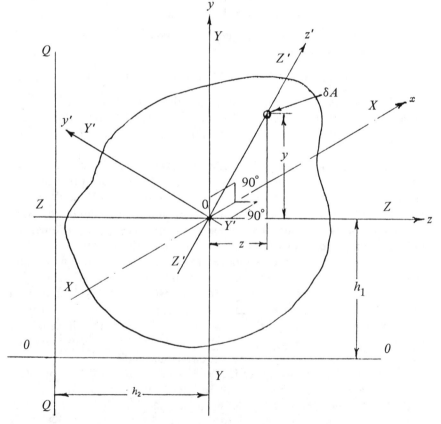

Fig. 1.14 Transformation of coordinates

The coordinates y, z and z' are related by the **theorem of Pythagoras**, namely:

$$z'^2 = y^2 + z^2 , \qquad (1.41)$$

thus it is clear that: $\Sigma z'^2 \delta A = \Sigma y^2 \delta A + \Sigma z^2 \delta A$ and hence:

$$I_{XX} = I_{YY} + I_{ZZ}. \qquad (1.42)$$

This is known as the **theorem of perpendicular axes**.

Similarly, the radii of gyration are related by the equation:

$$k^2{}_{XX} = k^2{}_{YY} + k^2{}_{ZZ}. \qquad (1.43)$$

For bending and torsion problems, the x,y,z axes are centred at the centroid of the lamina, as just discussed [equation (1.42)]. It is of interest and importance to observe that a consequence of equation (1.42) is that the sum of the second moments of area of a lamina about any two perpendicular axes, lying in the plane of the lamina and possessing the same origin, is constant.

The **product moment of area** of the lamina about the $z-$, $y-$axes which both pass through the centroid is defined as:

$$I_{ZY} = \Sigma zy \delta A. \qquad (1.44)$$

The product moment of area of the lamina shown in Fig. 1.14 about a parallel set of axes passing through another origin in the plane of the lamina, denoted by OO and QQ, will therefore be given by:

$$I_{OQ} = \Sigma(y+h_1)(z+h_2)\delta A = \Sigma(yz + h_1 z + h_2 y + h_1 h_2)\delta A =$$

$$\Sigma yz \delta A + h_1 \Sigma z \delta A + h_2 \Sigma y \delta A + h_1 h_2 \Sigma \delta A.$$

Now $\Sigma z \delta A = \Sigma y \delta A = 0$, since these are the first moments of the total area about the $y-$ and $z-$axes, which pass through the centroid. Hence:

$$I_{OQ} = I_{YZ} + Ah_1 h_2. \qquad (1.45)$$

This is the analogue of the theorem of parallel axes for second moments of area.

It is important to note that the product moment of area given by equation (1.44) for a **symmetrical** section about any of its axes of symmetry must be zero, since for any positive value of the product $zy \delta A$ there will be an equal and opposite value to cancel it out in the summation $\Sigma zy \delta A$. The second moments of area about these axes are then the maximum and minimum values for the lamina and these axes are called the **principal axes**, as will now be discussed.

Referring back to the general case shown in Fig. 1.14, it is necessary to determine how the second moments of area of the lamina about any pair of coordinate axes such as the primed set Z'Z', Y'Y' are related to those about the unprimed set ZZ, YY, both sets passing through the centroid. In other words the **equations of rotational transformation** for the second moments of area. To do this it is necessary to relate the z', y' coordinates to the z, y coordinates through any angular rotation θ about the centroid G, as illustrated in Fig. 1.15. It can be seen from the trigonometry of that diagram, with the help of the broken construction lines shown, that:

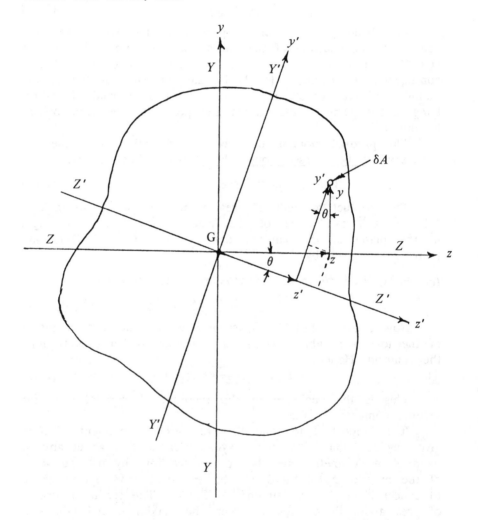

Fig. 1.15 Transformation of coordinates about centroid

$$z' = z\cos\,\theta - y\sin\,\theta, \text{ and} \qquad y' = y\cos\,\theta + z\sin\,\theta. \qquad (1.46)$$

Thus, the second moment of area of the lamina about the Z'Z' axis will be:

$$I_{Z'Z'} = \Sigma y'^2\delta A = \Sigma\{y^2\cos^2\theta + 2yz\sin\,\theta\cos\,\theta + z^2\sin^2\theta\}\delta A$$

$$= \frac{I_{ZZ} + I_{YY}}{2} + \frac{I_{ZZ} - I_{YY}}{2}\cos\,2\theta + I_{YZ}\sin\,2\theta. \qquad (1.47)$$

Likewise, its product moment about the Z'Y' axes will be defined here as:

$$I_{Z'Y'} =$$

$$\Sigma y'z'\delta A = \Sigma\{z^2\sin\,\theta\cos\,\theta + yz(\cos^2\theta - \sin^2\theta) - y^2\sin\,\theta\cos\,\theta\}\delta A$$

$$= -\frac{I_{ZZ} - I_{YY}}{2}\sin\,2\theta + I_{ZY}\cos\,2\theta. \qquad (1.48)$$

Comparison of equations (1.47) and (1.48) with equations (3.6) and (3.7) of Chapter 3 (p.68) of the companion volume reveals that there is a complete analogy between the equations of transformation for second moments and for stress. Second moments or product moments of area (or moments of inertia) are tensors of the second order, obeying the tensor equations of transformation. The second moment of area I_{ZZ} of this lamina 'rotating' about the ZZ axis is analogous to σ_{zz}, the direct stress acting in the z-direction normal to a y-plane whilst the product moment I_{ZY} about the z-and y-axes is analogous to the shear stress τ_{zy}. Thus, Mohr's circle can conveniently be used to represent second and product moments of area (or inertia), and this brings out the fact that maximum and minimum values of $I_{Z'Z'}$ exist, corresponding with zero product moments $I_{Z'Y'}$, as illustrated in Fig. 1.16 for a rectangular lamina. The **Prager pole** can also be introduced, in which case it is convenient to define the Prager Pole for the second and product moment plane illustrated in Fig. 1.16(b) as the point on the Mohr circle from which a line parallel to the *axis* of interest cuts the circle at the point representing the second and product moment about that *axis.* The product moment I_{ZY} is negative for the z-axis shown because the quadrant areas of the rectangular lamina shown shaded in which either z or y is negative (but not both), are obviously greater in extent than the remaining quadrants in which both z and y are of the same sign. Thus I_{YZ} (the product moment when the y-axis becomes the z'-axis) becomes positive. Analogous to the shear stress, the product moment of area attains stationary values about axes at 45˚ to the z'- and y'-axes which are the principal axes already referred to.

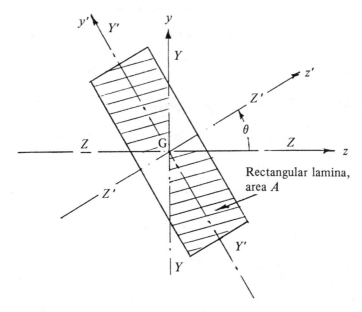

(a) Physical plane

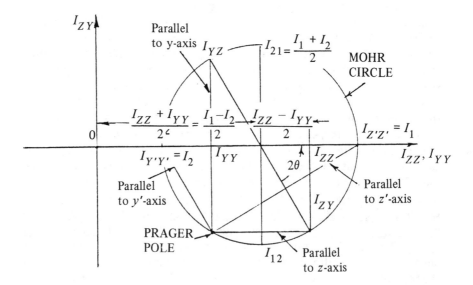

(b) Second and product moment plane

Fig. 1.16 Mohr's circle representation of
second and product moments of area

For simple symmetrical sections such as the rectangular lamina of Fig. 1.16(a), θ is easily determined by inspection. The principal second moments of area I_1 and I_2, say, may be determined readily from the following equations, easily derived from the Mohr circle:

$$I_1 = \frac{I_{ZZ}+I_{YY}}{2} + \frac{I_{ZZ}-I_{YY}}{2}\sec 2\theta$$

and

$$I_2 = \frac{I_{ZZ}+I_{YY}}{2} - \frac{I_{ZZ}-I_{YY}}{2}\sec 2\theta. \tag{1.49}$$

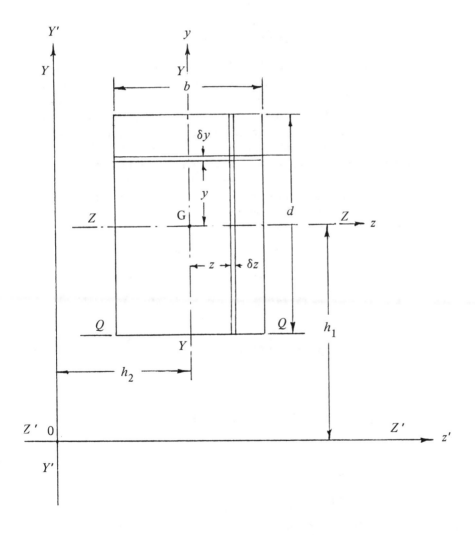

Fig. 1.17 A rectangular lamina

Similarly, the maximum product moment of area I_{12} is:

$$I_{12} = \pm\frac{I_{ZZ}+I_{YY}}{2}\sec 2\theta = \pm\frac{I_1-I_2}{2}. \qquad (1.50)$$

Conversely, if I_1 and I_2 are known for a symmetrical section and it is required to find the second and product moments about any axis inclined at an angle θ to the major principal axis {the z'–axis of Fig. 1.16(a)} then it is seen from the trigonometry of the Mohr circle that:

$$I_{ZZ} = \frac{I_1+I_2}{2} + \frac{I_1-I_2}{2}\cos 2\theta \qquad (1.51)$$

$$I_{ZY} = \pm\frac{I_1-I_2}{2}\sin 2\theta. \qquad (1.52)$$

In many text books an **ellipse of inertia** (or ellipse of second moment of area) and a **momental ellipse** of radii of gyration are introduced but these concepts are of limited usefulness. The Mohr circle construction gives, in the simplest possible way, all the information necessary for the solution of problems of unsymmetrical bending, as will be discussed later.

Some simple examples of second and product moments of area for typical sectional shapes will now be discussed. A rectangular lamina is shown in Fig. 1.17, with the origin of the z– and y–axes at the centroid G of the rectangle, its centre of symmetry. It can readily be seen that:

$$I_{ZZ} = b\int_{-d/2}^{+d/2} y^2 dy = b\left[\frac{y^3}{3}\right]_{-d/2}^{+d/2} = \frac{bd^3}{12} \qquad (1.53)$$

$$I_{YY} = d\int_{-b/2}^{+b/2} z^2 dz = d\left[\frac{z^3}{3}\right]_{-b/2}^{+b/2} = \frac{db^3}{12} \qquad (1.54)$$

$$I_{ZY} = \int_{-b/2}^{+b/2}\int_{-d/2}^{+d/2} zy\,dy\,dz = 0, \text{ by symmetry.} \qquad (1.55)$$

$$I_{Z'Z'} = b\int_{-b/2}^{+b/2}(y+h_1)^2 dy = b\int_{-b/2}^{+b/2}(y^2+2yh_1+h_1)^2 dy$$

$$= I_{ZZ}+Ah_1^2, \qquad (1.56)$$

the result already obtained as the **theorem of parallel axes**, cf. equation (1.37). Hence:

$$I_{Z'Z'} = bd^3/12 + bdh_1{}^2. \tag{1.57}$$

If the axis Z'Z' coincides with the lower edge QQ of the rectangle, then $h_1 = d/2$ and:

$$I_{QQ} = \frac{bd^3}{12} + \frac{bd^3}{4} = \frac{bd^3}{3} \tag{1.58}$$

$$I_{Z'Y'} = \int_{-b/2}^{+b/2}(z+h_2)\int_{-d/2}^{+d/2}(y+h_1)\,dy\,dz = \int_{-b/2}^{+b/2}\left[\frac{y^2}{2}+h_1 y\right]_{-d/2}^{+d/2}(z+h_2)\,dz$$

$$= h_1 d\left[\frac{z^2}{2}+h_2 z\right]_{-b/2}^{+b/2} = bdh_1 h_2. \tag{1.59}$$

This result could have been obtained directly from the theorem of parallel axes for product moments of area. Equation (1.45) was $I_{QQ} = I_{YZ}+Ah_1 h_2$. In this case $I_{QQ} = I_{Z'Y'}$, $A = bd$ and $I_{YZ} = 0$, by symmetry. Substitution of these values in equation (1.45) leads again to equation (1.59), as expected.

The various second and product moments of area for the rectangle just discussed enable the calculation of slightly more complicated sections such as those illustrated in Fig. 1.18. The centroid of each section is indicated by the letter G. In the case of the channel section of Fig. 1.18(b) the position of the centroid must be determined by taking first moments of area about any point in the section. The sections shown are all typical approximations to **rolled steel structural sections** which have been developed over the years for the various engineering industries. **Webs** and **flanges** are defined as illustrated. Their actual profiles are conditioned by requirements of the rolling process and their fatigue strength in use. Thus, wherever possible, there are no re-entrant sharp corners in the cross-section, but smooth rounded corners of large radius. Also, certain faces of the prismatic section have a tapered profile to enable the section to be rolled efficiently without sticking in the grooves. The actual typical profile of the industrial sections are shown with broken lines in Fig. 1.18 and it can be accepted that the sharp rectangular profiles of these approximating sections are sufficiently close to the real profiles to enable the second moment of area to be estimated reasonably well from those approximations.

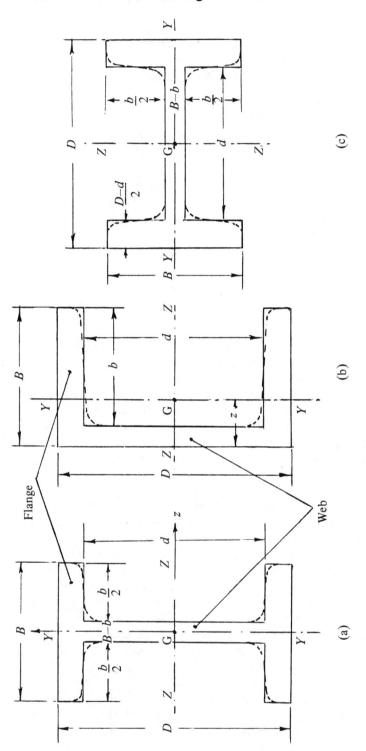

Fig. 1.18 Typical industrial structural sections

For the I-section beam of Fig. 1.18(a) and the channel of Fig. 1.18(b):-

$$I_{ZZ} = (BD^3 - bd^3)/12. \qquad (1.60)$$

If the I-section is turned through 90˙ as shown in Fig. 1.18(c), it is clear that:

$$I_{YY} = \{(D-d)B^3 + d(B-b)^3\}/12. \qquad (1.61)$$

For unsymmetrical sections such as the channel of Fig. 1.18(b) or the angle shown in Fig. 1.19, the theorems of parallel axes for both second and product moments of area provide a simple method for determining the required values. To simplify the algebra a typical unequal angle cross-section of specific dimensions has been shown in Fig. 1.19 and its approximating rectangular profile has been split into two simple rectangles shown by their diagonals which define their own centroids by their intersection and which can be used to find easily all the required properties of the section.

To find the position of the centroid G it is convenient to take first moments of the section about the long edges of the angle. The total area A is:

$$\begin{aligned}
A &= 8\times1+4\times1 &&= 12, \text{ thus:} \\
A\bar{z}' &= 12\bar{z}' &&= 8\times1\times\tfrac{1}{2}+4\times1\times3 = 4+12 = 16, \text{ hence:} \\
\bar{z}' &= 16/12 &&= 4/3 \text{ , as indicated. Similarly} \\
A\bar{y}' &= 12\bar{y}' &&= 8\times1\times4+4\times1\times\tfrac{1}{2} = 32+2 = 34 \text{ hence:} \\
\bar{y}' &= 34/12 &&= 17/6.
\end{aligned}$$

Using the theorems of parallel axes:

$$I_{z'z'} = \frac{1\times8^3}{12} + \frac{4\times1^3}{12} + 1\times8\times\left[1\tfrac{1}{6}\right]^2 + 4\times1\times\left[2\tfrac{1}{3}\right]^2 = 75\tfrac{2}{3}$$

$$I_{Y'Y'} = \frac{8\times1^3}{12} + \frac{1\times4^3}{12} + 8\times1\times\left[\tfrac{5}{6}\right]^2 + 1\times4\times\left[1\tfrac{2}{3}\right]^2 = 22\tfrac{2}{3}$$

$$I_{Z'Y'} = 1\times8\times\left[-\tfrac{5}{6}\right]\times1\tfrac{1}{6} + 1\times4\times\tfrac{5}{3}\times\left[-2\tfrac{1}{3}\right] = -23\tfrac{1}{3}$$

(note that the <u>sign</u> of the product moment depends on correctly designating the signs of the $z'-$ and $y'-$coordinates).

The Mohr's circle for this cross-section is shown on the stress plane of Fig. 1.19(b), with the Prager pole. Details of the construction have been indicated on the diagram. The maximum principal second moment of area $I_{11} = 84.48$. I_{11} is referred to the 11-axis which is inclined at an angle of 20.68˙ to the z'-axis shown. The minimum principal second moment of area is $I_{22} = 13.86$ about the 22-axis.

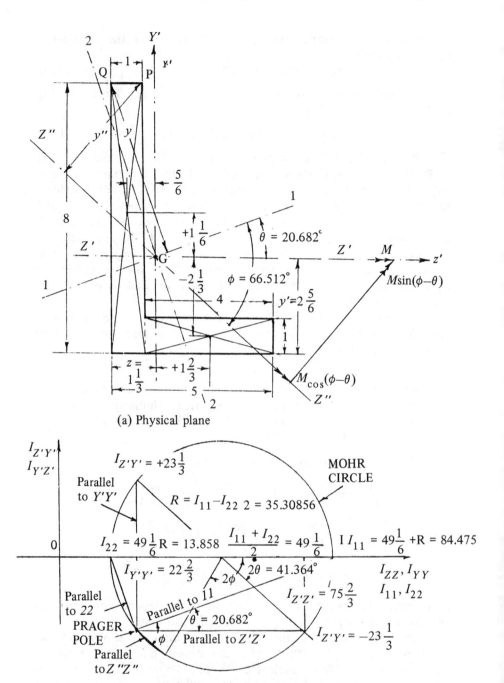

(a) Physical plane

(b) Second and product moment plane

Fig. 1.19 Typical angle section

The main purpose for determining these principal axes and second moments of area is in connection with the bending of such unsymmetrical sections, as will be discussed later. It may be noted here that our 'hands on' experience of the behaviour of a prismatic angular section of the type shown, in the bending mode, would lead us to expect that its weakest plane of bending would be in the mode which would have the 22–axis as the neutral axis. In practice the plane of bending is usually parallel to either the $z'-$ or $y'-$axis shown, so it is necessary to resolve the applied bending moments into their components in the principal directions, about axes 11 and 22.

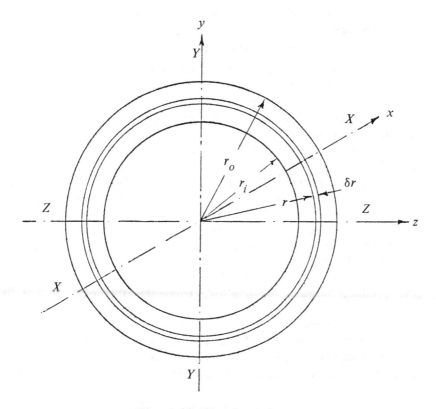

Fig. 1.20 Circular tube

One of the most efficient engineering cross–sections, both in bending and in torsion, is the **circular tube**, illustrated in Fig. 1.20. For both bending and torsion theories it is necessary to determine I_{XX}, I_{YY} and I_{ZZ}. The polar second moment of area I_{XX} (as shown in Fig. 1.20) is often denoted by the letter J.

To determine these three second moments of area it is simplest to make use of the theorem of perpendicular axes, i.e.: $$I_{XX} = I_{YY} + I_{ZZ} ,$$ (1.42)

noting that, in the case of all-round symmetry,

$$I_{YY} = I_{ZZ} = \tfrac{1}{2} I_{XX}, \text{ from equation (1.42).}$$ (1.62)

The polar second moment of area I_{XX} (or J), is most easily determined from the integral:

$$I_{XX} = J = \int_{r_i}^{r_0} 2\pi r^3 dr = \frac{\pi}{2}(r_0{}^4 - r_i{}^4) =$$
$$\pi(r_0{}^2 - r_i{}^2) \frac{r_0{}^2 + r_0{}^2}{2}$$ (1.63)

thus: $I_{XX} = J = Ak^2{}_{XX} = A(r_0{}^2 + r_i{}^2)/2$, since $A = \pi(r_0{}^2 - r_i{}^2)$, and therefore:

$$k_{XX} = \{(r_0{}^2 + r_i{}^2)/2\}^{\frac{1}{2}}$$ (1.64)

is the **polar radius of gyration** of the tubular cross-section.

Finally, for cross-sectional shapes which are very irregular and complicated, the second, product, and polar moments of area can only be determined numerically. A tabular method has already been described for the determination of the second moment of area about an axis through the centroid (Fig. 1.13 and Table 1.1).

A more general tabular method, suitable for the modern digital computer, is illustrated in Fig. 1.21 and Table 1.2. An arbitrary set of axes z' and y' with origin 0 is chosen and a table prepared as previously described (except that the breadth of a generic strip is here denoted by β instead of z). The second moment of area of any strip about the $Z'Z'$ axis is approximately $\beta y'^2 \delta y'$; about the $Y'Y'$ axis it is $(\beta^3/12 + \beta z'^2)\delta y'$; and the product moment is $\beta y'z'\delta y'$, from the theorems of parallel axes. Thus the tabular method indicated in Table 1.2 can be used to determine the position of the centroid G, by virtue of having found $\bar{z}$ and $\bar{y}$. Then the various theorems of parallel and perpendicular axes may be used to determine the second, product and polar moments as follows:

$$\left. \begin{array}{l} I_{ZZ} = I_{Z'Z'} - A\bar{y}^2 \\ I_{YY} = I_{Y'Y'} - A\bar{z}^2 \\ I_{ZY} = I_{Z'Y'} - A\bar{z}\bar{y} \\ I_{XX} = I_{YY} + I_{ZZ}. \end{array} \right\}$$ (1.65)

As previously mentioned, it is convenient to keep δy constant for all strips so that the data which has to be determined from the given shape of cross-section reduces to three values, namely β, y' and z'.

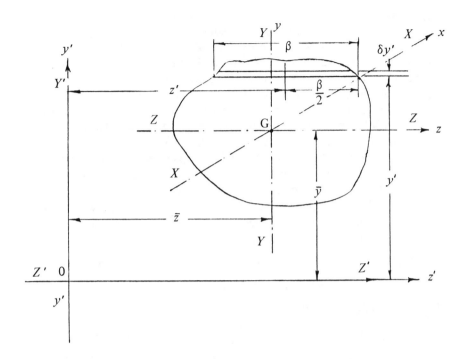

Fig. 1.21 Lamina − notation for Table 1.2

TABLE 1.2

INPUT DATA
(δy constant)

z'_1	y'_1	β_1	$\beta_1\delta y$	$\beta_1 y'_1 \delta y$	$\beta_1 z'_1 \delta y$	$\beta_1 y'^2_1 \delta y$	$\left[\dfrac{\beta_1{}^3}{12}+\beta_1 z'^2_1\right]\delta y$	$\beta_1 y'_1 z'_1{}^2\delta y$
---	---	--	----	-------	-------	--------	----------------	-----------
---	---	--	----	-------	-------	--------	----------------	-----------
z'_n	y'_n	β_n	$\beta_n\delta y$	$\beta_n y'_n \delta y$	$\beta_n z'_n \delta y$	$\beta_n y'^2_n \delta y$	$\left[\dfrac{\beta_n{}^3}{12}+\beta_n z'^2_n\right]\delta y$	$\beta_n y'_n z'_n{}^2\delta y$
Σ			A	$A\bar{y}$	$A\bar{z}$	$I_{Z'Z'}$	$I_{Y'Y'}$	$I_{Z'Y'}$

The tabular procedure is then as previously described but with the additional summation required to determine $\bar{z}$, $I_{Y'Y'}$, and $I_{Z'Y'}$ as indicated in Table 1.2. Again, it is best to have an odd number of values to be summed in each column, so that **Simpson's rule** can be used. A computer program written in Fortran follows, entitled Program 1.1, the notation used in it broadly corresponding with that given on Fig. 1.21. An example solution (Solution 1.1) is also given, for the equilateral triangle shown in Fig. 1.22. As shown in that Figure, the triangle has been split into 12 horizontal strips separated by 13 horizontal ordinates, β ranging from zero at the apex to 36 at the base of the triangle. This simple triangle has been chosen because accurate values of the various second moments can be calculated for comparison with the values computed in Solution 1.1. For the triangle shown:

$A = h(b_1+b_2)/2 = 6\times36 = 216; \quad I_{ZZ} = Ah^2/18 = 1728;$

$I_{YY} = A(b_1{}^2+b_1b_2+b_2{}^2)/18 = 11664;$

$I_{XX} = I_{YY}+I_{ZZ} = 13392;$

$I_{Z'Z'} = I_{ZZ}+A\bar{y}^2 = 23328; \qquad I_{Y'Y'} = I_{YY}+A\bar{z}^2 = 206064;$

$I_{X'X'} = I_{Z'Z'}+I_{Y'Y'} = 229392; \qquad I_{YZ} = 0 \text{ (by symmetry)};$

$I_{Y'Z'} = I_{YZ}+A\bar{y}\bar{z} = 64800.$

As can be seen from Solution 1.1, the computer program gives the correct values for all these parameters.

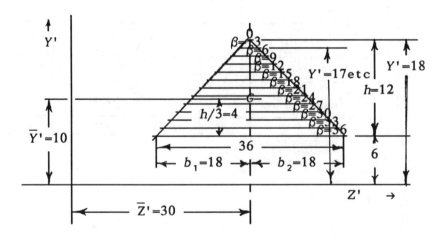

Fig. 1.22 Example lamina used for computer Solution 1.1

PROGRAM 1.1 Second moments of area

```
 1:*           PROGRAM SECMOM
 2:            DIMENSION BETA(100),YP(100),ZP(100),ALPHA(100)
 3:        &   ,GAMMA(100),DELTA(100),ZETA(100),ETA(100),THETA(100)
 4:            WRITE (*,1)
 5: 1          FORMAT(' "SECMOM" CALCULATES SECOND AND PRODUCT MOMENTS',/1X,
 6:        &   ' OF AREA (IYY,IZZ,IYZ AND IXX).Z IS HORIZONTAL,Y IS',/1X,
 7:        &   ' VERTICAL AXIS.ZDASH(=ZP IN PROGRAM)IS HORIZONTAL',/1X,
 8:        &   ' DISTANCE FROM YP-AXIS TO CENTRE OF STRIP,YDASH=YP IS',/1X,
 9:        &   ' VERTICAL DISTANCE FROM ZP-AXIS TO ORDINATE,N IS NUMBER',/1X
10:        &   ' OF ORDINATES DOWN THE LAMINA AND MUST BE ODD FOR',/1X,
11:        &   ' SIMPSONS RULE.BETA IS WIDTH OF HORIZONTAL STRIP, ',/1X,
12:        &   ' BREADTH DY.  INPUT N,THEN BETA,YP,ZP FOR EACH ORDINATE.'//)
13:            WRITE (*,2)
14: 2          FORMAT(' TYPE IN N ')
15:            READ  (*,*) N
16:            DO 6 I=1,N
17:            WRITE (*,4)I
18: 4          FORMAT(' TYPE IN BETA,YP,ZP FOR ORDINATE ',I4, )
19:            READ  (*,*) BETA(I),YP(I),ZP(I)
20: 6          CONTINUE
21:            WRITE (*,7)
22: 7          FORMAT(//14X,1HI,5X,7HBETA(I),5X,5HY'(I),7X,5HZ'(I)/)
23:            DO 9 I=1,N
24:            WRITE (*,8) I,BETA(I),YP(I),ZP(I)
25: 8          FORMAT(12X,I3,3(2X,E10.5))
26: 9          CONTINUE
27:            DY=ABS(YP(1)-YP(2))
28:            DO 10 I=1,N
29:            ALPHA(I)=BETA(I)*DY
30:            GAMMA(I)=ALPHA(I)*YP(I)
31:            DELTA(I)=ALPHA(I)*ZP(I)
32:            ZETA (I)=((BETA(I)**3)*DY/12.0+DELTA(I)*ZP(I))
33:            ETA  (I)=GAMMA(I)*YP(I)
34:            THETA(I)=GAMMA(I)*ZP(I)
35: 10         CONTINUE
36:            CALL SIMPSN (ALPHA,A,N)
37:            CALL SIMPSN (GAMMA,YB,N)
38:            YBAR=YB/A
39:            CALL SIMPSN (DELTA,ZB,N)
40:            ZBAR=ZB/A
41:            CALL SIMPSN (ZETA,AIYYP,N)
42:            CALL SIMPSN (ETA,AIZZP,N)
43:            CALL SIMPSN (THETA,AIYZP,N)
44:            AIYY =AIYYP-A*(ZBAR**2)
45:            AIZZ =AIZZP-A*(YBAR**2)
46:            AIYZ =AIYZP-A*YBAR*ZBAR
47:            AIXX =AIYY+AIZZ
48:            AIXXP=AIYYP+AIZZP
49:            WRITE (*,11)A,YBAR,ZBAR
50: 11         FORMAT(/5X,5HAREA=,E10.5,5X,5HYBAR=,E10.5,5X,5HZBAR=,E10.5)
51:            WRITE (*,12)AIYYP,AIZZP,AIYZP
52: 12         FORMAT(/5X,5HIYY'=,E10.5,5X,5HIZZ'=,E10.5,5X,5HIYZ'=,E10.5)
53:            WRITE (*,13)AIYY,AIZZ,AIYZ
54: 13         FORMAT(/5X,5HIYY =,E10.5,5X,5HIZZ =,E10.5,5X,5HIYZ =,E10.5)
55:            WRITE (*,14)AIXXP,AIXX
56: 14         FORMAT(/5X,5HIXX'=,E10.5,5X,5HIXX =,E10.5)
57:            STOP
58:            END
59:
60:            SUBROUTINE SIMPSN (P,Q,N)
61:            DIMENSION P(N)
62: C          USE SIMPSON'S RULE FOR INTEGRATION
63:            EVEN=0.0
64:            ODD =0.0
65:            DO 1 I=2,N-3,2
66:            EVEN=EVEN+P(I)
67:            ODD =ODD+P(I+1)
68: 1          CONTINUE
69:            Q=(1.0/3.0)*(P(1)+4.0*(EVEN+P(N-1))+2.0*ODD+P(N))
70:            RETURN
71:            END
```

SOLUTION 1.1

```
secmom
"SECMOM" CALCULATES SECOND AND PRODUCT MOMENTS
OF AREA (IYY,IZZ,IYZ AND IXX).Z IS HORIZONTAL,Y IS
VERTICAL AXIS.ZDASH(=ZP IN PROGRAM)IS HORIZONTAL
DISTANCE FROM YP-AXIS TO CENTRE OF STRIP,YDASH=YP IS
VERTICAL DISTANCE FROM ZP-AXIS TO ORDINATE,N IS NUMBER
OF ORDINATES DOWN THE LAMINA AND MUST BE ODD FOR
SIMPSONS RULE.BETA IS WIDTH OF HORIZONTAL STRIP,
BREADTH DY.  INPUT N,THEN BETA,YP,ZP FOR EACH ORDINATE.


TYPE IN N
13
TYPE IN BETA,YP,ZP FOR ORDINATE    1
   0.   18.   30.
TYPE IN BETA,YP,ZP FOR ORDINATE    2
   3.   17.   30.
TYPE IN BETA,YP,ZP FOR ORDINATE    3
   6.   16.   30.
TYPE IN BETA,YP,ZP FOR ORDINATE    4
   9.   15.   30.
TYPE IN BETA,YP,ZP FOR ORDINATE    5
  12.   14.   30.
TYPE IN BETA,YP,ZP FOR ORDINATE    6
  15.   13.   30.
TYPE IN BETA,YP,ZP FOR ORDINATE    7
  18.   12.   30.
TYPE IN BETA,YP,ZP FOR ORDINATE    8
  21.   11.   30.
TYPE IN BETA,YP,ZP FOR ORDINATE    9
  24.   10.   30.
TYPE IN BETA,YP,ZP FOR ORDINATE   10
  27.    9.   30.
TYPE IN BETA,YP,ZP FOR ORDINATE   11
  30.    8.   30.
TYPE IN BETA,YP,ZP FOR ORDINATE   12
  33.    7.   30.
TYPE IN BETA,YP,ZP FOR ORDINATE   13
  36.    6.   30.
```

I	BETA(I)	Y'(I)	Z'(I)
1	.00000E+00	.18000E+02	.30000E+02
2	.30000E+01	.17000E+02	.30000E+02
3	.60000E+01	.16000E+02	.30000E+02
4	.90000E+01	.15000E+02	.30000E+02
5	.12000E+02	.14000E+02	.30000E+02
6	.15000E+02	.13000E+02	.30000E+02
7	.18000E+02	.12000E+02	.30000E+02
8	.21000E+02	.11000E+02	.30000E+02
9	.24000E+02	.10000E+02	.30000E+02
10	.27000E+02	.90000E+01	.30000E+02
11	.30000E+02	.80000E+01	.30000E+02
12	.33000E+02	.70000E+01	.30000E+02
13	.36000E+02	.60000E+01	.30000E+02

```
    AREA=.21600E+03     YBAR=.10000E+02     ZBAR=.30000E+02

  IYY'=.20606E+06     IZZ'=.23328E+05     IYZ'=.64800E+05

  IYY =.11664E+05     IZZ =.17280E+04     IYZ =.00000E+00

  IXX'=.22939E+06     IXX =.13392E+05
Stop - Program terminated.
```

1.7 SIMPLE EXAMPLES, UNSYMMETRICAL BENDING

A **steel channel section** of the form shown in Fig. 1.18(b) is subjected to a pure bending moment so that the neutral axis of bending is ZZ. The dimensions of the beam are $D = 200$ mm, $B = 100$ mm, $d = 160$ mm and $b = 85$ mm. If the allow– able maximum stress is 150 MPa, determine the magnitude of the maximum bending moment which can be applied. Compare this with the maximum bending moment which could be applied if the channel were orientated so that the neutral axis of bending was YY. Also, compare the latter value with that for an I–section beam of the same dimensions, such as that shown in Fig. 1.18(c), being bent about axis YY.

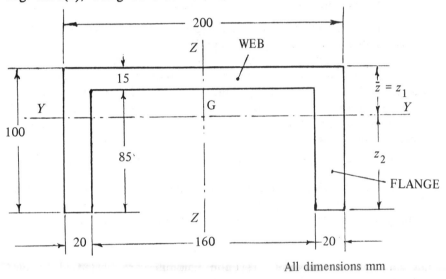

Fig. 1.23 Example channel section

For the channel section, from equation (1.60):

$$I_{ZZ} = (BD^3 - bd^3)/12 = 100 \times 200^3 - 85 \times 160^3/12 = 37.65 \times 10^6 \text{ mm}^4,$$

$$Z = I/y_{max} = 37.65 \times 10^6/100 = 37.65 \times 10^4 \text{ mm}^3,$$

$$M_{max} = \sigma_{max}Z = 150 \times 37.65 \times 10^4 \text{ Nmm} = 56.48 \times 10^6 \text{ Nmm, or}$$

<u>56.48 kNm.</u>

If the channel is rotated through 90° with respect to the plane of bending, so that YY is the neutral surface, as shown in Fig. 1.23, the following calculations apply:

Taking first moments of area about the top edge shown,

$$200 \times 100 \times 50 - 160 \times 85 \times (15 + 85/2) = (200 \times 100 - 160 \times 85)\bar{z}, \text{ hence:}$$

$\bar{z}$ =34.06 mm = z_1, 100-$\bar{z}$ = 65.94 mm = $-z_2$, $\bar{z}$-15=19.06mm.

Dividing the cross-section into rectangles, each with one edge along the YY axis, and remembering that the second moment of area of a rectangle about its edge is $bd^3/3$,

I_{YY} = $(40 \times 65.94^3 + 200 \times 34.06^3 - 160 \times 19.06^3)/3 \triangleq 6.088 \times 10^6$mm^4

Z_1 = I/z_1 = $6.088 \times 10^6/34.06 \triangleq 0.1787 \times 10^6$ mm^3 (= Z_{max})

Z_2 = I/z_2 = $6.088 \times 10^6/-65.94 \triangleq -0.0923 \times 10^6$ mm^3.

Hence: M = $\sigma_1 Z_1$ = $150 \times 0.1787 \times 10^6 \triangleq 26.81 \times 10^6$ Nmm

$$\triangleq \underline{26.8 \text{ kNm}}, \text{ if } \sigma_1 = 150 \text{ MPa}.$$

Alternatively, M = $\sigma_2 Z_2 \triangleq \underline{13.85 \text{ kNm}}$, if σ_2 = -150 MPa.

Therefore, this latter value represents the maximum bending moment which can actually be applied.

This example illustrates a danger of using the concept of two section moduli. It might be thought from the above calculations that a bending moment of 26.8 kNm could be applied to the beam without the stress exceeding ±150 MPa. If that bending moment is applied then the stress in the extreme lower fibres, where $z = z_2 = -65.938$ mm, will be:-

$$\sigma_2 = M/z_2 = 26.8/-0.0923 \triangleq -290 \text{ MPa},$$

which greatly exceeds the allowable stress of ±150 MPa. Perhaps it is better not to use the concept of section moduli at all but simply to use the basic equations of simple bending, viz.:

$$\frac{M}{I} = \frac{\sigma}{y} = \frac{E}{R}$$

at all times. However, section moduli are listed in many published tables of properties of structural steel sections. The important thing to remember is that it is the material furthest away from the neutral surface which is most heavily stressed.

If the web of the channel of Fig. 1.23 were at the centre of the section, thereby converting it into the geometry of the I-beam shown in Fig 1.18(c), the second moment of area would be given by equation (1.61). Inserting the appropriate values for this example,

I_{YY} = $[(200-160) \times 100^3 + 160(100-85)^3]/12 \triangleq 3.378 \times 10^6$ mm^4.

The maximum bending moment which could be applied to this section would be M_{max} = $\sigma_{max} I_{YY}/y_{max}$ = $150 \times 3.378 \times 10^6/50$ $\triangleq 10.14 \times 10^6$ Nmm, or 10.14 kNm. Thus the channel section is superior to the I-section when they are both bent in their weakest mode.

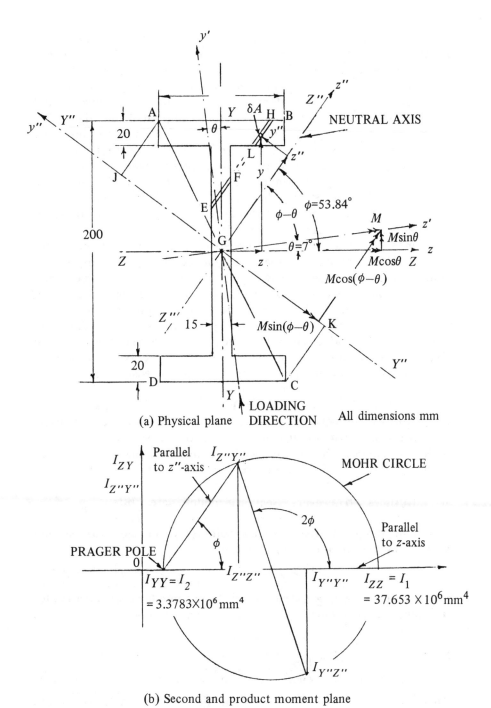

(a) Physical plane   LOADING DIRECTION   All dimensions mm

(b) Second and product moment plane

Fig. 1.24 Example I-beam section

It is interesting to investigate what happens to the stress distribution in the I–section beam, for example, if the neutral surface of bending does not coincide with the principal axis ZZ. The direction of loading and hence the **plane of bending** is assumed to be in the y'–direction through the centroid and acting at an angle θ to the y–axis, as indicated in Fig. 1.24. The double–headed arrow in the z'–direction may be imagined as a vector representing the applied bending moment M. The direction of the double–headed arrow may be taken as the vector representing a **hogging** bending moment which would introduce positive tensile stresses in the upper part of the beam shown. The dimensions of the web and flanges are the same as those of the channel and I–beam just considered.

Since the plane of the applied bending moment does not now coincide with one of the principal axes of the beam, it will have components causing bending about <u>both</u> principal axes and the beam will therefore also bend about its weakest axis (YY). Thus the problem resolves itself into that of summing the effects of the bending about the two principal axes. The component of bending moment $M\cos\theta$ will cause hogging about axis ZZ and $M\sin\theta$ will cause hogging about axis YY. At both corners A and B the component $M\cos\theta$ will introduce tensile stresses, but the component $M\sin\theta$ will introduce tensile stress at corner A and compressive stress at corner B. In determining the sense of these stresses the bending moment vectors have been drawn using the **right–hand corkscrew** rule. Using the basic equation $\sigma = My/I$ the stresses thereby induced on the elemental area δA in terms of the z and y axes are as follows, from each of the bending components:

$$_a\sigma_{xx} = \frac{My}{I_{ZZ}}\cos\theta \quad \text{and:} \quad _b\sigma_{xx} = -\frac{Mz}{I_{YY}}\sin\theta.$$

Thus the total stress at any point (z,y) in the cross–section is given by the equation:

$$\sigma = {}_a\sigma_{xx} + {}_b\sigma_{xx} = M\left[\frac{y}{I_{ZZ}}\cos\theta - \frac{z}{I_{YY}}\sin\theta\right] \quad (1.66)$$

and the stress will be zero at all those points with coordinates satisfying the equation:

$$\frac{y}{I_{ZZ}}\cos\theta = \frac{z}{I_{YY}}\sin\theta, \quad \text{or}$$

$$\frac{y}{z} = \frac{I_{ZZ}}{I_{YY}}\tan\theta = \tan\varphi, \quad (1.67)$$

which is the equation of a straight line at an angle φ to the z–axis, where:

$$\varphi = \tan^{-1}\left[\frac{I_{ZZ}}{I_{YY}}\tan\theta\right]. \quad (1.68)$$

This angle φ defines the neutral axis for unsymmetrical bending. It might have been thought at first sight that the neutral axis would have coincided with the Z'Z' axis about which the bending moment is applied but this is not the case. For the specific case under discussion, if $\theta = 7°$, for example, then:

$$\varphi = \tan^{-1}\left[\frac{37.65\times10^6}{3.378\times10^6}\tan 7°\right] \doteq 53.84°.$$

The neutral axis is indicated by Z"Z" in Fig. 1.24, along the z"-axis. Stresses are therefore constant along lines such as EFLH, parallel to the neutral axis Z"Z", and it is interesting to discover that the stress at any point along the strip EFLH is given by the simple basic equation $\sigma = My/I$ in which the bending moment M is replaced by $M\cos(\varphi-\theta)$, y by y'' and I by $I_{Z"Z"}$. Thus:

$$\sigma = M\left[\frac{y}{I_{ZZ}}\cos\theta - \frac{z}{I_{YY}}\sin\theta\right] = \frac{My''}{I_{Z"Z"}}\cos(\varphi-\theta). \qquad (1.69)$$

The proof of this is rarely given in text books and is as follows, using the geometry of the physical plane and second moment plane of Fig. 1.24.

$$y'' = y\cos\varphi - z\sin\varphi, \qquad (1.70)$$

$$I_{Z"Z"} = \frac{I_{ZZ}+I_{YY}}{2} + \frac{I_{ZZ}-I_{YY}}{2}\cos2\varphi = I_{ZZ}\cos^2\varphi + I_{YY}\sin^2\varphi.$$

From equation (1.69), therefore,

$$\frac{y''\cos(\theta-\varphi)}{I_{Z"Z"}} = \frac{(y\cos\varphi - z\sin\varphi)(\cos\varphi\cos\theta + \sin\varphi\sin\theta)}{I_{ZZ}\cos^2\varphi + I_{YY}\sin^2\varphi}$$

$$= \frac{(y - z\tan\varphi)(\cos\theta + \sin\theta\tan\varphi)}{I_{ZZ} + I_{YY}\tan^2\varphi}$$

$$= \frac{\left[y - z\dfrac{I_{ZZ}}{I_{YY}}\tan\theta\right]\left[\cos\theta + \dfrac{I_{ZZ}}{I_{YY}}\sin\theta\tan\theta\right]}{I_{ZZ} + \dfrac{I^2_{ZZ}}{I_{YY}}\tan^2\theta}$$

[using equation (1.67)]

$$= \frac{\left[y\cos\theta - z\dfrac{I_{ZZ}}{I_{YY}}\sin\theta\right]\left[1 + \dfrac{I_{ZZ}}{I_{YY}}\tan^2\theta\right]}{I_{ZZ}\left[1 + \dfrac{I_{ZZ}}{I_{YY}}\tan^2\theta\right]}.$$

Hence:

$$\frac{\sigma}{M} = \frac{y}{I_{ZZ}}\cos\theta - \frac{z}{I_{YY}}\sin\theta = \frac{y''}{I_{Z"Z"}}\cos(\varphi-\theta) \quad \text{Q.E.D.}$$

Considering the components of M, $M\cos\theta$ and $M\sin\theta$, about axes ZZ and YY, the total stress at area δA was found by summing the effects of these two components as:

$$\sigma = \frac{My}{I_{ZZ}}\cos\,\theta - \frac{Mz}{I_{YY}}\sin\,\theta.$$

It might therefore reasonably be supposed that the correct expression for the stress on δA considering the components of M, [namely: $M\cos(\varphi-\theta)$ and $M\sin(\varphi-\theta)$], about the axes Z"Z" and Y"Y", would be:

$$\sigma = \frac{My''}{I_{Z''Z''}}\cos(\varphi-\theta) + \underline{\frac{Mz''}{I_{Y''Y''}}\sin(\varphi-\theta)}.$$

This is <u>not</u> the case, however. The term underlined should <u>not</u> be included, as has just been proved analytically. The bending moment component $M\sin(\varphi-\theta)$ is produced by the asymmetrical stress distribution about the Y"Y" axis (in spite of the fact that stresses are constant along lines parallel to the neutral axis Z"Z"). This may be proved as follows. Referring again to Fig. 1.24, the bending moment produced by the stresses given by equation (1.69) about the Y"Y" axis is:

$$M_{Y''Y''} = \Sigma\sigma z''\delta A = \frac{M\Sigma y''z''\delta A}{I_{Z''Z''}}\cos(\varphi-\theta)$$

$$= M\frac{I_{Z''Y''}}{I_{Z''Z''}}\cos(\varphi-\theta)$$

$$= M\frac{(I_{ZZ}-I_{YY})\sin\,2\varphi}{(I_{ZZ}+I_{YY}) + (I_{ZZ}-I_{YY})\cos\,2\varphi}\cos(\varphi-\theta) =$$

$$M\frac{(I_{ZZ}-I_{YY})\tan\,\varphi}{I_{ZZ}+I_{YY}\tan^2\varphi}\cos(\varphi-\theta) = M\frac{\tan\,\varphi - \tan\,\theta}{1 + \tan\,\varphi\tan\,\theta}\cos(\varphi-\theta)$$

$$= M\tan(\varphi-\theta)\cos(\varphi-\theta) = M\sin(\varphi-\theta). \qquad Q.E.D.$$

For the given angular orientation θ of the applied bending moment M, the maximum stresses will occur at those extreme points of the cross-section defined by lines JA and KC drawn parallel to the neutral axis Z"Z". In the case of the symmetrical I-section shown in Fig. 1.24, the maximum stresses will occur at A and C, of equal and opposite magnitude (tensile at A, compressive at C). As the angle θ is increased, more and more of the applied bending moment M is apportioned to the weaker axis YY of the beam and, if the magnitude of M remains constant, the maximum stress will increase by a very great amount. It is of interest to specify a maximum allowable stress to which the beam may be subjected and investigate how the magnitude of the permissible bending moment decreases as θ is increased from 0 to 90°. The result is shown in Table 1.3, for a maximum stress of 150 MPa at A, where $z = -50$ mm and $y = +100$ mm.

TABLE 1.3

$\theta°$	0	5	10	15	20	40	60	75	79.83	85	90.00
$\varphi°$	0	44.28	63.03	71.49	76.15	83.90	87.03	88.62	89.08	89.55	90.00
M	56.48	38.11	28.93	23.45	19.85	12.99	10.60	10.01	9.976	10.02	10.14

It can be seen from this table that, contrary to what might be expected, for a constant value of σ_{max} the minimum value of M which can be applied does not occur for $\theta = 90°$ but for $\theta = 79.83°$. From equation (1.66):

$$\frac{\sigma}{M} = \frac{y\cos \theta}{I_{ZZ}} - \frac{z\sin \theta}{I_{YY}} \ .$$

For given values of z, y, I_{ZZ} and I_{YY}:

$$\frac{d}{d\theta}\left[\frac{\sigma}{M}\right] = -\frac{y}{I_{ZZ}}\sin \theta - \frac{z}{I_{YY}}\cos \theta .$$

For stationary σ/M this must be zero, at a value of:

$$\tan \theta = -\frac{I_{ZZ}}{I_{YY}}.\frac{z}{y} \ . \tag{1.71}$$

For the values given,

$$\theta = \tan^{-1}\left[-\frac{-50}{100}.\frac{37.65}{3.378}\right] \triangleq 79.83°$$

and thus: $\varphi \triangleq 89.08°$, for minimum M.

Hence this beam is very slightly stronger when the plane of the applied bending moment coincides with the minor principal axis YY than when it is being subjected to a bending moment in a plane at about $10°$ to that axis. This fact could be significant if such a beam is used as a strut and subject to buckling. In practice, beams of the simple I, channel, or angle cross-sections so far discussed are generally loaded in directions normal to their outer faces. Under such circumstances the angle cross-section shown in Fig 1.19, for example, would be subjected to unsymmetrical bending and it is sections of that type which are the most important and therefore most often dealt with in text books. Another standard unsymmetrical section is the Z-section, often encountered in structures. Considering the angle section illustrated in Fig. 1.19 as an example of unsymmetrical bending, it should be noted that the principal second moments of area

were denoted by I_{11} = 84.48 cm^4 (if the unit dimension is taken
as 1 cm) and I_{22} = 13.86 cm^4. The z'- and y'-axes in that
example were taken as the axes parallel to the sides of the
right-angled angle section because they are the axes about which
the bending moment is generally applied and therefore correspond with the z'- and y'-axes of the I-section beam just
discussed. In this case, therefore, θ = -20.68˚.

As for the I-section beam, the neutral surface will be at an
angle φ to the principal 11 and 22 axes, where:

$$\varphi = \tan^{-1}\frac{I_{11}}{I_{22}}\tan\theta = \tan^{-1}\frac{84.47}{13.86}\tan-20.682˚ = -66.51˚$$

as shown in Fig. 1.19(a). If a hogging moment of amount M is
applied about the Z'Z' axis then the maximum stress will be
tensile at the point P due to the component $M\cos(\varphi-\theta)$ about
the neutral axis Z'Z', given by the previously derived expression:

$$\sigma = \frac{My''}{I_{z''z''}}\cos(\varphi-\theta). \tag{1.69}$$

In this case, y'' = $y'\cos(\varphi-\theta)-z'\sin(\varphi-\theta)$, from equation
(1.70) with reference to the z'- and y'-axes, thus:

$$y'' = 5\frac{1}{6}\cos(-66.51˚+20.68˚) - \left[-\frac{1}{3}\right]\sin(-45.83˚) =$$

3.361 cm and:

$$I_{z''z''} = \frac{I_{11}+I_{22}}{2} + \frac{I_{11}-I_{22}}{2}\cos 2\varphi = 49\frac{1}{6} + 35.39\cos(-133˚)$$
$$\triangleq 25.08 \text{ cm}^4.$$

If an allowable stress of 150 MPa is again assumed, the
permissible bending moment in this case would be:

$$M = \frac{\sigma I_{z''z''}}{y''\cos(\varphi-\theta)} = \frac{150\times25.08\times10^4}{33.61\cos(-45.83˚)} \triangleq 1.606\times10^6 \text{ Nmm,}$$
$$\text{i.e. } M \triangleq 1.606 \text{ kNm.}$$

The maximum bending moment which this angle section
could accept would occur if the beam were bent about its major
principal axis 11, in which case the maximum stress would be
tensile at point Q. The magnitude of this bending moment, with
σ = 150 MPa is given by the equation:

$$M = \frac{\sigma I_{11}}{y}, \text{ where: } y = y'\cos\theta - z'\sin\theta,$$

from equation (1.70),

thus: $y = 5\frac{1}{6}\cos(-20.68˚)-\left[-1\frac{1}{3}\right]\sin(-20.68˚) = 5.305\text{cm}$
Hence:

$$M = \frac{150\times84.48\times10^4}{53.05} \triangleq 2.389\times10^6 \text{ Nmm or } \underline{2.389 \text{ kNm.}}$$

It should be mentioned here that all beams, particularly those having unsymmetrical cross-sections such as the angle under discussion, may not only bend but also <u>twist</u> due to induced shear stresses, unless the resultant line of action of the load passes through a certain point known as the <u>shear centre</u>. This effect will be discussed later.

1.8. COMPOSITE BEAMS, SIMPLE EXAMPLES

Composite beams are defined as beams comprising different materials. The best-known present day form of composite beam is the **reinforced concrete beam**. It is also well known that concrete is very strong in compression but has virtually no strength in tension (although modern cements are being developed which give the concrete improved tensile properties). In reinforced concrete, steel is cast into the concrete, wherever possible in those positions at which tensile stresses would otherwise exist. The steel is assumed to be firmly embedded and "stuck" to the concrete so that the reinforced concrete beam, for example, would bend so that **plane sections remain plane**. The steel often takes the form of reinforcing rods which have very rough or serrated surfaces to aid the transmission of the shear stresses necessary to achieve the condition of plane sections remaining plane. **Pre-stressing** of the rods is sometimes carried out; the concrete is either cast around bars which are stretched a certain amount or the bars are screwed at their ends and fitted with end plates and nuts so that the nuts can be tightened so as to give an initial pre-tensioning to the bars. This **pre-tensioning** introduces compression into the concrete beam thereby enhancing its strength against subsequent bending.

The earliest type of composite beam was the **flitch** beam in which a timber beam of rectangular cross-section was reinforced by nailing flat steel strips along certain of its faces. These were generally either on the two vertical or the two horizontal faces as shown in Figs. 1.25(a) and (b) but a central core of steel embraced by wooden beams as illustrated in Fig. 1.25(c) was also used, mainly to give the steel some protection against rusting. The typical reinforced concrete beam cross-section is shown in (d), although the steel can be in both the top and bottom sections of the beam. In general, for symmetrical composite beams of the type shown in Figs. 1.25(a), (b) and (c) and reinforced concrete beams with steel reinforcement symmetrically disposed about the central axis, the following simple analysis can be applied for simple bending.

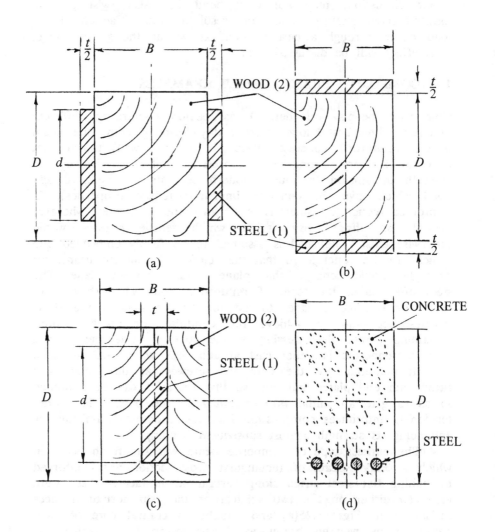

Fig. 1.25 Typical composite beam sections

If the two different materials are denoted by subscripts 1 and 2 as shown in Fig. 1.25, the bending moment which is carried by material 1 can be denoted M_1, that by material 2,

M_2. Then the total bending moment carried by the composite beam is $M = M_1 + M_2$. Also, the radius of curvature of the composite beam is approximately the same for both its elements, i.e. $R = R_1 = R_2$. From equations (1.33), $R = E_1 I_1 / M_1 = E_2 I_2 / M_2$ and hence the total moment of resistance of the composite beam must be:

$$M = M_1 + M_2 = E_1 I_1 / R + E_2 I_2 / R = (E_1 I_1 + E_2 I_2)/R. \quad (1.72)$$

Again, from the basic equations (1.33), $\sigma_1 / y_1 = E_1 / R_1$ and $\sigma_2 / y_2 = E_2 / R_2$, hence:

$$1/R = 1/R_1 = 1/R_2 = \sigma_1/(E_1 y_1) = \sigma_2/(E_2 y_2). \quad (1.73)$$

Combining equations (1.72) and (1.73),

$$M = \frac{\sigma_1}{y_1}\left[I_1 + \frac{E_2}{E_1}I_2\right] \text{ or } \frac{\sigma_2}{y_2}\left[I_2 + \frac{E_1}{E_2}I_1\right]. \quad (1.74)$$

If the stiffer material is denoted by the suffix 1, then the **modular ratio** is defined as:

$$m = E_1/E_2 \quad (1.75)$$

and equations (1.74) may be written as:

$$\frac{\sigma_1}{y_1} = \frac{M}{I_1 + I_2/m} \text{ or } \frac{\sigma_2}{y_2} = \frac{M}{I_2 + mI_1}. \quad (1.76)$$

Comparing equations (1.76) with equations (1.33), it is evident that the composite beam may be regarded either as a beam made entirely of the stiffer material, with its second moment of area augmented by a small fraction of that of the less stiff material (e.g. for steel and wood, $m = E_1/E_2 \cong 20$); or as being made entirely from the less stiff material with its second moment of area augmented by a large multiple of the stiffer material's second moment of area. The flitch beam shown in Fig. 1.25(b) may therefore be regarded in one of two ways, as illustrated in Figs. 1.26(b) and (c), since the second moment of area of a rectangular cross-section about its neutral axis is directly proportional to its breadth. The historical development of the steel I-section girder from this type of flitch beam is implicitly indicated by Fig. 1.26(b).

The analysis of the unsymmetrically reinforced concrete beam (the usual case, for reasons of economy) is more complicated, as will now be described. Normal concrete has a very high strength in compression but is unable to withstand virtually any tensile stress. Thus, in a reinforced concrete beam, the concrete is assumed to take all the compressive stress and the steel all the tension.

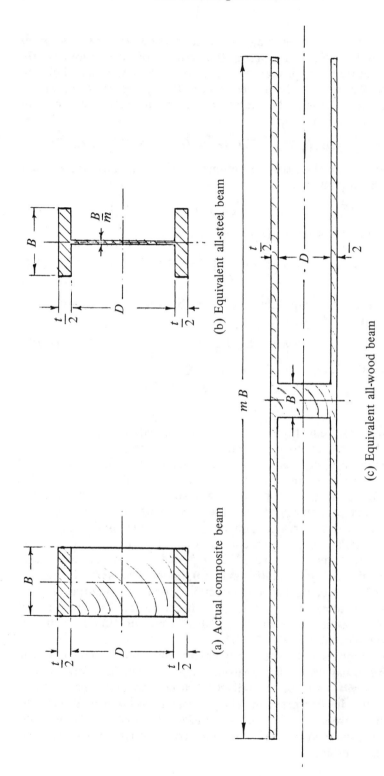

(a) Actual composite beam

(b) Equivalent all-steel beam

(c) Equivalent all-wood beam

Fig. 1.26 A composite beam and its equivalents

The typical cross-section of a **reinforced concrete beam** is shown again in Fig. 1.27(a) together with the linear distribution of strain assumed in the simple theory of bending in Fig.1.27(b). The resulting distributions of stress in the concrete and steel are shown in Fig. 1.27(c) and the tensile stresses induced in the concrete, shown by the broken lines, are usually neglected altogether. The neutral axis is denoted by NA and its position is determined as follows:

It can be seen from the geometry of the strain distribution that the maximum strain ϵ_c in the concrete is related to the strain in the steel, ϵ_s, by the equation:

$$\frac{\epsilon_c}{h} = -\frac{\epsilon_s}{d-h}. \qquad (1.77)$$

Thus, if E_c and E_s are the Young's moduli of the concrete and steel respectively, σ_c the maximum (compressive) stress in the concrete and σ_s the (tensile) stress in the steel,

$$\frac{\sigma_s}{\sigma_c} = \frac{E_s\,\epsilon_s}{E_c\,\epsilon_c} = -\frac{E_s}{E_c}\cdot\frac{d-h}{h} = -\frac{m(d-h)}{h} \qquad (1.78)$$

where $m = E_s/E_c$ is the **modular ratio**, generally about 16 for these materials.

Under a pure applied bending moment M, the resultant forces in the concrete and steel acting longitudinally on the upper and lower portion of the cross-section are $P_c = \frac{1}{2}\sigma_c Bh$ (compressive) and $P_s = \sigma_s A_s$ (tensile), where A_s is the total cross sectional area of the **steel reinforcement** and $\frac{1}{2}\sigma_c$ is the mean stress in the concrete. Thus, since $P_c = -P_s$ for equilibrium:–

$$\tfrac{1}{2}\sigma_c Bh = -\sigma_s A_s. \qquad (1.79)$$

The applied bending moment M must then be equal to the couple generated by these two equal and opposite forces and, since P_c acts at a distance of $(2/3)h$ from the neutral axis:

$$M = P_c(2h/3 + d - h) = P_c(d - h/3), \text{ hence:}$$

$$M = -\sigma_s A_s(d - h/3) = \tfrac{1}{2}\,\sigma_c Bh(d - h/3). \qquad (1.80)$$

To be consistent with the convention of sagging moments being negative, the sign of M will be negative for loading applied to the top surface of the beam illustrated, which would produce compression in the concrete and tension in the steel, as required. In general, the values of σ_s and σ_c which are permissible can be specified, in which case the value of h can be found from equation (1.78), since the modular ratio will be known. (Note that σ_s/σ_c will always be <u>negative</u>, by convention.)

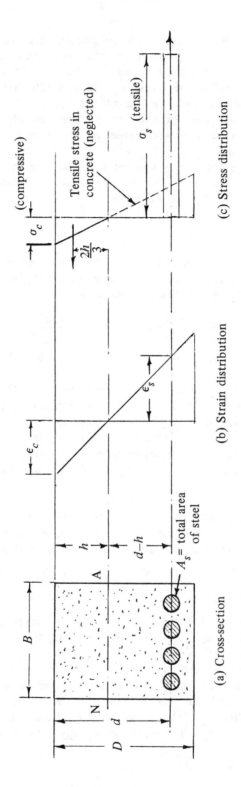

Fig. 1.27 Typical reinforced concrete beam section

The appropriate algebra gives, in fact,

$$h = \frac{d}{1 - (1/m)(\sigma_s/\sigma_c)}. \qquad (1.78a)$$

Then, the total area of steel required (A_s) can be found from equation (1.79) and the resisting moment M from equation (1.80). This section is known as the **economic section**, since the maximum allowable stresses in both steel and concrete are realized.

On the other hand, if the dimensions of the beam and the modular ratio are given, as is often the case, the ratio σ_s/σ_c can be eliminated between equations (1.78) and (1.79) to give a quadratic equation for h, with the solution:

$$h = \alpha\{\sqrt{(1 + 2d/\alpha)} - 1\}, \text{ where } \alpha = mA_s/B. \qquad (1.81)$$

Considering now some simple examples of composite beams, first consider the three flitched beams of wood and steel having the geometries previously illustrated in Fig. 1.25(a), (b), and (c), with dimensions of the wood $B = 150$ mm, $D = 200$ mm and of the steel $t = 12$ mm and $d = 150$ mm. Assume that Young's modulus for the steel is $E_1 = 200$ GPa, for the wood $E_2 = 10$ GPa and that the maximum allowable stresses are $\sigma_1 = 120$ MPa in the steel and $\sigma_2 = 30$ MPa in the wood. Compare the maximum bending moments which can be accepted by the three beams and consider how the steel should be attached to the wood in each case.

Referring to the three cases as (a), (b), and (c), (of Fig. 1.25), the following calculations apply, with $m = E_1/E_2 = 200/10 = 20$, for all cases:

(a) $I_1 = td^3/12 = 12\times150^3/12 = 3.375\times10^6$ mm^4.
$I_2 = BD^3/12 = 150\times200^3/12 = 100\times10^6$ mm^4.

From equations (1.74):

$$M = \frac{\sigma_1}{y_1}\left[I_1 + \frac{I_2}{m}\right] = \frac{\sigma_1}{75}\left[3\cdot375 + \frac{100}{20}\right]\times10^6$$

$$= 0.1117\times10^6\sigma_1 \text{ Nmm or:}$$

$$M = \frac{\sigma_2}{y_2}(I_2 + mI_1) = \frac{\sigma_2}{100}(100 + 20\times3.375)\times10^6$$

$$= 1.675\times10^6 \sigma_2 \text{ Nmm.}$$

Thus $\sigma_1/\sigma_2 = 1.675/0.1117 = 15$, for the chosen geometry and modular ratio. Since the maximum permissible stress in the steel is 120 N/mm^2, the accompanying maximum stress in the wood will be only 120/15 = 8 N/mm^2, which is considerably less than could be accepted by the wood. The total moment of resistance of this section is therefore:

$M = 0.1117 \times 120 = 1 \cdot 675 \times 8 = \underline{13.4 \text{ kNm.}}$

The contribution from the steel is $M_1 = \sigma_1 I_1 / y_1 = 120 \times 3.375/75 = 5.4$ kNm and from the wood $M_2 = \sigma_2 I_2 / y_2 = 8 \times 100/100 = 8$ kNm, as can be seen from equations (1.72) and (1.73). For the steel to be able to contribute this resisting moment of 5.4 kNm it must bolted or otherwise attached to the wooden beam so that it will be bent to the same curvature as the wood and stretched to give the same strain at the interface. The size and number of bolts required can readily be calculated to provide transmission of this moment.

(b) $I_1 = (B/12)\{(D + t)^3 - D^3\} = (150/12)(212^3 - 200^3) = 19.1 \times 10^6 \text{ mm}^4$, $y_1 = 106$ mm.
$I_2 = BD^3/12 = 100 \times 10^6 \text{ mm}^4$, $y_2 = 100$ mm. Thus:
$M = (\sigma_1/106)(19.1 + 100/20) \times 10^6 = 0.2274 \times 10^6 \sigma_1$ Nmm,
$= (\sigma_2/100)(100 + 20 \times 19.1) \times 10^6 = 4.82 \times 10^6 \sigma_2$ Nmm.

Hence $\sigma_1/\sigma_2 = 4.82/0.2274 \triangleq 21.2$ for this geometry and modular ratio. Again, the steel is the limiting factor and the moment of resistance of this section is much higher in this case, namely:

$M = 0.2274 \times 120 = \underline{27 \cdot 29 \text{ kNm}}$.

The maximum stress in the wood is thus $27 \cdot 29 \div 4.82 \triangleq 5 \cdot 66$ N/mm^2 for this case. Evidently this is a much more efficient way in which to employ the steel. However, it should be noted that the steel must be securely attached to the wood. If the top surface of the composite beam shown in Fig. 1.25(b) is in tension, for example, the whole of the top steel plate is in tension, its lower surface suffering the same tensile strain as the upper surface of the wood. Knowing the mean tensile strain in the steel plate it is a simple matter to calculate the tensile force which has to be transmitted to stretch the top steel plate (or compress the lower steel plate) and hence design the method of attachment (bolts, nails, or adhesive) required to form the composite beam.

(c) $I_1 = td^3/12 = 3 \cdot 375 \times 10^6 \text{ mm}^4$, $I_2 = (BD^3 - td^3)/12$
$= (100 - 3 \cdot 375) \times 10^6 = 96 \cdot 63 \times 10^6 \text{ mm}^4$.
$y_1 = 75$ mm, $y_2 = 100$ mm. Thus:
$M = (\sigma_1/75)(3.375 + 96.63/20) \times 10^6 = 0.1094 \times 10^6 \sigma_1$ Nmm,

or

$M = (\sigma_2/100)(96.63 + 20 \times 3.375) \times 10^6 = 1.641 \times 10^6 \sigma_2$ Nmm,
Hence $\sigma_1/\sigma_2 = 1.641/0.1094 = 15$ for this geometry and

modular ratio.

Thus, for this case, $M = 0.1094 \times 120 = 13.13$ kNm and the maximum stress in the wood is $13.13 \div 1.6413 = 8$ N/mm². This arrangement is less efficient than either (a) or (b), but the steel strip is obviously effectively constrained to bend to the same radius of curvature as the wood.

As an example of a reinforced concrete beam of the geometry previously illustrated in Fig. 1.25(d) and Fig. 1.27, assume the dimensions of the concrete are $B = 150$ mm and $D = 200$ mm and that the diameter of the round steel reinforcing rods will be about 20 mm, so that they will have to be placed at a depth d in the concrete of not more than 160 mm, to ensure that they are adequately supported. What then will be the economic section and the moment of resistance of the composite beam? The Young's modulus of the steel is $E_s = 208$ kN/mm² and that of the concrete $E_c = 13$ kN/mm²; the maximum allowable stress in the steel is $\sigma_s = 120$ N/mm² and in the concrete -20 N/mm². Thus $m = 16$ and $\sigma_s/\sigma_c = -6$.

From equation (1.78), $h = d/\{1 - (1/m)(\sigma_s/\sigma_c)\} = 160 \div (1 + 6/16)$

$= 116.4$ mm.

From equation (1.79), the total area of steel required is $A_s = -\frac{1}{2}Bh(\sigma_c/\sigma_s) = 150 \times 116.4/(2 \times 6) = 1455$ mm².

If there are 4 steel rods, each will have an area of $1455/4 = 363.6$ mm², and diameter 21.52 mm.

From equation (1.80), moment of resistance of the beam is
$M = -120 \times 1455(160 - 116.4/3) = 21.16$ kNm.

Alternatively, starting with the given dimensions of the beam and the modular ratio of $m = 16$, with the total area of steel $A_s = 1455$ mm², then the value of h would be given by equation (1.81), with $\alpha = mA_s/B = 16 \times 1455/150 = 155.2$, as:

$h = \alpha\{\sqrt{(1 + 2d/\alpha)} - 1\} =$
$155.2\{\sqrt{(1 + 320/155.2)} - 1\} = 116.4$ mm, as
before, with $M = 21.16$ kNm.

Quite often, generally for practical reasons, it is more convenient to have the reinforcement in both the tensile and compressive portions of a reinforced concrete beam. In that case, it is assumed that the steel reinforcement which is in compression does contribute to the moment of resistance whereas the concrete in tension makes no contribution. A linear distribution of strain across the section is assumed and the position of the neutral axis and the moment of resistance calculated using the methods already described.

1.9   COMBINED BENDING AND DIRECT STRESSES

The general case of a short column subjected to an axial load W which is **eccentric** to both principal $y-$ and $z-$axes of the cross–section of the column is illustrated in Fig. 1.28. The line of action of the load W is parallel with the $x-$axis. It is evident from Fig. 1.28 that the stresses induced on the elemental area δA will be all compressive, comprising a direct stress W/A and bending stresses Wby/I_{ZZ} and Waz/I_{YY}. Thus, the total stress at any point in the cross–section will be given by the expression (tensile stress positive)

$$\sigma = -W(1/A + by/I_{ZZ} + az/I_{YY}). \qquad (1.82)$$

The stress is zero along the neutral axis, which is therefore defined by the equation:

$$1/A + by/I_{ZZ} + az/I_{YY} = 0. \qquad (1.83)$$

This is an equation in y and z which represents a straight line, for given values of a and b. In the case of a rectangular cross–section of the type shown in Fig. 1.28, $A = BD$, $I_{ZZ} = BD^3/12$ and $I_{YY} = DB^3/12$. Hence equation (1.83) becomes:

$$1 + 12by/D^2 + 12az/B^2 = 0. \qquad (1.84)$$

When $y = 0$, $z = -B^2/12a$ and when $z = 0$, $y = -D^2/12b$. In Fig. 1.28 the neutral axis NA shown has been drawn for the load W with $a = B/6$ and $b = D/6$, so that equation (1.84) becomes $z/B + y/D = -\frac{1}{2}$ with intercepts on the axes of $z = -B/2$ and $y = -D/2$. The area shown shaded is clearly subjected to tensile stress. If $a = B/12$ and $b = D/12$, equation (1.84) becomes $z/B + y/D = -1$, and the neutral axis is the straight line N´A´ so that the whole section is subjected to compression only.

Short columns are often constructed from materials such as cast iron or concrete which have little or no tensile strength and it is desirable that no tensile stresses be induced in them. It is possible to define a **central core** or **kernel** for any given cross–section within which the line of action of the load must fall to ensure that no tensile stresses occur. For the rectangular section shown, with a and b both positive, the point at which tension is most likely to occur is $(z,y) = (-B/2, -D/2)$ at the lower left–hand corner of the cross–section as drawn. Substituting these values into equation (1.83) for zero stress gives:

$$a/b + b/D - 1/6 = 0 \qquad (1.85)$$

which represents a straight line for the variables a and b with intercepts on the $z-$ and $y-$axes of $B/6$ and $D/6$ respectively, as shown by the broken line in the positive quadrant. Repeating the analysis for the other corners of the rectangle leads to the conclusion that the line of action of the load must lie within the central diamond shaped core shown at the centre of the figure. This is often termed the **middle third rule for rectangular sections** (but note the diamond geometry).

A similar analysis applies to solid circular cross–sections, as illustrated in Fig. 1.29. In that case, if R is the radius of the circle, $A = \pi R^2$, $I_{ZZ} = I_{YY} = \pi R^4/4$ and substitution of these values into equation (1.82) gives the stress at any point in the cross–section as:

$$\sigma = -\frac{4W}{\pi R^2}\left[\frac{1}{4} + \frac{by}{R^2} + \frac{az}{R^2}\right]. \qquad (1.86)$$

Since a circular cross–section has all–round symmetry, the analysis may be simplified by choosing the $z-$axis to pass through the line of action of the load W so that $b = 0$ and the stress is a function of z only. In that case equation (1.86) becomes:

$$\sigma = -\frac{4W}{\pi R^2}\left[\frac{1}{4} + \frac{az}{R^2}\right] \qquad (1.86a)$$

from which it is readily determined that the stress is zero when:

$$z = -\frac{R^2}{4a}. \qquad (1.87)$$

The straight line defined by this value of z defines the neutral axis NA as indicated in Fig. 1.29 for the case of $a = R/2$, in which case $z = -R/2$ defines NA.

To ensure that the neutral axis lies outside the cross–section requires the stress to be zero at values of z numerically greater than $-R$. From equation (1.87), therefore, it is clear that when $z = -R$, a must be equal to $R/4$. Since the circle has all–round symmetry the central core within which the line of action of the load W must fall to ensure compressive stress only has a radius of $R/4$. This is known as the **middle quarter rule for circular sections** as shown shaded in Fig. 1.29.

For a hollow circular cross–section of external radius R and internal radius r the stress along any line parallel to the $y-$axis is given by a similar analysis, since $A = \pi(R^2 - r^2)$ and $I_{ZZ} = I_{YY} = \pi(R^4 - r^4)/4$, as:

$$\sigma = -\frac{4W}{\pi(R^2 - r^2)}\left[\frac{1}{4} + \frac{az}{R^2 + r^2}\right]. \qquad (1.86b)$$

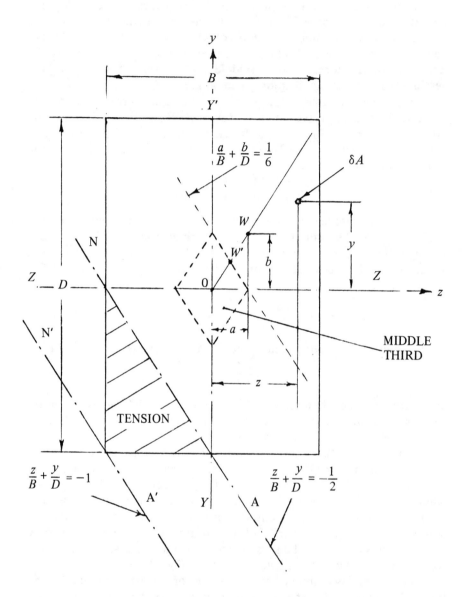

Fig. 1.28 Rectangular cross-section of a column

The neutral axis is the line:

$$z = -\frac{R^2 + r^2}{4a}. \qquad (1.87a)$$

and the central kernel is defined by a radius of:

$$a_k = \frac{R^2 + r^2}{4R}. \qquad (1.88)$$

Equation (1.86b) reduces to equation (1.86a), equation (1.87a) reduces to equation (1.87), and equation (1.88) gives a value of $a_k = R/4$ when $r = 0$, as required. For a very thin-walled cylindrical column, $r \to R$ and $a_k \to R/2$ which could properly be termed **the middle half rule for thin walled circular columns.**

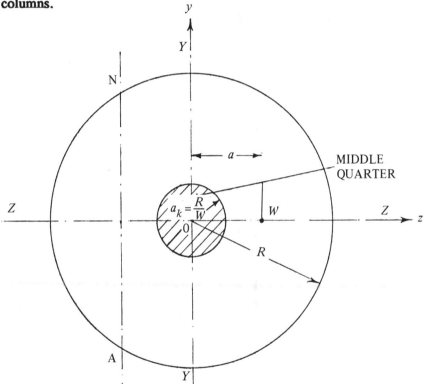

Fig. 1.29 Solid circular cross—section of a column

The core or kernel of more complicated cross—sectional shapes must be determined from equation (1.83) by substituting the extreme values of y and z for the boundaries of the section and establishing the corresponding equation between a and b which gives zero stress at any particular boundary. In the case of a **symmetrical I—section column,** as illustrated diagrammatically in Fig. 1.30, the situation is similar to that for the simple rectangular section just described.

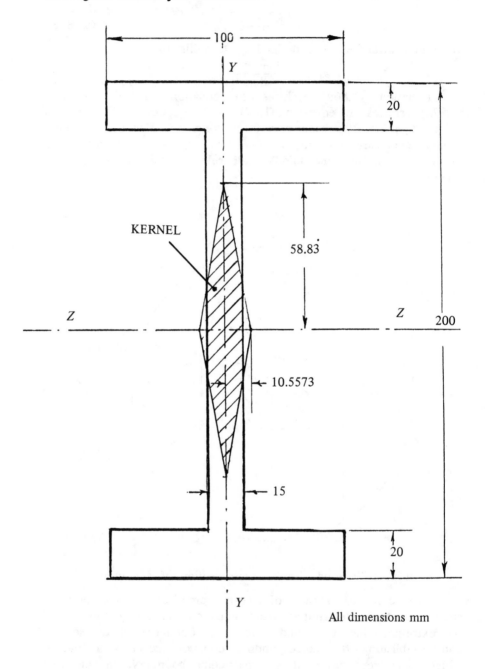

Fig. 1.30 Diagrammatical cross–section of an I–beam column

If B and D are the overall breadth and depth respectively of the I–section then the required equation between a and b within the positive $z - y$ quadrant will be given by substituting $z = -B/2$ and $y = -D/2$ into equation (1.83), to give the equations:

$$\frac{1}{A} - \frac{aB}{2I_{YY}} - \frac{bD}{2I_{ZZ}} = 0, \text{ or:} \tag{1.89}$$

$$\frac{aB}{k^2_{YY}} + \frac{bD}{k^2_{ZZ}} = 2, \tag{1.90}$$

where k_{YY} and k_{ZZ} are the radii of gyration about the principal axes.

As an example, the I–section of Fig. 1.24 has been reproduced again in Fig. 1.30 with the central core of rhombic or diamond shape indicated thereon. The required dimensions of the I–section are $B = 100$ mm, $D = 200$ mm, $A = 6400$ mm^2, $I_{ZZ} = 37.653 \times 10^6$ mm^4, $I_{YY} = 3.3783 \times 10^6$ mm^4, $k^2_{ZZ} = 5883.3$ mm^2, $k^2_{YY} = 527.86$ mm^2. From equation (1.86), when $b = 0$, $a = 2k^2_{YY}/B = 10.557$ mm and when $a = 0$, $b = 2k^2_{ZZ}/D = 58.833$ mm, as indicated.

1.10 ANTICLASTIC CURVATURE, BENDING INTO THE PLASTIC RANGE, RESIDUAL STRESSES, SIMPLE EXAMPLES

Within the elastic range the application of a pure bending moment to any beam leads to linear distributions of both normal strain and stress, under the assumptions made in the simple theory of bending, both strain and stress being zero at the neutral surface. The maximum strains and stresses occur at the extreme fibres of the beam, one side being tensile, the other compressive. If the beam is narrow, as illustrated diagram–matically in Fig. 1.31, and a pure hogging bending moment is applied as shown, the top fibres of the beam will effectively be under pure tensile stress and will therefore be subjected to lateral contraction, due to the **Poisson effect**. Since there is a linear increase of stress from the neutral surface to the outer fibres the whole beam will be subjected to this effect and a **transverse curvature** will take place. Bending a rubber eraser of rectangular cross–section demonstrates the effect spectacularly. This transverse curvature, known as the **anticlastic curvature,** is indicated with radius ρ in Fig. 1.31. For any layer distant y from the neutral surface the longitudinal strain is $\epsilon_x = y/R$, so that the transverse strain is $\epsilon_z = -\nu\epsilon_x = -y/\rho$. Thus:

$$\frac{\epsilon_x}{y} = \frac{1}{R} = \frac{1}{\nu\rho} \ , \ \text{ or } \ \ \rho = \frac{R}{\nu} \tag{1.91}$$

where R is the radius of curvature of bending and ν is Poisson's ratio. Since ν is generally less than 0.5, ρ is generally greater than twice R.

If the beam is very wide it will become a **plate**. **Anticlastic curvature** will be suppressed due to **membrane stresses** which will be induced in the edges of the plate and the theory of plates must be used, which involves taking into account transverse bending moments, as developed and described by Timoshenko.

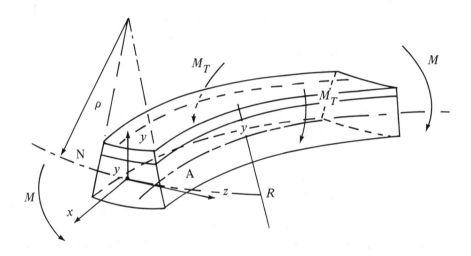

Fig. 1.31 Illustrating anticlastic curvature

If anticlastic curvature is suppressed the wide beam will be bent into a cylindrical surface by the pure bending moment and the transverse strain will be zero everywhere. Thus bending will occur under **plane strain conditions** and transverse bending moment will be induced. It is unnecessary to refer to the **theory of plates** to realize that the stresses induced are in this case, from equations (3.30) and (3.31) of the companion volume with σ_1 and ϵ_1 the stress and strain in any fibre in the bending direction, σ_2 the transverse stress and $\epsilon_2 = 0$,

$$\sigma_1 = E\epsilon_1/(1 - \nu^2) \tag{1.92}$$

$$\sigma_2 = E\nu\epsilon_1/(1 - \nu^2) = \nu\sigma_1. \qquad (1.93)$$

Thus, in order to suppress anticlastic curvature and produce a **cylindrical neutral surface,** the beam must be wide enough to generate a **transverse bending moment** $M_T = \nu M$. It can be seen from Fig. 1.31 that M_T would have to be a bending moment of the same sign, i.e. so as to cause hogging as drawn in the figure. Equation (1.92) also indicates that, for a given strain (or curvature of the beam), a higher stress (or bending moment M per unit width) must be applied to a very wide beam than if it were narrow, because Young's Modulus E is replaced by the higher effective modulus $E/(1 - \nu^2)$.

Beams in the context of this chapter are narrow and will exhibit anticlastic curvature. As the pure bending moment is increased the distributions of strain and stress will remain linear across the beam until the stresses at one or both extreme fibres attain the yield stress of the material, at which point the material will begin to deform plastically at the outer fibres. In pure bending the deformation is strain controlled and the beam will continue to bend in such a way that the curvature is constant along its length. It is reasonable, therefore, to assume that the distribution of <u>strain</u> will remain linear across the depth of the beam even if the distribution of stress does not. In fact, the distribution of stress will simply be a raised image of the stress–strain curve in uniaxial tension in the tensile part of the (narrow) beam and of the uniaxial compressive stress–strain curve in the compressive part, with the assumption of a linear strain distribution.

On this basis it is possible to obtain a good estimate of the real behaviour of narrow beams bent into the plastic range and a similar analysis can be used for wide beams also. For most engineering materials the stress–strain curve in compression is a mirror image of that in tension, at least for reasonably small strains. The simplest stress–strain curve is that of an **elastic–perfectly plastic material,** in which the flow stress (true stress) remains constant after yielding has occurred and the flow stresses in tension and compression are equal and opposite.

With these assumptions, as the bending moment is increased from M_1 to M_4, say, the distributions of strain and stress across a beam of simple rectangular section will be as sketched in Fig. 1.32, since the neutral axis will remain at the centre of the cross–section. In general, for <u>any</u> shape of cross–section, the distribution of stress across the beam will be as illustrated in Fig. 1.33. In both these figures Y is the yield or flow stress of the material. The forces acting on any cross–section are also shown in Fig. 1.33 and, for equilibrium,

$$F_1 + F_2 = F_3 + F_4 \tag{1.94}$$

where F_1, F_2, F_3 and F_4 are the forces acting on those portions of the cross–section associated with the stress distributions, which are either linear or constant as shown, and must be found numerically.

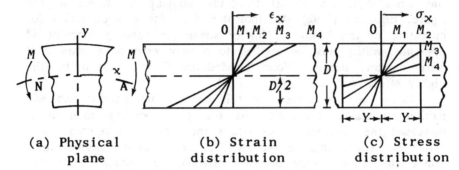

(a) Physical   (b) Strain   (c) Stress
  plane   distribution   distribution

Fig. 1.32 Stress and strain distributions across an
elastic–perfectly plastic beam

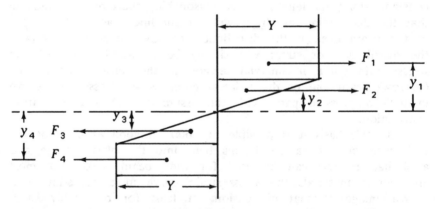

Fig. 1.33 Stress distribution across elastic–perfectly plastic beam
of any cross–section

The position of the neutral surface is then determined by trial, using equation (1.94). The moment of resistance is:

$$M = F_1 y_1 + F_2 y_2 + F_3 y_3 + F_4 y_4 \tag{1.95}$$

where y_1, y_2, y_3, and y_4 are the distances of the centres of gravity of the forces F_1, F_2, F_3, and F_4 from the neutral axis and must again be found numerically.

The stress distribution shown in Fig. 1.34 is for a simple rectangular section of dimensions $B{\times}D$, where $a/2$ is the value of y at the **plastic–elastic interface**, from which it can be seen that

$F_1 = F_4 = B(D - a)Y/2$, $F_2 = F_3 = BaY/4$ and that $y_1 = y_4$ $= (D + a)/4$, $y_2 = y_3 = a/3$. Hence the **moment of resistance** is, from equation (1.95):

$$M = YB(3D^2 - a^2)/12. \qquad (1.96)$$

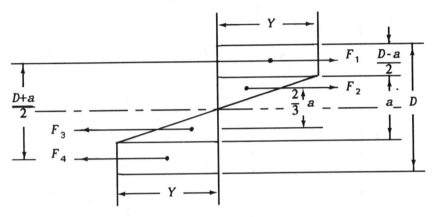

Fig. 1.34 Stress distribution across elastic–perfectly plastic beam of rectangular cross–section

At **initial yielding** of the outer fibres of the beam, $a=D$ and the moment of resistance is:

$$M = M_Y = YB(2D^2)/12 = YBD^2/6 \qquad (1.97)$$

as can readily be checked from the basic elastic equation:

$$M/I = \sigma/y, \text{ with } I = BD^3/12, \ y = D/2 \text{ and } \sigma = Y.$$

As bending of the beam is continued, more and more fibres yield, the **plastic zone** spreads inwards and a decreases until only a **vestigial elastic layer** remains at the neutral surface. In the limit, the **fully plastic bending moment**, M_P for this beam of rectangular cross–section of **elastic–perfectly plastic material** is given by equation (1.96) with $a = 0$, as:

$$M_P = YBD^2/4. \qquad (1.98)$$

The basic equations of elasticity will continue to apply to the **central elastic core** as a decreases from D to zero so that the curvature $1/R$ of the beam is related to the inverse of a by the equation:

$$1/R = 2Y/(Ea), \qquad (1.99)$$

since, when $y = a/2$, $\sigma = Y$. The variation of bending moment with curvature is illustrated in Fig. 1.35, where initial yielding occurs at a curvature given by equation (1.99) with $a = D$, as:

$$1/R = 2Y/(ED), \qquad (1.100)$$

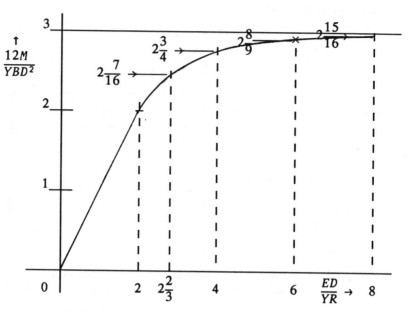

Fig 1.35 Moment–curvature relationship for an
elastic–perfectly plastic beam

The graph is drawn for specific values of $12m/(YB)$ versus $ED/(YR)$, corresponding with values of a of D, $\frac{3}{4}D$, $\frac{1}{2}D$, $\frac{1}{4}D$, and 0.

The ratio M_P/M_Y is called the **shape factor** S, i.e.

$$S = M_P/M_Y. \qquad (1.101)$$

Thus $S = 1.5$ for a rectangular cross–section.

The elastic–plastic analysis for more complicated shapes of cross–section, even for the elastic–perfectly plastic material assumed here, is much more difficult. However, it is usually only the fully plastic moment M_P which is of interest and this is easily determined, even for complicated shapes, as pointed out by Johnson and Mellor. A general lamina is shown in Fig. 1.36(a) with the fully plastic distribution of stress illustrated in Fig. 1.36(b). The total load on the top (tensile) part of the section above the neutral axis NA is:

$$\int_0^{y_1} Y\beta dy = Y\int_0^{y_1}\beta dy = YA_1,$$

where A_1 is the area of the top of the section. Likewise the total compressive load on A_2, the area below the neutral axis, is YA_2. Therefore, since these loads must be equal for equilibrium, the position of the neutral axis must be such as to ensure that $A_1\bar{y}_1 = A_2\bar{y}_2$. (If the shape of the lamina is cut out of a piece of cardboard, the location of the neutral axis can be found by balancing it on a knife edge, approximately.)

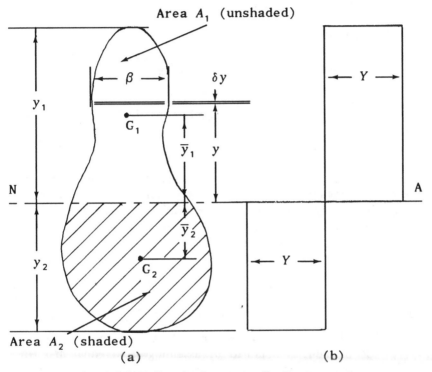

Fig. 1.36 Fully plastic stress distribution over
a beam of general cross–section

The fully plastic moment of resistance M_P is therefore given by the equation:

$$M_P = Y\left[\int_0^{y_1}\beta y\,dy + \int_0^{y_2}\beta y\,dy \right] = Y(A_1\bar{y}_1 + A_2\bar{y}_2),$$

hence:
$$M_P = YA(\bar{y}_1 + \bar{y}_2)/2 \qquad (1.102)$$

where A is the total area of the lamina and $\bar{y}_1 + \bar{y}_2 =$ the distance between the centroids G_1 and G_2 of the equal areas A_1 and A_2 perpendicular to the neutral axis NA.

Applying equation (1.102) to the I–beam and channel beam

cross–sections shown in Figs. 1.18(a) and (b), it can be seen that:

$A = BD - bd,$ and $\bar{y}_1 = \bar{y}_2 = (BD^2 - bd^2)/\{4(BD-bd)\}.$
    Substituting these values into equation (1.102) gives:

$$M_P = Y(BD^2 - bd^2)/4. \tag{1.103}$$

Also, since $I = (BD^3 - bd^3)/12,$ $y = D/2$ and $\sigma = Y,$

$$M_Y = \sigma I/y = Y(BD^2 - bd^3/D)/6. \tag{1.104}$$

Thus, for these sections:

$$S = \frac{M_P}{M_Y} = 1.5\,\frac{(1 - bd^2/BD^2)}{(1 - bd^3/BD^3)}. \tag{1.105}$$

For typical I–beam sections, $b/B = d/D = 0.82$ and $S \doteq 1.228.$ If $b/B = d/D = \lambda,$ the variation of S with λ is given by the equation:

$$S = 1.5(1 - \lambda^3)/(1 - \lambda^4) \tag{1.106}$$

and is illustrated in Fig. 1.37. This indicates that the thinner the I–beam or channel, the better the estimate of **fully plastic moment** by simply assuming it as the **initial yield moment** M_Y.
    For a circular cross–section of radius $r,$ $A = \pi r^2,$ $\bar{y}_1 = \bar{y}_2 = 4r/(3\pi),$ $I = \pi r^4/4,$ $y = r,$ hence:

$$M_P = 4Yr^3/3 \tag{1.107}$$

$$M_Y = \pi Yr^3/4 \tag{1.108}$$

and therefore: $S = 16/(3\pi) \doteq 1.698.$ (1.109)

With such a large **shape factor** only a poor estimate of the fully plastic moment is obtained by assuming it to equal M_Y.
    In modern design practice **factors of safety** are generally based upon the **fully plastic moment** M_P which can be estimated reliably and conservatively by the methods described above.
    Having bent a beam into the plastic range, removing the bending moment required to achieve the plastic state will result generally in elastic recovery of the beam. Considering again the simplest rectangular cross–section, if the beam has been bent so that $a = D/3$ in Fig. 1.34, the bending moment required to have achieved this is given by equation (1.96), with $a = D/3$ as:

$$M' = 2\frac{8}{9}\left[\frac{YBD^2}{12}\right] \tag{1.110}$$

at a curvature, from equation (1.100), of $1/R = 6Y/(ED).$ This is the point shown by a cross on Fig. 1.35.
    The extreme fibre stress associated with an elastic

distribution of stress, to give the value of M' in equation (1.110) would be: $\sigma = 6M'/(BD^2) = 13Y/9.$ (1.111)

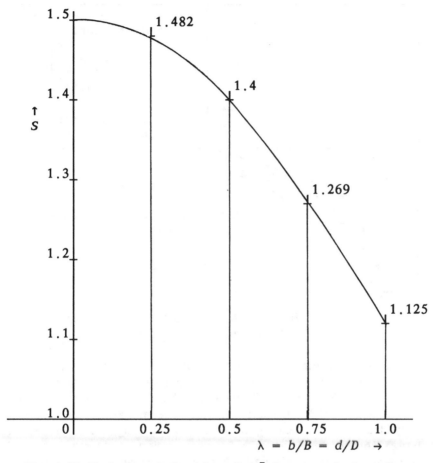

Fig. 1.37 Variation of S with λ for I-beams and channels

Since this stress is less than $2Y$ (the maximum <u>change</u> of stress permissible without causing yielding of the outer fibres in the reverse direction), it can be subtracted from the stress Y introduced into the outer fibres by the application of the bending moment M'. The linear elastic stress distribution introduced by applying the bending moment $-M'$ (to give zero moment) is shown superimposed on the original elastic–plastic stress distribution associated with $+M'$ in Fig. 1.38.

The value of M_P associated with the fully plastic case is given by equation (1.98) and the elastic outer fibre stress associated with M_P would be given by:

$$\sigma = 6M_P/BD^2 = 3Y/2 \qquad (1.112)$$

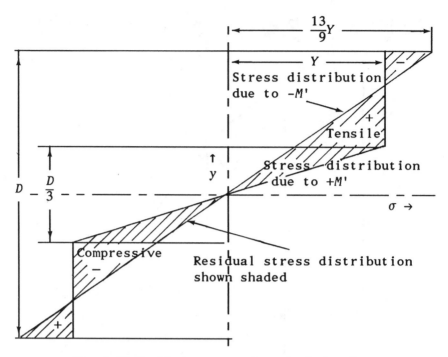

Fig. 1.38 Residual stresses after plastic bending
of elastic–perfectly plastic beam

as indicated in Fig. 1.39. In this case, large residual stresses are
introduced at the centre of the beam, equal to Y in fact.

Figs. 1.38 and 1.39 indicate the type of residual stress
distribution which is left in any beam after having been
subjected to pure bending into the plastic range. The analysis of
more complicated shapes of cross–section and more realistic
stress–strain curves generally has to be carried out numerically,
but the essential principles and features of the behaviour of the
material are as described above. In practice, bending is generally
not pure and hence shear stresses will also be present across any
transverse section as will be discussed later.

1.11 MOMENT–CURVATURE RELATIONSHIP, SLOPE AND DEFLECTION BY INTEGRATION

Returning to the case of the pure elastic bending of beams,
consider how the deflection of any beam may be determined.
The basic equations of the simple theory of bending, equations
(1.33), were:

$$\frac{M}{I} = \frac{\sigma}{y} = \frac{E}{R}$$

$$(1.33)$$

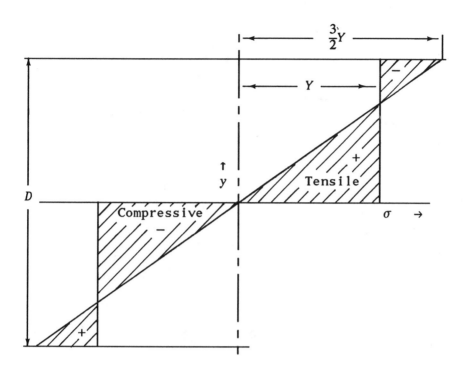

Fig. 1.39 Residual stresses after plastic
bending of perfectly plastic beam

and it may be observed that the curvature of the beam, viz.
$1/R$, is one of the parameters involved. Thus, it should be
possible to find the deformed shape of any elastically loaded
beam, by working from one of the supports along the beam,
since its curvature at any point is known from equations (1.33).

In considering the deflected shape of a beam, the <u>downward</u>
deflection at any section distant x from the left–hand supported
end is usually denoted by positive y.

The general relationship between the curvature $1/R$ and the
derivatives of y with respect to x can be determined as
illustrated in Fig. 1.40 with y positive <u>upwards</u>.

The elemental arc length:–

$$\delta s = R\delta\theta = \sqrt{(\delta x^2 + \delta y^2)} = \delta x \sqrt{\{1 + (\delta y/\delta x)^2\}} \qquad (1.113)$$

and the slope of the beam at any point is given by:

$$\frac{dy}{dx} = \tan\theta. \qquad (1.114)$$

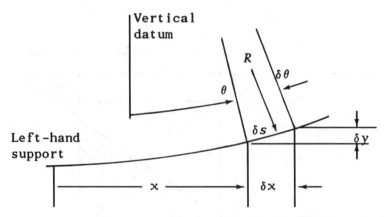

Fig. 1.40 Notation for beam element

Differentiating equation (1.114):

$$\frac{d^2y}{dx^2} = \sec^2\theta\frac{d\theta}{dx} = (1 + \tan^2\theta)\frac{d\theta}{dx} = \left\{1 + \left[\frac{dy}{dx}\right]^2\right\}\frac{d\theta}{dx}. \quad (1.115)$$

From equation (1.113), in the limit:

$$\frac{1}{R} = \frac{d\theta}{ds} = \frac{d\theta}{dx/\{1 + (dy/dx)^2\}}. \qquad (1.113a)$$

But:

$$\frac{d\theta}{dx} = \frac{\dfrac{d^2y}{dx^2}}{1 + \left[\dfrac{dy}{dx}\right]^2},$$

from equation (1.115), and substituting this into equation (1.113a) gives the **well-known mathematical relationship for curvature:**

$$\frac{1}{R} = \frac{\dfrac{d^2y}{dx^2}}{\left\{1 + \left[\dfrac{dy}{dx}\right]^2\right\}^{\frac{3}{2}}}. \qquad (1.116)$$

Since the elastic deflections of an initially straight beam will be fairly small, the slope $i = dy/dx = \tan\theta$ will also be small and the quantity $\{1 + (dy/dx)^2\}^{3/2}$ can be replaced by unity. Thus, for small deflections:

$$\frac{1}{R} = \frac{d^2y}{dx^2} \qquad (1.117)$$

and, from equation (1.31), the bending moment:

$$M = \frac{EI}{R} = EI\frac{d^2y}{dx^2}. \qquad (1.118)$$

This is the fundamental equation of the **theory of simple**

bending from which the slope and deflection of the beam at any point can be found, generally by successive integration. In fact successive integration, beginning with the load distribution, will give the distributions of shear force, bending moment, slope and deflection, respectively.

A note of caution is necessary here about the interpretation of the signs of the various quantities in the diagrams involved. Consider the case of pure positive (hogging) bending shown in Fig. 1.41 for a prismatic cantilevered beam subjected to equal and opposite couples, the left–hand end being fixed horizontally, the right–hand end being subjected to an applied couple or moment M.

The shear force $F = dM/dx$ is zero everywhere as is the load $w = dF/dx$. The bending moment is positive and equal to M everywhere. Thus, integrating equation (1.118) (with EI constant) gives:

$$EIdy/dx = Mx + A, \tag{1.119}$$

where A is the constant of integration. The given boundary condition relevant to this part of the problem is that at $x = 0$, $dy/dx = 0$, hence $A = 0$.

Integrating equation (1.119) gives $EIy = Mx^2/2 + B$, where B is another constant of integration also equal to zero, because of the boundary condition that, at $x = 0$, $y = 0$. Thus $y = Mx^2/(2EI)$ and $dy/dx = i = Mx/EI$ for this problem.

It should be observed that positive deflection is usually indicated <u>downwards</u> by the nature of these problems but the positive slope will be drawn upwards, on the i–diagram.

In this case the slope of the i–diagram is constant and positive, equal to M/EI, in fact. This is a consequence of assuming that 'hogging' moments are positive and, since deflections of loaded beams are generally downwards it seems a sensible convention to adopt. The nature of the problem usually indicates the correct deflected shape of any beam, in any case.

The simplest beam is a prismatic beam, simply supported, carrying a uniformly distributed load, as illustrated in Fig. 1.42. The bending moment at any section distant x from the left–hand end is $-wlx/2$ (sagging) $+ wx.x/2$ (hogging), i.e.

$$M = EId^2y/dx^2 = -wlx/2 + wx^2/2. \tag{1.120}$$

Integrating this twice gives the expressions:

$$EIdy/dx = -wlx^2/4 + wx^3/6 + A \tag{1.121}$$

and $$EIy = -wlx^3/12 + wx^4/24 + Ax + B \tag{1.122}$$

where A and B are constants of integration.

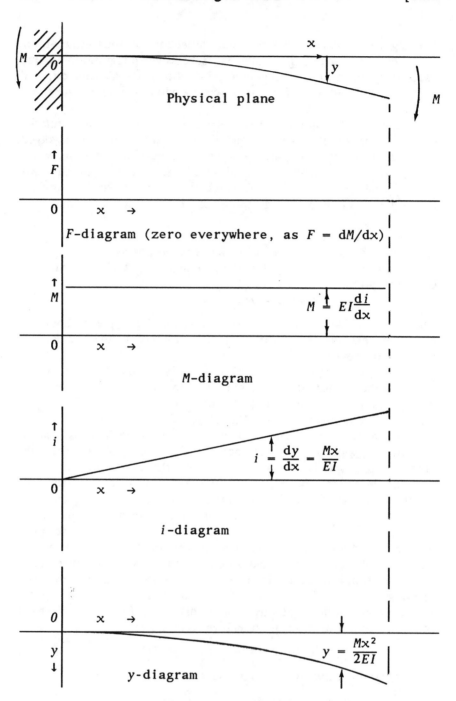

Fig. 1.41 Cantilevered beam subjected to
an applied couple at its free end

For this particular problem the boundary conditions are that, at $x = 0$ and $x = l$, $y = 0$. Substituting these values into equation (1.122) gives $B = 0$ and $A = wl^3/24$. Hence:

$$EIdy/dx = -wlx^2/4 + wx^3/6 + wl^3/24 \qquad (1.121a)$$

$$EIy = -wlx^3/12 + wx^4/24 + wl^3x/24. \qquad (1.122a)$$

At $x = l/2$, $dy/dx = 0$, as expected (by symmetry) and therefore $EIy = 5wl^4/384$. Again the deflection is positive, therefore <u>downwards</u>, as shown in Fig. 1.42. The slope varies from $+wl^3/(24EI)$ at $x = 0$ to $-wl^3/(24EI)$ at $x = l$, as illustrated in Fig. 1.42, but with positive slope <u>upwards</u>. It will be seen from these five diagrams that, with the adoption of the downwards positive convention for y and upwards positive convention for i there is consistency in the sign convention for all the diagrams as drawn. Thus, the initial slope of the y–diagram is positive (although apparently negative due to the convention adopted of y being positive downwards) and the initial value of the i–diagram is therefore shown positive (upwards).

The initial slope of the i–diagram is zero then negative as x increases, so that the initial value of the M–diagram is zero then negative. The initial slope of the M–diagram is negative as is the initial value of the F–diagram. The slope of the F–diagram is constant and positive as is the value of the load.

To summarize, with hogging bending moment positive, deflection y must be indicated as positive downwards and slope i as positive upwards in order for the diagrams to be correctly inter–related. In practice, there will generally be shear force and hence shear stresses induced across transverse sections which will lead to additional deflection, as will be discussed later.

1.12 MACAULAY'S METHOD, AREA–MOMENT METHODS, CLAPEYRON'S EQUATION, BEAM ON AN ELASTIC FOUNDATION

The introduction of discontinuous functions in order to give mathematical equations describing the variation of quantities along a beam was discussed in Section 1.2. **Macaulay's Method** utilizes these discontinuous functions when it is required to find the slope and deflection of beams carrying discontinuous loads. Considering again the beam illustrated in Fig. 1.1, the equation for the bending moment along the beam was:

$$M = -R_0 x + W[x - a], \qquad (1.14)$$

where $[x - a] = 0$ if $x < a$.

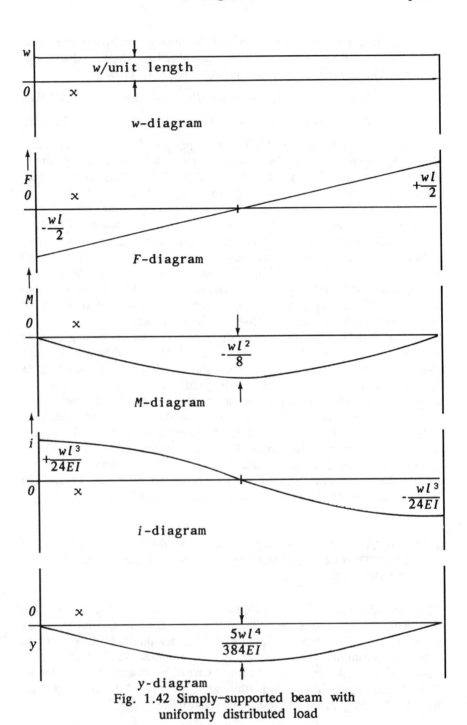

Fig. 1.42 Simply–supported beam with
uniformly distributed load

To determine the slopes and deflections along this beam
equation (1.12) may be written as:

$$M = EI d^2y/dx^2 = -R_0 x + W[x - a], \qquad (1.123)$$

Integrating equation (1.123) (assuming EI is constant along the beam) gives:

$$EI dy/dx = -R_0 x^2/2 + W[x - a]^2/2 + A, \qquad (1.124)$$

$$EI y = -R_0 x^3/6 + W[x - a]^3/6 + Ax + B, \qquad (1.125)$$

R_0 is known, since this is a statically determinate problem, from moment equilibrium as already discussed, viz:

$$R_0 = W(1 - a/l). \qquad (1.6)$$

The constants of integration A and B are determined from the boundary conditions on the deflection y, namely that $y = 0$ at $x = 0$ and $x = l$. Thus, from equation (1.125):

$$0 = -W(1 - a/l)l^3/6 + W[l - a]^3/6 + Al, \text{ hence:}$$

$$A = Wa(l - a)(2l - a)/(6l) \qquad (1.126)$$

and
$$B = 0. \qquad (1.127)$$

Substituting equations (1.6), (1.126) and (1.127) into equations (1.124) and (1.125) gives the slope and deflection everywhere along the beam.

The deflection under the load is given by substituting $x = a$ into equation (1.125) and is found to be:

$$y = Wa^2(l - a)^2/3EIl. \qquad (1.128)$$

If the load is at the centre, $a = l/2$ and the deflection is given by the well-known expression:

$$y = \frac{Wl^3}{48EI}. \qquad (1.129)$$

In Fig. 1.43 is shown the same system of loads as was discussed in relation to Fig. 1.1, but the ends of the beam are **encastré** i.e. 'encased' so that they remain horizontal, with zero slope. **Fixing moments** M_0 and M_l are introduced because of this, as well as reactions R_0 and R_l, and the values of these four unknown quantities can now only be found by introducing the known boundary conditions on both slope and displacement, i.e. $y = dy/dx = 0$ at $x = 0$ and $x = l$. This is a **statically indeterminate** problem, unlike the problem shown in Fig. 1.1 which was **statically determinate** so that the unknown reactions could be found simply by considering the equilibrium of the applied forces and moments, as was described in equations (1.3) to (1.10), for example.

The solution by Macaulay's method begins by setting up the

bending moment equation as follows:

$$M = EI d^2y/dx^2 = M_0 - R_0x + W[x - a] \qquad (1.130)$$

where $[x - a] = 0$ when $x < a$. As before [cf. equations (1.116) to (1.118)], this equation can be integrated to give:

$$EI dy/dx = M_0x - R_0x^2/2 + W[x - a]^2/2 + A. \qquad (1.131)$$

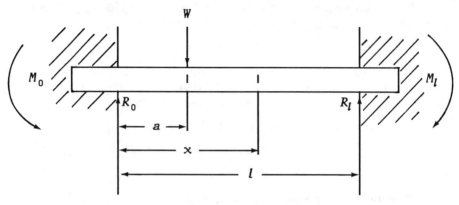

Fig. 1.43 Encastré beam with point load

The constant of integration A is zero, since when $x = 0$, $dy/dx = 0$, hence:

$$EI y = M_0x^2/2 - R_0x^3/6 + W[x - a]^3/6 + B \qquad (1.132)$$

and the constant of integration B is zero, since when $x = 0$, $y = 0$.

The values of M_0 and R_0 can be found by using the other boundary conditions, viz $y = dy/dx = 0$ at $x = l$.

Thus

$$0 = M_0l - R_0l^2/2 + W(l - a)^2/2 \qquad (1.131a)$$

and

$$0 = M_0l^2/2 - R_0l^3/6 + W(l - a)^3/6. \qquad (1.132a)$$

Multiplying equation (1.131a) by $l/3$ gives:

$$0 = M_0l^2/3 - R_0l^3/6 + W(l-a)^2l/6. \qquad (1.131b)$$

Subtracting equation (1.131b) from (1.132a) gives:

$$M_0 = W(l - a)^2a/l^2. \qquad (1.133)$$

From equation (1.131a): $R_0 = W(l - a)^2(2a + l)/l^3.$ $\qquad (1.134)$

By equilibrium, $R_l = W - R_0 = Wa^2(3l - 2a)/l^3.$ $\qquad (1.135)$

From equation (1.130): $M_l = M_0 - R_0l + W(l - a),$

hence: $M_l = W(l - a)a^2/l^2.$ $\qquad (1.136)$

By interchanging a and $(l - a)$ in equations (1.133) and (1.136) and in equations (1.134) and (1.135) these equations are seen to be correctly symmetrical. Substituting $a = l/2$ gives $R_0 = R_l = W/2$ and $M_0 = M_l = Wl/8$, for a central load.

Substitution of these values of M_0 and R_0 in equations (1.130) to (1.132) gives the solution to this problem. For example, the deflection under the load is given by substituting $x = a$ in equation (1.132) which gives:

$$EIy = M_0 a^2/2 - R_0 a^3/6 \qquad (1.137)$$

hence: $$y = Wa^3(l - a)^3/(6EIl^3). \qquad (1.138)$$

When $a = l/2$, for a central load,

$$y = Wl^3/(192EI), \text{ another well-known result.} \qquad (1.139)$$

The maximum deflection occurs at the point at which $dy/dx = 0$. If $a > l/2$, then this will occur on the left-hand side of the load, in the portion of the beam for which $x < a$.

From equation (1.131), $dy/dx=0$ when $0=M_0 x - R_0 x^2/2$, if $x<a$.

Hence: $$x = 2al/(2a + l) \qquad (1.140)$$

and, from equation (1.132),

$$y_{max} = 2Wa^3(l - a)^2/\{(3EI(2a + l)^2\}. \qquad (1.141)$$

Again, this reduces to $Wl^3/(192EI)$ when $a = l/2$, as required.

The procedure described above can be applied to any beam problem of the kind illustrated in Figs. 1.1 and 1.2 to determine slope and deflection. An example of a more complicated problem is illustrated in Fig. 1.44. The bending moment equation for this problem, using Macaulay's method, is as follows:

$$M = EId^2y/dx^2 =$$

$$M_0 - R_0 x + W_1[x - a] + w[x{-}b]^2/2 - w[x - c]^2/2 +$$

$$W_2[x - d]+W_3[x - e]. \qquad (1.142)$$

This can be integrated twice, as before, to give the slope and deflection equations. Again, the two constants of integration and the two unknowns M_0 and R_0 can be found by using the four known boundary conditions on slope and deflection, viz. $dy/dx = y = 0$ at $x = 0$ and $x = l$. Such lengthy equations are best solved numerically rather than algebraically, by substituting all the given values of loads and dimensions immediately into equation (1.142).

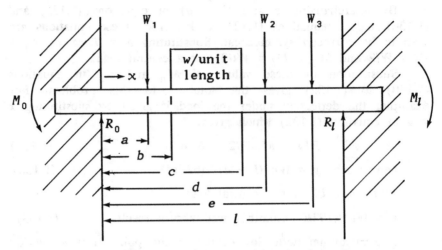

Fig. 1.44 Encastré beam with mixed loading

An alternative, semi–graphical, method of determining slope and deflection from the bending moment diagram is occasionally used, known as the **area moment method.** It can be seen from equation (1.118) viz:

$$M = \frac{EI}{R} = EI\frac{d^2y}{dx^2} \qquad (1.118)$$

that:

$$\frac{d^2y}{dx^2} = \frac{M}{EI} = \frac{di}{dx},$$

where i = slope. Integrating this equation, with EI constant along the beam, gives:

$$i_2 - i_1 = \frac{1}{EI}\int_{x_1}^{x_2} M dx. \qquad (1.143)$$

This equation states that the change in slope of the beam between any two points situated at x_1 and x_2 equals the area of the bending moment diagram between those two points, divided by EI. Also:

$$x\frac{d^2y}{dx^2} = \frac{Mx}{EI},$$

and, integrating by parts:

$$\left[xi - y\right]_{x_1}^{x_2} = \frac{1}{EI}\int_{x_1}^{x_2} Mx dx. \qquad (1.144)$$

Thus the change in $(xi - y)$ between points x_1 and x_2 equals the **moment of area** of the bending moment diagram about $x = 0$. In many cases it is possible to choose x_1 and x_2 so that xi

is zero at each point and equation (1.144) then gives the change in deflection between x_1 and x_2. In fact, when applying this method it is usual to define the integration limits at locations which have either x or i zero, so that the deflection can be determined directly.

A useful application of the area–moment method just described occurs in the case of continuous beams which have many supports and therefore have a high degree of redundancy. The area–moment method can give equations relating the three fixing moments at the supports of any two neighbouring spans along the beam, as follows.

Fig. 1.45 illustrates the problem, and represents any two spans of a continuous beam having more than three supports. The 'fixing moments' at each support are M_a, M_b and M_c. The actual M–diagram can conveniently be regarded as the sum of the (negative) bending moment diagram for the given loading as if it were applied to a freely supported span, plus the (positive) bending moment diagram for the fixing moments applied alone at the ends of the span concerned, as shown.

The 'free' M–diagrams have areas A_1 and A_2 and their centroids are distant $\bar{x}_1$ and $\bar{x}_2$ from the left– and right–hand supports respectively, as shown. The 'fixed' M–diagrams are trapezoidal and have their areas split into two triangles with centroids positioned at 1/3 and 2/3 along each span. In the analysis which follows the left–hand support is taken as the datum level, with its deflection $\delta_a = 0$, the deflections of the central support and right–hand support being δ_b and δ_c respectively.

Using the area–moment method, equation (1.144) can be applied to each of these spans, x being taken positive from the left–hand support and negative (leftwards) from the right–hand support, to provide two equations for the slope i_b at the middle support, Young's Modulus E being assumed the same for each span, as follows:

$$(l_1 i_b - \delta_b) - (0 - 0) = (1/EI_1)[M_a(l_1/2)(l_1/3) +$$
$$M_b(l_1/2)(2l_1/3) - A_1\bar{x}_1]$$

$$\{-l_2(-i_b) - (-\delta_b)\} - \{0 - (-\delta_c)\} =$$
$$(1/EI_2)[M_b(l_2/2)(-2l_2/3) + M_c(l_2/2)(-l_2/3) - (-A_2\bar{x}_2)].$$

Hence:– $i_b = (1/EI_1)(M_a l_1/6 + 2M_b l_1/6 - A_1\bar{x}_1/l_1) + \delta_b/l_1 =$
$$(1/EI_2)(-2M_b l_2/6 - M_c l_2/6 + A_2\bar{x}_2/l_2) - (\delta_b - \delta_c)/l_2.$$

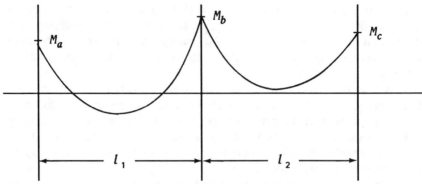

Actual M-diagram

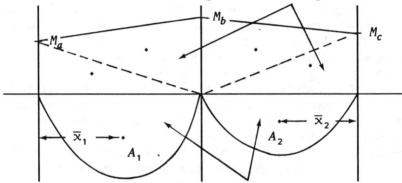

M-diagram for fixing moments only

M-diagram for loads on 'free' supports

Separated M-diagrams

Fig. 1.45 Neighbouring spans in a continuous beam

Therefore: $M_a l_1/I_1 + 2M_b(l_1/I_1 + l_2/I_2) + M_c l_2/I_2 =$

$6(A_1\bar{x}_1/I_1l_1 + A_2\bar{x}_2/I_2l_2) + 6E[\delta b/l_1 + (\delta_b-\delta_c)/l_2]$.   (1.145a)

This is the most general form of the **theorem of three moments** or **Clapeyron's equation**.

If $I_1 = I_2 = I$, often the case, equation (1.145a) becomes:

$$M_a l_1 + 2M_b(l_1 + l_2) + M_c l_2 =$$

$6(A_1\bar{x}_1/l_1 + A_2\bar{x}_2/l_2) + 6EI[\delta b/l_1 + (\delta_b - \delta_c)/l_2]$.   (1.145b)

If the supports are all at the same level:

$$M_a l_1 + 2M_b(l_1 + l_2) + M_c l_2 = 6(A_1\bar{x}_1/l_1 + A_2\bar{x}_2/l_2)$$ (1.145c)

and if the ends are simply supported ($M_a = M_c = 0$):

$$M_b(l_1 + l_2) = 3(A_1\bar{x}_1/l_1 + A_2\bar{x}_2/l_2).$$ (1.145d)

Equation (1.145c) is most often used in practice. $A_1\bar{x}_1$, $A_2\bar{x}_2$, l_1, l_2, δ_a, δ_b, and δ_c are usually known, so that an equation can be found for M_a, M_b and M_c for any three neighbouring supports. Generally the fixing moments are known or can be found at the ends of the whole continuous beam so that a set of simultaneous equations for all the fixing moments at every support can be found and the problem solved by applying Clapeyron's equation successively to each neighbouring pair of spans along the beam.

Sometimes the continuously supported beam has an infinite number of supports, becoming a beam on a (usually) elastic foundation. The local reaction of this elastic support is proportional to its displacement which is equal to the local deflection y of the beam. Recalling one of the basic equations of the simple bending theory, viz:

$$M = EI d^2y/dx^2 \qquad (1.118)$$

and noting again the relationships between w, F and M, viz:

$$w = dF/dx \ (= d^2M/dx^2) \qquad (1.22)$$

and: $\qquad\qquad F = dM/dx, \qquad\qquad (1.23)$

then, if E and I are both constant along the length of the beam (often the case) from equations (1.118), (1.22) and (1.23):

$$w = EI d^4y/dx^4. \qquad (1.146)$$

If the upward load on the beam is proportional to its deflection at that point, then the upward loading can be represented by the expression $w = -ky$, k having the dimensions force/(length)^2.

Consider the simple case shown in Fig. 1.46 for a central concentrated load W on a uniform beam supported on an elastic foundation. Since the loading is symmetrical, it is necessary to consider only half of the beam, of length l, say, with the origin of x at the left-hand end.

Then, the loading diagram for the whole of the left half of the beam up to the load W may be written as:

$$EI d^4y/dx^4 = -ky, \text{ or: } d^4y/dx^4 = -4\alpha^4 y, \qquad (1.147)$$

$$\text{where } \alpha^4 = k/4EI.$$

The general solution to equation (1.147) is:

$y =$
$(A\sin \alpha x + B\cos \alpha x)\sinh \alpha x + (C\sin \alpha x + D\cos \alpha x)\cosh \alpha x.$

$$(1.148)$$

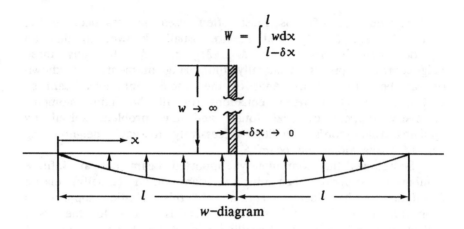

$$W = \int_{l-\delta x}^{l} w\,dx$$

$w \to \infty$

$\delta x \to 0$

x

l l

w–diagram

Fig. 1.46 Beam on an elastic foundation

At $x = 0$, $y = 0$. Substituting $x = 0$ in equation (1.148) gives $D = 0$. Also, at $x = 0$, $M = EI\,d^2y/dx^2 = 0$. Successive differentiation of equation (1.148) and substitution of $x = 0$ gives $A = 0$. Likewise, at $x = 0$, $F = EI\,d^3y/dx^3 = 0$, and $B = C$.

Thus, equation (1.148) reduces to:

$$y = B(\cos\,\alpha x.\sinh\,\alpha x + \sin\,\alpha x.\cosh\,\alpha x). \qquad (1.148a)$$

At $x = l$, $i = dy/dx = 0$, hence $B\alpha\cos\,\alpha l.\cosh\,\alpha l = 0$. The least solution of this is for $\alpha l = \pi/2$, which determines the minimum length l for which this solution applies, namely:

$$l_{min} = \pi/2\alpha = \pi/\{2(k/4EI)^{\frac{1}{4}}\} \propto k^{-\frac{1}{4}}.$$

It can be seen from this that the stiffer the elastic foundation, (i.e. the larger is k), the smaller the value of l_{min}.

For this minimum length, at the centre, $F = W/2 = EI\,d^3y/dx^3$, and: $W/2 = EI(4\alpha^3\sinh\,\alpha l.B\sin\alpha l)$, or $B = W/(8EI\alpha^3\sinh\,\pi/2) = W\alpha/(2k\sinh\,\pi/2)$. At the centre: $y_{max} = (W\alpha/2k).\coth\,\pi/2$ and $M_{max} = -(W/4a).\coth\,\pi/2$.

For $l < l_{min}$, in which the whole beam sinks into the elastic foundation, another solution applies, of the same form as equation (1.148a), but with an additional term $D\cos\,\alpha x.\cosh\,\alpha x$, so that: $y = B(\cos\,\alpha x.\sinh\,\alpha x + \sin\,\alpha x.\cosh\,\alpha x) + D\cos\,\alpha x.\cosh\,\alpha x$

$$(1.148b)$$

where:

$$B = -\,\frac{W\alpha}{k}\,\frac{\sin\,\alpha l.\cosh\,\alpha l - \cos\,\alpha l.\sinh\,\alpha l}{\sin\,2\alpha l + \sinh\,2\alpha l}$$

and:

$$D = -\frac{2W\alpha}{k}\frac{\cos\alpha l.\cosh\alpha l}{\sin 2\alpha l + \sinh 2\alpha l}.$$

As before, y_{max} and M_{max} occur under the load W.

1.13 MAXWELL'S GENERAL RECIPROCAL THEOREM

Maxwell's general reciprocal theorem can be derived by considering the behaviour of the elastic body illustrated in Fig. 1.47. In that figure forces F_a and F_b are applied at points A and B in arbitrary directions as shown.

Let:

δ_{aa} =	deflection of	A in the direction	of	F_a,	due to	F_a		
δ_{ab} =	"	" A " "	"	"	"	F_a	"	" F_b
δ_{bb} =	"	" B " "	"	"	"	F_b	"	" F_b
δ_{ba} =	"	" B " "	"	"	"	F_b	"	" F_a

In general F_a and F_b need not be in the same plane, so that the total deflections at A and B will contain components perpendicular to their lines of action. Since no work will be done by these components, they do not enter into the theorem.

If F_a is applied first, gradually, the work done = $\frac{1}{2}F_a\delta_{aa}$. The force F_b is then gradually applied so that the work done at the point B will be $\frac{1}{2}F_b\delta_{bb}$. However, there will be an additional deflection of the point A due to this force, of amount δ_{ab}. This will involve the movement of the now constant force F_a, so that the work done due to the application of F_b will be $\frac{1}{2}F_b\delta_{bb}+F_a\delta_{ab}$ and the total work done on the body will be:

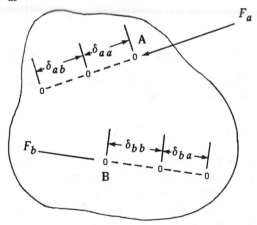

Fig. 1.47 An elastic body subjected to arbitrary forces

$$\tfrac{1}{2}F_a\delta_{aa} + \tfrac{1}{2}F_b\delta_{bb} + F_a\delta_{ab}.$$

If the loads were to have been applied in the reverse order, the total work done would have been:

$$\tfrac{1}{2}F_b\delta_{bb} + \tfrac{1}{2}F_a\delta_{aa} + F_b\delta_{ba}.$$

The total work done must be the same, for an elastic body, irrespective of the order in which the loads are applied, thus:

$$F_a\delta_{ab} = F_b\delta_{ba}. \qquad (1.149)$$

If $F_b = F_a$ then $\delta_{ab} = \delta_{ba}$. That is, the deflection at A due to a load at B is the same as the deflection at B due to the _same_ load at A, the deflections being in the direction of the loads.

This simple version of Maxwell's reciprocal theorem is sometimes useful in beam deflection problems, for example, the deflection at the centre of the simply supported beam due to the load W in Fig. 1.48 is equal to the deflection at the point A if the load were moved to the centre.

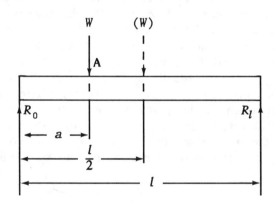

Fig. 1.48 Simply supported beam with point load

1.14 SHEAR STRESSES IN BEAMS, I–BEAM WEBS AND FLANGES, SHEAR CENTRE

1.14.1 _Shear Stresses_

The discussion so far in this chapter has concentrated on the _direct_ stresses induced over the cross–sections of beams subjected to bending, as discussed in Section 1.5 and illustrated in Fig. 1.12 for the case of pure bending, when shear forces and therefore shear stresses are absent.

In general, shear forces do exist in beams and hence shear stresses will act across transverse sections, thereby inducing

complementary shear stresses acting along longitudinal sections of the beam, which will lead to **warping** of cross–sections, as already mentioned. A simple estimate of the distribution of these shear stresses and the resultant additional deflection of the beam due to shear can be made by neglecting this warping, as will now be described.

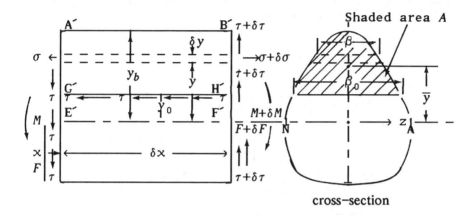

cross–section

Fig. 1.49 Free body diagram showing shear stresses

The situation assumed is illustrated in Fig. 1.49, which is based on Figs. 1.3, 1.12 and 1.13. The shear force F acting across any transverse cross–section of a beam will induce shear stresses τ there, as shown in Fig. 1.49. In the simple bending theory it is assumed that all stresses are uniform <u>across</u> the beam, i.e. in the z–direction, and also that these induced stresses will not affect the simple linear distribution of the direct stress σ already assumed. This cannot be true; these stresses must vary with y and this will cause some distortion (warping) of the transverse planes, as already mentioned, so that these transverse planes will not remain plane. However, the simple theory gives a reasonable approximation to the actual situation.

Due to the shear stresses τ acting in the y–direction there will be complementary shear stresses τ induced on longitudinal planes such as G´H´ in Fig. 1.49, distant y_0 from the neutral surface E´F´. Assume that the width of the generic cross–section at a distance of y_0 from the neutral axis NA is β_0, and that the cross–sectional area of the beam above that line is A, with its centre of gravity distant $\bar{y}$ from NA. Since there are no shear stresses acting at the top surface A´B´ of the beam, the section A´B´H´G´ of the beam above the longitudinal surface of area $\beta_0 \delta x$ will be in equilibrium provided that the following equilibrium equation is satisfied:

$$\tau \beta_0 \delta x = \int_{y_0}^{y_b} \delta\sigma . \beta \delta y. \qquad (1.150)$$

From equation (1.33), $\sigma = M.y/I$, hence $\delta\sigma = \delta M.y/I$.

Substituting that value into equation (1.150), and noting that $\delta M/I$ is constant over the section:

$$\tau \beta_0 \delta x = (\delta M/I) \int_{y_0}^{y_b} y\beta\delta y = (\delta M/I) A\bar{y}.$$

Thus:

$$\tau = \frac{1}{I\beta_0} \frac{dM}{dx} A\bar{y} = \frac{F}{I\beta_0} A\bar{y} \qquad (1.151)$$

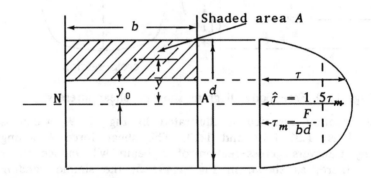

Fig. 1.50 Shear stress distribution over rectangular cross-section

since $F = dM/dx$. Note that β_0 is the actual width of the longitudinal section where the shear stress τ is acting, $A\bar{y}$ is the first moment of area of the area above β_0 about NA, I is the second moment of area of the <u>whole</u> area about NA and F is the shear force acting across the section.

For the rectangular cross-section shown in Fig. 1.50,

$$I = bd^3/12, \quad A = b(d/2-y_0),$$

$\bar{y} = y_0+(d/2-y_0)/2 = (d/2+y_0)/2, \quad \beta_0 = b$, and hence:

$$\tau = \frac{6F(d^2/4 - y_0^2)}{bd^3}. \qquad (1.152)$$

Thus at the surface, where $y_0 = d/2$, $\tau = 0$. At the centre, where $y_0 = 0$, $\tau = 6Fd^2/(4bd^3) = 1.5F/bd = 1.5\tau_m$ $\qquad (1.153)$

where: τ_m = the mean shear stress = F/bd. $\qquad (1.154)$

1.14.2 *Shear Stresses in Webs and Flanges of I-Beams*

One of the more important cross–sectional shapes is that of the I-beam, as discussed previously in Section 1.7 and shown in Fig. 1.18(a), repeated below for convenience, as Fig. 1.51.

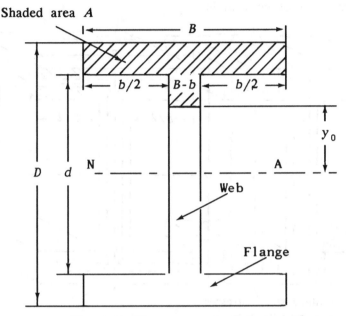

Fig. 1.51 Shear stress in I-beam

Firstly, considering the shear stresses in the web, their values will be given by equation (1.151), with:

$$A\bar{y} = B\left[\frac{D-d}{2}\right]\left[\frac{d}{2} + \frac{D-d}{4}\right] = B\left[\frac{D-d}{2}\right]\left[\frac{D+d}{4}\right],$$

for the flange part of A, and:

$$A\bar{y} = (B-b)\left[\frac{d}{2} - y_0\right]\left[\frac{d}{2} - \left[\frac{d}{2} - y_0\right]/2\right] =$$

$$(B-b)\left[\frac{d}{2} - y_0\right]\left[\frac{d}{4} + \frac{y_0}{2}\right]$$

for the web part of A, and $\beta_0 = B - b$. Hence:

$$\tau_w = \frac{F}{I(B-b)}\left[\frac{B}{8}(D^2 - d^2) + \left[\frac{B-b}{2}\right]\left[\frac{d^2}{4} - y_0{}^2\right]\right]. \quad (1.155)$$

The maximum value of τ_w occurs when $y_0 = 0$ and is given by the expression:

$$\tau_{wmax} = \frac{F}{8I(P-b)}(BD^2 - bd^2). \quad (1.156)$$

At the top of the web, $y_0 = d/2$ and τ_w will be given by:

$$\tau_{wd} = \frac{FB}{8I(B - b)}(D^2 - d^2).$$ (1.157)

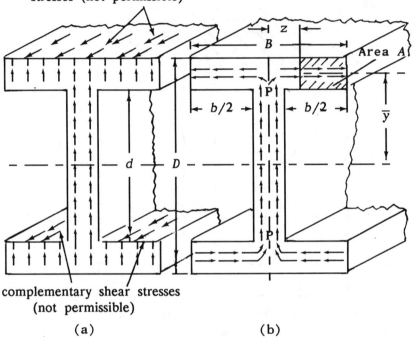

Complementary shear
stresses (not permissible)

complementary shear stresses
(not permissible)

(a)   (b)

Fig. 1.52 Shear stresses in flanges

When considering the flanges it is necessary to realize that
the direction of the shear stresses acting across them cannot be
vertical, as shown previously in Fig. 1.49. If the shear stresses
were vertical in the flanges, as sketched in Fig. 1.52(a), then
complementary shear stresses would have to be invoked on the
surfaces of the flanges, as illustrated.

It follows, and this is generally true, that shear stresses
cannot act vertically across flanges but will act tangentially to
their horizontal surfaces, as shown in Fig. 1.52(b). The "width"
β_0 will not be the width B of the flange shown in Fig. 1.51 but
its thickness, namely $(D-d)/2$, the area moment $A\bar{y}$ being given
by the parameters sketched in Fig. 1.52(b).

Hence, for the flanges:

$$A\bar{y} = \left[\frac{B}{2} - z\right]\left[\frac{D - d}{2}\right]\left[\frac{D + d}{4}\right], \quad \beta_0 = \frac{D - d}{2}$$

and the shear stress in the flange will be τ_f, acting horizontally, given by:

$$\tau_f = \frac{F(D + d)}{4I}\left[\frac{B}{2} - z\right]. \qquad (1.158)$$

It can be seen from equation (1.158) that τ_f will vary from a maximum value at the centre of the beam, where $z = 0$, of:

$$\tau_{fmax} = \frac{F(D + d)B}{8I} \qquad (1.159)$$

to a value of zero at the tips of the flanges, where $z = B/2$.

On the centre line of the web at the point P shown in Fig. 1.52(b), where $y_0 = d/2$, the shear stress in the web, t_{wd}, is given by equation (1.157), whilst the value of the shear stress in the flange, τ_{fmax}, is given by equation (1.159). The ratio of these two values is:

$$\tau_{wd}/\tau_{fmax} = (D - d)/(B - b) \qquad (1.160)$$

i.e. $2 \times$ (Thickness of Flange)/(Thickness of Web).

Therefore, the shear stress in the web will be double that in the flange at point P when their thicknesses are equal. Normally, the flange thickness is greater than that of the web, so that the shear stress in the web is always much greater than that in the flange, and most of the shearing force is carried by the web, the shear stresses in the flanges being negligible. Also, the variation of stress over the web is comparatively small and it is convenient for design purposes to assume that the web carries all the shear force, uniformly distributed over it to give a constant shear stress, only in the web. Likewise, it can be reasonably assumed that the flanges take all the bending moment, manifested as constant tensile and compressive stresses in opposite flanges.

Although the maximum direct stresses occur at the outer fibres of any I-beam, it is generally found that the maximum principal stresses (found by taking the shear stresses into account) occur at the transition point between web and flange {point P in Fig. 1.52(b)}.

The bending stress at point P in an I-beam bent elastically will be, from equation (1.33):

$$\sigma_p = \frac{My}{I} = \frac{Md}{2I}. \qquad (1.161)$$

The shear stress at P is given by equation (1.157), namely:

$$\tau_{wp} = \frac{FB}{8I(B - b)}(D^2 - d^2). \qquad (1.157)$$

Assuming that the vertical direct stresses acting through this horizontal beam are zero, then the maximum principal stresses will be:

$$\sigma_1 = \frac{\sigma_p}{2} + \sqrt{\left[\left(\frac{\sigma_p}{2}\right)^2 + \tau_{wp}^2\right]} \quad \text{and}$$

$$\sigma_2 = \frac{\sigma_p}{2} - \sqrt{\left[\left(\frac{\sigma_p}{2}\right)^2 + \tau_{wp}^2\right]} \quad (1.162)$$

as seen most easily from the Mohr's circle shown in Fig. 1.53.

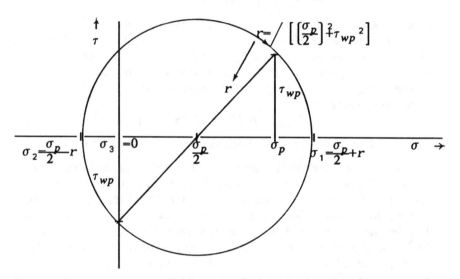

Fig. 1.53 Mohr's circle for beam stresses

If it is assumed that the transverse direct stresses (acting normal to the web) are zero, with zero shear stress on those planes (parallel to the web surface), then $\sigma_3 = 0$ and the **yield stress**, according to the **yield criterion of von Mises** (see Chapter 5), will be: –

$$\bar{\sigma} = (1/\sqrt{2})[(\sigma_1 - \sigma_2)^2 + (\sigma_2 - \sigma_3)^2 + (\sigma_3 - \sigma_1)^2]$$

$$= \sqrt{(\sigma_1^2 - \sigma_1\sigma_2 + \sigma_2^2)}. \quad (1.163)$$

Typically, in practice, $\tau_{wp} = \sigma_p/2$. Substituting this value into equation (1.162) gives:

$$\sigma_1 = \sigma_p[(1/2) + (1/\sqrt{2})] \triangleq 1.2071\sigma_p,$$

and $\sigma_2 = \sigma_p[(1/2) - (1/\sqrt{2})] \triangleq 0.2071\sigma_p.$

Substituting those values into equation (1.163) gives:

$$\bar{\sigma} = \sigma_p\sqrt{(7/4)} = 1.323\sigma_p.$$

This yield stress (or effective stress) is therefore seen to be greater in magnitude than the maximum principal stress and is the value which should be used when estimating the static yield strength of the beam. If straining of the beam is continued beyond this point the outer fibres will yield plastically and the behaviour of the beam will be described approximately by the analysis previously given in Sec. 1.10.

Returning to a consideration of the shear stress in the flanges of the I-beam illustrated in Fig. 1.53, the fact that the shear stresses in such flanges act in a direction tangential to the surfaces of the flanges can cause twisting in unsymmetrical sections. To avoid this, the line of action of the shear force at the section must pass through the **shear centre**, will now be illustrated using a channel section, as shown in Fig. 1.54.

As for the I-beam, the shear stresses in the flanges will be directed <u>along</u> the flanges as shown by the arrows in Fig. 1.54 and the shear stress at any section distant z from the centre line of the web will be given as before by:

$$\tau_f = \frac{F}{I\beta_0} A\bar{y},$$

where $A\bar{y} = \left[\frac{B + b}{2} - z\right]\left[\frac{D - d}{2}\right]\left[\frac{D + d}{4}\right]$, and $\beta_0 = \frac{D - d}{2}$

hence

$$\tau_f = \frac{F(D + d)}{4I}\left[\frac{B + b}{2} - z\right] \tag{1.164}$$

which varies linearly with z from a maximum of $F(D + d)(B + b)/(8I)$ at $z = 0$ to a value of zero at $z = (B + b)/2$, with a mean value of:

$$\tau_{fm} = \frac{F(D + d)(B + b)}{16I}. \tag{1.165}$$

Thus, the total force R in each flange is τ_{fm} multiplied by the flange area, viz:

$$R = \tau_{fm}\frac{D - d}{2} \times \frac{B + b}{2} = \frac{F(D + d)(D - d)(B + b)^2}{64I}. \tag{1.166}$$

These forces act to produce a clockwise twisting couple which must be balanced by applying the external shear force F through the shear centre S, distant h from the centre of the web as shown, where: –

$$Fh = R\left[D - \frac{D - d}{2}\right] = R\left[\frac{D + d}{2}\right]$$

hence:

$$h = R\left[\frac{D + d}{2F}\right]. \tag{1.167}$$

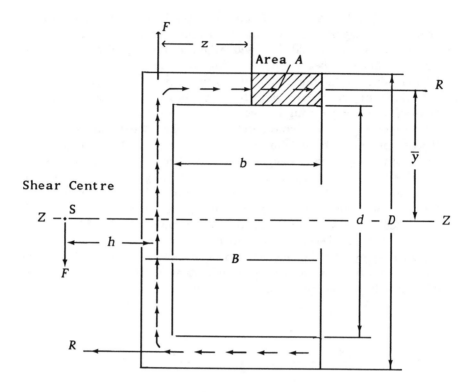

Fig. 1.54 Shear stresses in a beam of channel section

Substituting the value of R given by equation (1.166) into equation (1.167) gives:

$$h = \frac{(D - d)(D + d)^2(B + b)^2}{128I}. \qquad (1.168)$$

To obtain some idea of the actual magnitude of this dimension by which the supplied force must be offset to avoid twisting a channel section, consider the channel section which was shown in Fig. 1.23 on p.43.

The values of the dimensions are as follows:

D = 200 mm, d = 160 mm, B = 100 mm, b = 85 mm.

Substituting these values into equation (1.168) and recalling that I_{zz} for that section was calculated to be 37.65×10^6 mm^4, gives the value of h as:

$$h = \frac{40\times360^2\times185^2}{128\times37.65\times10^6} = 36.81 \text{ mm.}$$

This places the shear centre S at a point about twice the thickness of the web above the web as configured in Fig. 1.23.

If the lengths of the flanges were equal to the length of the web, i.e. $B = 200$ mm instead of 100 mm, then h would be:

$$h = \frac{40 \times 360^2 \times 385^2}{128 \times 37.65 \times 10^6} = 159.4 \text{ mm}.$$

These unduly long flanges would therefore increase the twisting moment to an extent which would make the offsetting of the applied load inconveniently large in practice, and is one reason for standard channel sections having $B \triangleq D/2$.

1.15 CURVED BEAMS

The elementary theory of beams relates to beams which are initially straight, as illustrated in Fig. 1.12. A beam with an **initial curvature** is shown in Fig 1.55, where R_0 is the initial radius of the neutral surface of the beam and the length of the generic fibre GH is $(R_0 + y)\theta$.

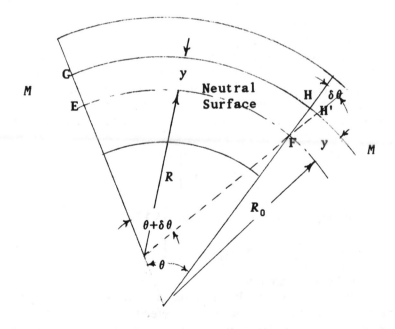

Fig. 1.55 Bending of a curved beam

After bending, the radius of the neutral surface EF will change to R, as sketched, and the length of the fibre GH will change to GH', given by $(R+y)(\theta+\delta\theta)$. The strain in that fibre will therefore be:

$$\epsilon = \frac{(R+y)(\theta+\delta\theta)-(R_0+y)\theta}{(R_0+y)\theta} = \frac{y\delta\theta}{(R_0+y)\theta} \tag{1.169}$$

where use has been made of the fact that the length EF along the neutral surface remains unchanged, i.e.:

$$R(\theta+\delta\theta) = R_0\theta. \tag{1.170}$$

For beams of small curvature, y may be neglected in comparison with R_0 in the denominator of equation (1.169) and $\delta\theta/\theta = (R_0/R-1)$, from equation (1.170), hence equation (1.169) becomes:

$$\epsilon = y\left[\frac{1}{R} - \frac{1}{R_0}\right] \tag{1.171}$$

and the bending stress is:

$$\sigma = Ey\left[\frac{1}{R} - \frac{1}{R_0}\right]. \tag{1.172}$$

For an initially straight beam $R_0 \to \infty$ and equation (1.172) reduces to equation (1.27).

As for the simple beam theory, neglecting any lateral stresses, the total force on the beam cross–section $= \int\sigma dA$ is zero, i.e. $E(1/R-1/R_0)\int ydA = 0$, so that the neutral surface must pass through the centroid of the section, as before. Likewise, the moment of resistance of the curved beam will be:

$$M = \int\sigma ydA = E(1/R-1/R_0)\int y^2dA$$

$$= EI\left[\frac{1}{R} - \frac{1}{R_0}\right]. \tag{1.173}$$

Combining equations (1.172) and (1.173),

$$\frac{\sigma}{y} = \frac{M}{I} = E\left[\frac{1}{R} - \frac{1}{R_0}\right] \tag{1.174}$$

which can be compared with equation (1.33).

If the beam is of **large initial curvature**, y may not be neglected in comparison with R. Referring to Fig. 1.55, the strain of the fibre GH will be given, as before, by the ratio HH'/GH, so that the stress will be given by:

$$\sigma = E.HH'/GH = \frac{Ey\delta\theta}{(R_0+y)\theta}. \tag{1.175}$$

For pure bending the total normal force on the cross-section must be zero, i.e.

$$\int \sigma dA = \frac{E\delta\theta}{\theta}\int\frac{ydA}{R_0+y} = 0. \qquad (1.176)$$

The moment of resistance $M = \int\sigma ydA$, which is, from equation (1.176):

$$M = \frac{E\delta\theta}{\theta}\int\frac{y^2dA}{R_0+y}. \qquad (1.177)$$

Now $\int\dfrac{y^2dA}{R_0+y} = \int\dfrac{\left[y(R_0+y)-R_0y\right]}{R_0+y}dA = \int ydA - R_0\int\dfrac{ydA}{R_0+y} = Ae-0,$

using equation (1.176). The variable e is the distance between the neutral surface EF and the principal axis through the centroid of the cross-section, which will not coincide with the neutral surface for beams of large initial curvature. Substituting this value for the integral in equation (1.177) gives the moment of resistance as:

$M = (E\delta\theta/\theta)Ae = [\sigma(R_0+y)/y]Ae$, [from equation (1.175)], hence

$$\sigma = \frac{My}{Ae(R_0+y)}. \qquad (1.178)$$

In this equation both e and y are positive measured outwards, a positive bending moment tending to increase the curvature of the beam.

As an example, consider the rectangular cross-section shown in Fig. 1.56.

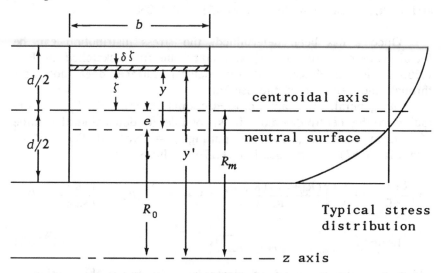

Fig. 1.56 Rectangular section beam of large initial curvature

Let $\zeta = y-e$ = distance of any elemental strip from the centroidal axis, the **mean radius of curvature** $= R_m = R_0+e$, and $\delta A = b\delta\zeta$.

Then, from equation (1.176), since $R_0+y = R_m+\zeta$ and $y = \zeta+e$

$$\int \frac{ydA}{R_0+y} = \int_{-d/2}^{d/2} \frac{(\zeta+e)}{R_m+\zeta}.bd\zeta = 0$$

hence:

$$\int_{-d/2}^{d/2} \frac{(R_m+\zeta)-(R_m-e)}{R_m+\zeta}.d\zeta = 0$$

i.e.

$$\int_{-d/2}^{d/2} d\zeta - (R_m-e)\int_{-d/2}^{d/2} \frac{d\zeta}{R_m+\zeta} = 0$$

or

$$d - (R_m-e)\ln\frac{R_m+d/2}{R_m-d/2} = 0.$$

Hence $$e = R_m - \frac{d}{\ln \frac{R_m+d/2}{R_m-d/2}} .$$ (1.179)

If $R_m = d$, for example, $e = d(1-1/\ln 3) = 0.08976d$, and if $R_m = 2d$, $e = 2d - d/\ln(5/3) = 0.04238d$.

Once e has been determined, the stress distribution can be found from equation (1.178) and takes the form sketched in Fig. 1.56. It is non-linear and the neutral surface does not pass through the centroid of the section.

In general, for any cross-sectional shape, if R_m is the radius to the centroidal axis, it is possible to determine the value of e from equation (1.176) by putting $y = \zeta+e$ and $R_0 = R_m-e$ (see Fig. 1.57). Then equation (1.176) reduces to:

$$\int \frac{\zeta+e}{R_m+\zeta}.dA = \int \frac{(R_m+\zeta)-(R_m-e)}{R_m+\zeta}.dA = A - (R_m-e)\int \frac{dA}{R_m+\zeta} = 0.$$

Hence: $$e = R_m - \frac{A}{\int \frac{\beta d\zeta}{R_m+\zeta}}$$ (1.180)

where β is the width of an elemental strip across the section as indicated in Fig. 1.57.

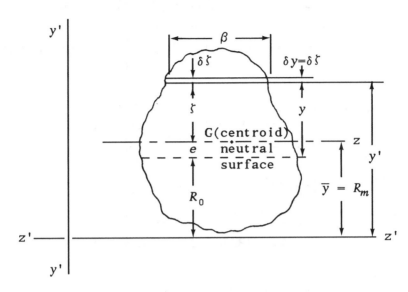

Fig 1.57 General section of beam of large initial curvature

The integral in Eq. 1.180 can be found for any section either graphically or numerically, giving the value of e. The distribution of σ can then be found from Eq. 1.178, for a given applied bending moment M. Alternatively, a computer program can be used and Program 1.2 which follows is based on Program 1.1 (see p.41) previously given for determining first and second moments of area. The input data required for the solution, either for a numerical tabular procedure or for this computer program, is set out in Table 1.4 which relates to Fig. 1.57 and, if an odd number of ordinates is chosen, Simpson's rule can again be used. Program 1.2 enables the determination of e and hence the distribution of σ/M for any section, Solution 1.2 giving a typical example.

The example chosen is for a rectangular section curved beam of the geometry shown in Fig. 1.56, with $R_m = b = d = 2$. Thus the theoretical value e would be, from Eq. 1.179, $e = R_m - d/\ln 3 = d - d/\ln 3 = 0.08976d = 0.08976 \times 2 = 0.1795$, as compared with 0.1797 for the numerical solution with 9 ordinates. If 5 ordinates are taken, the value of e becomes 0.18182, and for 7 ordinates $e = 0.18007$. Thus it would seem advisable to take at least 9 ordinates (N = 9) in this program.

TABLE 1.4 Input Data ($\delta y = \delta y' = \delta \zeta =$ constant)

β_1	y_1'	$\beta_1 \delta y$	$\beta_1 y_1' \delta y$	$\zeta_1 = y_1' - R_m$	$\beta_1 \delta \zeta / (R_m + \zeta_1)$	$y_1 = \zeta_1 + e$	$\dfrac{\sigma_1}{M} = \dfrac{y_1}{Ae(R_0 + y_1)}$
¦	¦	¦	¦	¦	¦	¦	¦
β_n	y_n'	$\beta_n \delta y$	$\beta_n y_n' \delta y$	$\zeta_n = y_n' - R_m$	$\beta_n \delta \zeta / (R_m + \zeta_n)$	$y_n = \zeta_n + e$	$\dfrac{\sigma_n}{M} = \dfrac{y_n}{Ae(R_0 + y_n)}$
SUMS $\rightarrow$	A	$A\bar{y}$ thus $R_m = \bar{y} = \dfrac{A\bar{y}}{A}$			$\Sigma = \int \dfrac{\beta \delta \zeta}{R_m + \zeta}$ thus $e = R_m - A/\Sigma$ $R_0 = R_m - e$		

PROGRAM 1.2 Curved beams

```
 1:*          PROGRAM CURBEA
 2:           DIMENSION ALPHA(100),BETA(100),GAMMA(100),DELTA(100),
 3:        &  SIGMAM(100),ZETA(100),Y(100),YP(100)
 4:           WRITE(*,1)
 5: 1         FORMAT(' "CURBEA" CALCULATES DISTANCE E BETWEEN ',/1X,
 6:        &  ' CENTROIDAL AND NEUTRAL SURFACES, ALSO SIGMA/M (SIGMAM)',/1X
 7:        &  ' IN A CURVED BEAM. RADII OF CENTROIDAL AND NEUTRAL ',/1X,
 8:        &  ' SURFACES ARE RM AND R0. YDASH=YP AND ZDASH=ZP IN ',/1X,
 9:        &  ' PROGRAM. N IS NUMBER OF ORDINATES DOWN THE LAMINA',/1X,
10:        &  ' AND MUST BE ODD FOR SIMPSONS RULE. BETA IS WIDTH,',/1X,
11:        &  ' OF HORIZONTAL ELEMENTAL STRIP, BREADTH DY. YP IS',/1X,
12:        &  ' VERTICAL DISTANCE FROM ZP AXIS TO ORDINATE. ',/1X,
13:        &  ' INPUT N AND THEN BETA, YP FOR EACH ORDINATE. '//)
14:           WRITE(*,2)
15: 2         FORMAT(' TYPE IN N ')
16:           READ(*,*) N
17:           DO 5 I=1,N
18:           WRITE(*,4) I
19: 4         FORMAT(' TYPE IN BETA, YP, FOR ORDINATE ',I3,)
20:           READ (*,*) BETA(I),YP(I)
21: 5         CONTINUE
22:           DY=ABS(YP(1)-YP(2))
23:           DO 6 I=1,N
24:           ALPHA(I)=BETA(I)*DY
25:           GAMMA(I)=ALPHA(I)*YP(I)
26: 6         CONTINUE
27:           CALL SIMPSN(ALPHA,A,N)
28:           CALL SIMPSN(GAMMA,YB,N)
29:           YBAR=YB/A
30:           DO 7 I=1,N
31:           ZETA(I)=YP(I)-YBAR
32:           DELTA(I)=BETA(I)*DY/(YBAR+ZETA(I))
33: 7         CONTINUE
34:           CALL SIMPSN(DELTA,EB,N)
35:           E=YBAR-A/EB
36:           R0=YBAR-E
37:           DO 8 I=1,N
38:           Y(I)=ZETA(I)+E
39:           SIGMAM(I)=Y(I)/(A*E*(R0+Y(I)))
40: 8         CONTINUE
41:           WRITE(*,9) A,E,R0,YBAR
42: 9         FORMAT(//4X,'A=',E10.5,2X,'E=',E10.5,2X,'R0=',E10.5,
43:        &  2X,'RM=',E10.5,/)
44:           WRITE(*,10)
45: 10        FORMAT(//10X,1HI,4X,7HBETA(I),6X,5HY'(I),7X,4HY(I),7X,
46:        &  9HSIGMAM(I))
47:           DO 12 I=1,N
48:           WRITE(*,11) I,BETA(I),YP(I),Y(I),SIGMAM(I)
49: 11        FORMAT(8X,I3,4E12.5)
50: 12        CONTINUE
51:           STOP
52:           END
```

```
53:
54:            SUBROUTINE SIMPSN(P,Q,N)
55:            DIMENSION P(N)
56: C          USE SIMPSON'S RULE FOR INTEGRATION
57:            EVEN=0.0
58:            ODD =0.0
59:            DO 1 I=2,N-3,2
60:            EVEN=EVEN+P(I)
61:            ODD =ODD+P(I+1)
62: 1          CONTINUE
63:            Q=(1.0/3.0)*(P(1)+4.0*(EVEN+P(N-1))+2.0*ODD+P(N))
64:            RETURN
65:            END
```

SOLUTION 1.2

curbea
"CURBEA" CALCULATES DISTANCE E BETWEEN
CENTROIDAL AND NEUTRAL SURFACES, ALSO SIGMA/M (SIGMAM)
IN A CURVED BEAM. RADII OF CENTROIDAL AND NEUTRAL
SURFACES ARE RM AND RO. YDASH=YP AND ZDASH=ZP IN
PROGRAM. N IS NUMBER OF ORDINATES DOWN THE LAMINA
AND MUST BE ODD FOR SIMPSONS RULE. BETA IS WIDTH,
OF HORIZONTAL ELEMENTAL STRIP, BREADTH DY. YP IS
VERTICAL DISTANCE FROM ZP AXIS TO ORDINATE.
INPUT N AND THEN BETA, YP FOR EACH ORDINATE.

TYPE IN N
9
TYPE IN BETA, YP, FOR ORDINATE   1
2. 3.
TYPE IN BETA, YP, FOR ORDINATE   2
2. 2.75
TYPE IN BETA, YP, FOR ORDINATE   3
2. 2.5
TYPE IN BETA, YP, FOR ORDINATE   4
2. 2.25
TYPE IN BETA, YP, FOR ORDINATE   5
2. 2.
TYPE IN BETA, YP, FOR ORDINATE   6
2, 1,75
TYPE IN BETA, YP, FOR ORDINATE   7
2. 1.5
TYPE IN BETA, YP, FOR ORDINATE   8
2. 1.25
TYPE IN BETA, YP, FOR ORDINATE   9
2. 1.
```

A=.40000E+01   E=.17971E+00   RO=.18203E+01   RM=.20000E+01
```

I	BETA(I)	Y'(I)	Y(I)	SIGMAM(I)
1	.20000E+01	.30000E+01	.11797E+01	.54705E+00
2	.20000E+01	.27500E+01	.92971E+00	.47031E+00
3	.20000E+01	.25000E+01	.67971E+00	.37823E+00
4	.20000E+01	.22500E+01	.42971E+00	.26568E+00
5	.20000E+01	.20000E+01	.17971E+00	.12500E+00
6	.20000E+01	.17500E+01	-.70291E-01	-.55877E-01
7	.20000E+01	.15000E+01	-.32029E+00	-.29705E+00
8	.20000E+01	.12500E+01	-.57029E+00	-.63468E+00
9	.20000E+01	.10000E+01	-.82029E+00	-.11411E+01

```
Stop - Program terminated.

A>
```

1.16 STRUTS, COLUMNS

A **strut** is generally understood to mean a long slender member of uniform cross–section subjected to compression along its length. A **column** is usually taken to mean a shorter, thicker member. A strut would fail by **buckling,** whereas a column would fail by **crushing** in compression. In practice combinations of these failure modes will occur.

The classical theory of the buckling of a strut is due to Euler. The simplest geometry of loading is that indicated in Fig. 1.58, which shows a thin strut of uniform cross–section subjected to an end load P applied axially through the centroid of the cross–section. The ends of the strut are free to rotate, usually termed **pin–ended,** and the end loads are constrained to move towards one another. The strut is assumed to deflect into the bowed shape (approximately sinusoidal) as indicated.

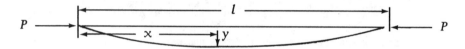

Fig. 1.58 Pin–ended strut

Assuming that the simple theory of the bending of beams will apply and adopting the usual sign conventions, at the section distant x from one end the bending moment will be $-Py$, the negative sign being associated with the sagging, positive downward deflection y shown in Fig. 1.58. For this situation the fundamental equation of the simple theory of bending [equation (1.118)] becomes:

$$EI\frac{d^2y}{dx^2} = M = -Py \tag{1.181}$$

which can be written as:

$$\frac{d^2y}{dx^2} + \frac{P}{EI}y = 0$$

or:

$$\frac{d^2y}{dx^2} + \lambda^2 y = 0, \text{ where } \lambda^2 = \frac{P}{EI}. \tag{1.182}$$

The well–known solution of this second order differential equation is:

$$y = A\sin \lambda x + B\cos \lambda x. \tag{1.183}$$

The boundary conditions are:

When $x = 0$, $y = 0$; $\therefore B = 0$.
When $x = l$, $y = 0$, $\therefore 0 = A\sin \lambda l$.

Therefore, either $A = 0$ or $\sin \lambda l = 0$, in which case $\lambda l = \pi$, 2π, 3π, etc. Taking the smallest value gives:

$$\lambda = \sqrt{\frac{P}{EI}} = \frac{\pi}{l} , \text{ hence } P = P_e = \frac{\pi^2 EI}{l^2} \qquad (1.184)$$

generally denoted as P_e, the **Euler buckling load**.

Substituting $\lambda = \pi/l$ into equation (1.183) gives the deflected shape of the strut as the half sine wave $y = A\sin(\pi/l)x$, from which the maximum deflection is A, occurring in the middle of the strut, where $x = l/2$. The interpretation of this solution is that, with an initially perfectly straight strut, there will be no deflection until the Euler load P_e is reached. At that value the strut will deflect into a half sine wave having an indeterminate maximum deflection B. With the higher values of $\lambda = 2\pi/l$, $3\pi/l$, etc., the buckling loads will be larger (by factors of 2^2, 3^2, etc.), and are therefore of no relevance for this problem. In practice, application of the Euler buckling load would be unstable since any slight increase in load would lead to collapse.

It is important to note that, since the cross-section of the strut can have any shape, the value of I used in equation (1.184) must clearly be <u>the minimum value for the section</u>. The plane in which buckling occurs will therefore be determined by the plane of minimum I.

The other end condition which can be prescribed is, as for beams, **fixed-ended** or **direction-fixed,** the term usually applied to struts. For the cases in which either one or both ends are fixed the basic Euler solution can be used by simply changing the length of the strut, as will now be described. In Fig. 1.59 is shown a strut with one end **free, or pin-ended,** the other direction-fixed.

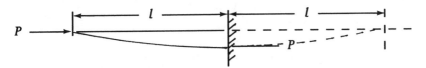

Fig. 1.58 Pin-ended, fixed-ended strut

This strut will deflect into a quarter sine wave and is therefore behaving as though it were one half of a strut of total length $2l$. Thus:

$$P = \frac{\pi^2 EI}{(2l)^2} = \frac{\pi^2 EI}{4l^2} \qquad (1.185)$$

i.e. only one quarter of the load if both ends of the strut were pin-ended.

If the two ends of the strut are fixed in both position and direction, i.e. **position-fixed** and **direction-fixed,** then the buckling load will be the same as that for a strut of only half the actual length, as indicated in Fig. 1.60.

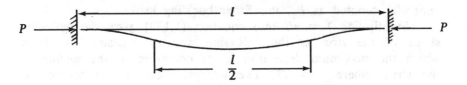

Fig. 1.60 Fixed-ended strut

Thus, in this case:

$$P = \frac{\pi^2 EI}{(l/2)^2} = \frac{4\pi^2 EI}{l^2} \qquad (1.186)$$

i.e. four times that of the free-ended strut.

From this it seems advisable to fix both ends of the strut, wherever possible. In practice it is difficult to fix the ends of any strut completely, however, and a great variety of intermediate end conditions is possible.

An intermediate condition between those shown in Figs. 1.59 and 1.60 is shown in Fig 1.61, in which a lateral force Q has to be introduced in order to maintain one end of the strut in line with the applied end load P. This end is termed **position-fixed,** whilst the other end is **direction-fixed.**

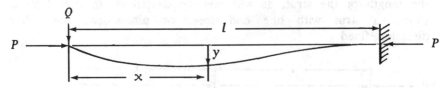

Fig. 1.61 Position-fixed, direction-fixed strut

In this case:

$$EI\frac{d^2y}{dx^2} = -Py + Qx, \text{ or } \frac{d^2y}{dx^2} + \lambda^2 y = \frac{Qx}{EI}, \qquad (1.187)$$

The solution to this equation is:

$$y = A\sin \lambda x + B\cos \lambda x + \frac{Qx}{P}. \qquad (1.188)$$

The boundary conditions are:

When $x = 0$, $y = 0$; $\therefore B = 0$.
When $x = l$, $y = 0$ and $dy/dx = 0$,
$\therefore 0 = A\sin \lambda l + Ql/P$, and $0 = \lambda A\cos \lambda l + Q/P$.

Eliminating Q/P from these last two equations gives the equation: $\lambda l = \tan \lambda l$, the least solution of which gives the value of $\lambda l = 4.4934$. Hence:

$$P = \lambda^2 EI = 4.4934^2 \frac{EI}{l^2} = 2.046 \frac{\pi^2 EI}{l^2}. \qquad (1.189)$$

In general, the Euler buckling load can be written as:

$$P_e = n\frac{\pi^2 EI}{l^2} = n\frac{\pi^2 EAk^2}{l^2} \qquad (1.190)$$

where n is a constant depending on the end fixing conditions, k is the least radius of gyration of the cross-section and A is the area of the cross-section (i.e. $I_{min} = Ak^2$). Note that the direct compressive stress in the Euler strut is:

$$\sigma_e = \frac{P_e}{A} = n\frac{\pi^2 Ek^2}{l^2} = n\frac{\pi^2 E}{(l/k)^2} \qquad (1.191)$$

and it must obviously not exceed the compressive yield stress σ_c. The ratio l/k is termed the **slenderness ratio** and is often denoted by the variable s. From equation (1.191) with $\sigma = \sigma_c$, P_e cannot possibly apply for values of s less than s_e given by:

$$s_e = \sqrt{\frac{n\pi^2 E}{\sigma_c}}. \qquad (1.192)$$

A column is a short strut, short enough not to buckle, and its collapse or yield load would be given by:

$$P_c = \sigma_c A. \qquad (1.193)$$

The variation of P_e with the slenderness ratio s is shown in Fig. 1.62, for given values of n, E, and A, from which it can be seen that s cannot be less than that associated with the yield load P_c.

To include both of the buckling and yielding failure modes, Rankine proposed a criterion of failure:

$$\frac{1}{P_R} = \frac{1}{P_c} + \frac{1}{P_e}. \qquad (1.194)$$

Rankine's formula becomes, after substitution from equations

(1.190) and (1.193):

$$P_R = \frac{\sigma_c A}{1+\frac{\sigma_c s^2}{n\pi^2 E}} = \frac{P_c}{1+as^2}. \tag{1.195}$$

It was originally suggested by Rankine that (with $n = 1$), the term $\sigma_c/(n\pi^2 E)$ in equation (1.195) should be replaced by a, and he gave the average values of σ_c and a for the metals shown in Table 1.5, based on experience. For example, with $n = 1$ and a value of $E = 207$ GPa for steel, the value of σ_c required to give $a = 1/7500$ would be 272 MPa, which is about halfway between yield and ultimate stress values for mild steel. Probably his figures are appropriate for structural steel.

TABLE 1.5

Material	σ_c MPa	a Pinned ends $(n=1)$	Fixed ends $(n=4)$
Mild steel	324	$0.1\dot{3} \times 10^{-3}$	$0.3\dot{3} \times 10^{-4}$
Wrought Iron	247	$0.1\dot{1} \times 10^{-3}$	$0.2\dot{7} \times 10^{-4}$
Cast Iron	556	0.625×10^{-3}	0.5625×10^{-3}

Other end conditions can be catered for by using an appropriate value for n, and a **safety factor** can be introduced by reducing the value of σ_c, which may then be regarded as the safe working stress.

As discussed in detail by Morley, initial curvature of the strut and eccentricity of the load can be combined to give an equivalent initial curvature of amplitude c. The total maximum compressive stress due both to the direct load and to bending is then:

$$\sigma_c = \frac{cPP_e}{P_e-P}\cdot\frac{r}{I} + \frac{P}{A}$$

where r is the maximum distance of any fibre from the neutral axis for the buckling mode associated with the minimum I, given by:

$$r = \frac{(\eta P/A)\sigma_e}{\sigma_e-P/A} + \frac{P}{A},$$

where $\eta = cr/k^2$ and is taken as $\eta = 0.003s$ in British Standards. Substituting $r = \eta k^2/c$ and $I = Ak^2$ into the first of the above two equations leads to the following quadratic in P/A:

$$(P/A)^2-[\sigma_c+(\eta+1)\sigma_e]P/A+\sigma_c\sigma_e = 0, \text{ with the least solution:}$$

$$P_P/A = (1/2)[\sigma_c+(\eta+1)\sigma_e] - (1/2)\sqrt{\{[\sigma_c+(\eta+1)\sigma_e]^2-4\sigma_c\sigma_e\}}. \tag{1.196}$$

This is the **Perry–Robertson Formula,** (with $P = P_P$), which is favoured in Britain, forming the basis of **B.S.S. 449 (1937).**

Other approximate formulae have been developed over the years. The simplest, often used in America, is the **straight–line formula**:–

$$P_A = \sigma_c A[1-\alpha s] = P_c[1-\alpha s] \qquad (1.196)$$

but this can only apply over a limited range of the variables. Typical values used for structural steel in a similar equation in British Standards are $\sigma_c = 140$ MPa (approximately half the **yield stress**), with $\alpha = 0.54 \times 10^{-2}$ for pinned ends and 0.36×10^{-2} for fixed ends.

An alternative approximation is **Johnson's parabolic formula**:

$$P_J = \sigma_c A[1-\beta s^2] = P_c[1-\beta s^2] \qquad (1.197)$$

with $\sigma_c = 290$ MPa for structural steel, $\beta = 0.3 \times 10^{-4}$ for pinned ends and 0.2×10^{-4} for fixed ends. It is interesting to note that Rankine's formula [equation (1.195)] can be expanded into a series, by division, to give $P_R/P_c = 1-as^2+a^2s^4-a^3s^6+a^4s^8-....$ etc. Probably, Johnson based his parabolic formula on the first two terms of that series, but the value of β recommended is not equal to a, however. For structural "mild" steel Rankine's theoretical value of a, with $n = 1$, is $a = \sigma_c/(\pi^2 E) = 324/(\pi^2 \times 207000) \triangleq 1.586 \times 10^{-4}$, which is slightly higher than the value 1.333×10^{-4} proposed by Rankine himself (see Table 1.5) and <u>much</u> higher than the value of 0.3×10^{-4} proposed by Johnson for his constant [β in equation (1.197)]. In fact, at a typically practical median value of $s = 100$, Rankine's formula gives the value of the ratio $P_R/P_c = 0.7/.3867 = 1.81$. This can be seen from the curves shown in Fig. 1.62. It is tempting to suggest that a better value for β would be that which would make $1/(1+100^2 a) = 1/2.589 = 1-100^2 \beta$, i.e. $\beta = 0.6133 \times 10^{-4}$. Three curves for P_J, with $\beta = 0.3 \times 10^{-4}$, 0.1586×10^{-4}, and 0.6133×10^{-4}, are shown in Fig. 1.62.

Computer Program 1.3 enables determination of P_c, P_e, P_R, P_P, P_A and P_J for values of the slenderness ratio s increasing in steps of 10 from 10 up to 200, Solution 1.3 giving a typical example for a mild steel I–beam of the section shown in Fig. 1.24 on p.45. For that beam, the cross–sectional area is $A = 0.0064$ m^2, the least value of I is $I_{min} = I_{YY} = 3.3783 \times 10^{-6}$ m^4, hence $k_{min} = \sqrt{(I_{min}/A)} \triangleq 23$ mm. Thus for an I–beam column of this cross–section, say 2 m long, the s ratio would be $2/0.023 \triangleq 87$, for which $P_c = 2.074$ MN, $P_e = 1.743$ MN, $P_R = $

0.943 MN, P_P = 1.159 MN, P_A = 1.099, and P_J = 1.601. In this computer program the <u>theoretical</u> value of the constant $a=\sigma_c/\{n\pi^2 E\}$ (AR in the program) has been used in Rankine's formula for the allowable load P_R.

The limiting column loads obtained from Solution 1.3 are sketched in Fig. 1.62. It is interesting to observe from these results that the buckling loads predicted by Rankine's formula are less but do not differ very greatly from those predicted by the Perry–Robertson formula, which forms the basis of the British Standards for columns. In general, it is more conservative (safer). Therefore, it is recommended that the Rankine buckling load P_R be used to obtain a reliable first estimate of the buckling load for any long column.

The Rankine formula is generally adequate for giving a reasonably accurate estimate of the strength of any column, over the whole range of slenderness ratios. It may also be observed that Johnson's parabolic formula with the <u>recommended</u> value of β (0.3E–4) gives a rather poor estimate of buckling loads. With the value β = 0.6133E–04 required to give correspondence with P_R at s = 100, however, Johnson's parabolic formula gives the best approximation of all up to s = 100, nearly corres– ponding with P_P in fact. Beyond s = 100 it is quite useless. The straight line approximation P_A applies up to s = 140. However, for designing a building or structure, the appropriate Codes of Practice (i.e. P_P in the U.K.) must be rigidly adhered to. It must always be remembered that <u>the minimum I value for the section should be chosen</u>, since that will be the most favourable mode for buckling.

PROGRAM 1.3 Column

```
 1:*       PROGRAM COLUMN
 2:        WRITE(*,1)
 3: 1      FORMAT(1X,' "COLUMN" CALCULATES ALLOWABLE COLUMN LOADS',/1X,
 4:       & ' PC,PE,PR,PP,PA AND PJ, WHERE PC=SC*A,PE=PI**2*E*A/S**2',/1X
 5:       & ' = EULER BUCKLING LOAD,PR=SC*A/(1.0+AR*S**2)=RANKINE ',/1X,
 6:       & ' ALLOWABLE LOAD,PP=A*(X/2.-((X**2-4.*SC*SE)**.5)/2.,',/1X,
 7:       & ' PJ=PC*(1.-BETA*S**2),SC=MAXIMUM ALLOWABLE COMPRESSIVE',/1X,
 8:       & ' =PERRY-ROBERTSON ALLOWABLE LOAD,PA=PC*(1.-ALPHA*S)',/1X,
 9:       & ' STRESS,SE=PE/A=EULER BUCKLING STRESS,A=AREA,',/1X,
10:       & ' AR=SC/(XN*PI**2*E)=CONSTANT IN RANKINE FORMULA,XN=1,',/1X,
11:       & ' ALPHA=.0054,BETA=.00003 FOR PINNED ENDS,XN=4,ALPHA=',/1X,
12:       & ' .0036,BETA=.00002,FOR FIXED ENDS ETC.,E=YOUNGS',/1X,
13:       & ' MODULUS,S=SLENDERNESS RATIO L/K OF COLUMN,',/1X,
14:       & ' X=SC+SE*(ETA+1.0),ETA=0.003*S,PI=3.14159. '//)
15:        WRITE(*,2)
16: 2      FORMAT(1X,' TYPE IN A, SC, E, XN, ALPHA, BETA '/)
17:        READ(*,*) A, SC, E, XN, ALPHA, BETA
18:        WRITE(*,3) A, SC, E, XN, ALPHA, BETA
19: 3      FORMAT(/1X,'A=',E10.5,' SC=',E10.5,' E=',E10.5,' XN=',E10.5,/
20:       & 1X,' ALPHA=',E10.5,' BETA=',E10.5,/)
21:        PI=3.141592654
22:        AR=SC/(XN*PI**2*E)
23:        S=10.
24:        PC=SC*A
25:        WRITE(*,4) PC
26: 4      FORMAT(1X,' PC=',E10.5,/)
27:        WRITE(*,7)
28: 7      FORMAT(/8X,1HS,9X,2HPE,9X,2HPR,9X,2HPP,9X,2HPA,9X,2HPJ/)
29: 5      ETA=0.003*S
30:        PE=PI**2*E*A/S**2
31:        SE=PE/A
32:        X=SC+SE*(ETA+1.0)
33:        PR=PC/(1.0+AR*S**2)
34:        PP=A*(X/2.0-((X**2-4.0*SC*SE)**0.5)/2.0)
35:        PA=PC*(1.-ALPHA*S)
36:        PJ=PC*(1.-BETA*S**2)
37:        WRITE(*,6) S, PE, PR, PP, PA, PJ
38: 6      FORMAT(2X,6E11.5)
39:        S=S+10.0
40:        IF(S.LT.201.0) GO TO 5
41:        STOP
42:        END
```

SOLUTION 1.3

```
column
"COLUMN" CALCULATES ALLOWABLE COLUMN LOADS
PC,PE,PR,PP,PA AND PJ, WHERE PC=SC*A,PE=PI**2*E*A/S**2
= EULER BUCKLING LOAD,PR=SC*A/(1.0+AR*S**2)=RANKINE
ALLOWABLE LOAD,PP=A*(X/2.-((X**2-4.*SC*SE)**.5)/2.,
=PERRY-ROBERTSON ALLOWABLE LOAD,PA=PC*(1.-ALPHA*S)
PJ=PC*(1.-BETA*S**2),SC=MAXIMUM ALLOWABLE COMPRESSIVE
STRESS,SE=PE/A=EULER BUCKLING STRESS,A=AREA,
AR=SC/(XN*PI**2*E)=CONSTANT IN RANKINE FORMULA,XN=1,
ALPHA=.0054,BETA=.00003 FOR PINNED ENDS,XN=4,ALPHA=
.0036,BETA=.00002,FOR FIXED ENDS ETC.,E=YOUNGS
MODULUS,S=SLENDERNESS RATIO L/K OF COLUMN,
X=SC+SE*(ETA+1.0),ETA=0.003*S,PI=3.14159.


TYPE IN A, SC, E, XN, ALPHA, BETA

.0064  324.  207000.  1.  .0054  .00003

A=.64000E-02 SC=.32400E+03 E=.20700E+06 XN=.10000E+01
  ALPHA=.54000E-02 BETA=.30000E-04

PC=.20736E+01
```

S	PE	PR	PP	PA	PJ
.10000E+02	.13075E+03	.20412E+01	.20123E+01	.19616E+01	.20674E+01
.20000E+02	.32688E+02	.19499E+01	.19492E+01	.18497E+01	.20487E+01
.30000E+02	.14528E+02	.18146E+01	.18793E+01	.17377E+01	.20176E+01
.40000E+02	.81720E+01	.16539E+01	.17971E+01	.16257E+01	.19741E+01
.50000E+02	.52301E+01	.14849E+01	.16968E+01	.15137E+01	.19181E+01
.60000E+02	.36320E+01	.13200E+01	.15737E+01	.14018E+01	.18497E+01
.70000E+02	.26684E+01	.11669E+01	.14282E+01	.12898E+01	.17688E+01
.80000E+02	.20430E+01	.10291E+01	.12693E+01	.11778E+01	.16755E+01
.90000E+02	.16142E+01	.90765E+00	.11111E+01	.10658E+01	.15697E+01
.10000E+03	.13075E+01	.80189E+00	.96565E+00	.95386E+00	.14515E+01
.11000E+03	.10806E+01	.71040E+00	.83851E+00	.84188E+00	.13209E+01
.12000E+03	.90800E+00	.63148E+00	.73029E+00	.72991E+00	.11778E+01
.13000E+03	.77368E+00	.56345E+00	.63922E+00	.61793E+00	.10223E+01
.14000E+03	.66710E+00	.50473E+00	.56275E+00	.50596E+00	.85432E+00
.15000E+03	.58112E+00	.45391E+00	.49838E+00	.39398E+00	.67392E+00
.16000E+03	.51075E+00	.40981E+00	.44396E+00	.28201E+00	.48108E+00
.17000E+03	.45243E+00	.37140E+00	.39768E+00	.17004E+00	.27579E+00
.18000E+03	.40356E+00	.33781E+00	.35807E+00	.58061E-01	.58061E-01
.19000E+03	.36220E+00	.30834E+00	.32397E+00	-.53914E-01	-.17211E+00
.20000E+03	.32688E+00	.28237E+00	.29443E+00	-.16589E+00	-.41472E+00

$\beta=.00006133$ $\beta=.0000943$ $\beta=.00015859$

PJ	PJ	PJ
20609E+01	.20540E+01	.20407E+01
.20227E+01	.19954E+01	.19421E+01
.19591E+01	.18976E+01	.17776E+01
.18701E+01	.17607E+01	.15474E+01
.17557E+01	.15847E+01	.12515E+01
.16158E+01	.13697E+01	.88973E+00
.14504E+01	.11155E+01	.46222E+00
.12597E+01	.82214E+00	-.31054E-01
.10435E+01	.48972E+00	-.59010E+00
.80186E+00	.11820E+00	-.12149E+01
.53480E+00	-.29244E+00	-.19055E+01
.24230E+00	-.74218E+00	-.26619E+01
-.75639E-01	-.12310E+01	-.34840E+01
-.41901E+00	-.17590E+01	-.43719E+01
-.78781E+00	-.23261E+01	-.53256E+01
-.11821E+01	-.29322E+01	-.63450E+01
-.16017E+01	-.35775E+01	-.74302E+01
-.20468E+01	-.42619E+01	-.85812E+01
-.25174E+01	-.49854E+01	-.97980E+01
-.30134E+01	-.57480E+01	-.11080E+02

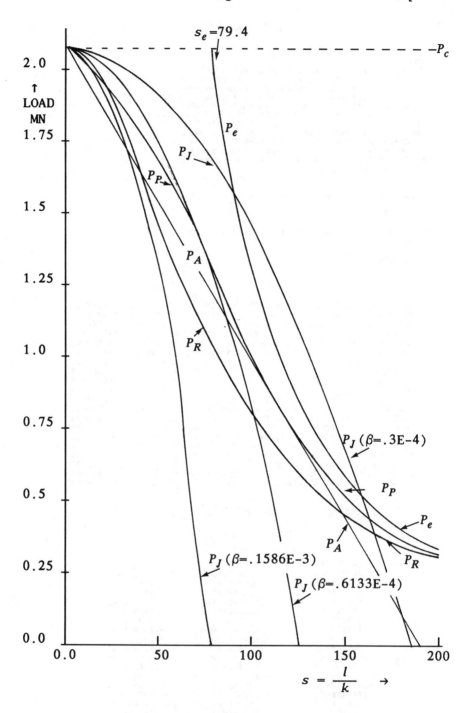

Fig. 1.62 Limiting column loads for various theories

Chapter 2

THE TORSION OF SHAFTS

2.1 TORSION OF CIRCULAR SHAFTS

This chapter deals with *twisting* or *torsion*. A machine element subjected to twisting is referred to as a *shaft*. This phenomenon is observed in applications like power transmissions, torsion springs, instrumentation using small torsion springs, structural members subject to eccentric loading etc. The primary interest here is the relationship between the applied twisting moment and the resulting stress and angular displacement of the element. Torsion in a machine member involves variable stress. In order to study this, the distribution of stress and strain throughout the machine member will have to be considered.

The steps given below are generally followed to derive the stress-strain relations for torsion. They are:

1. Determine the relations between stresses that are compatible with the deformations from the study of elastic deformations produced by a specific load and application of Hooke's law. These are known as *equations of compatibility*.

2. Apply the equilibrium conditions to a free body diagram to obtain relations between the stresses. These relations result from equilibrium considerations between external loads and internal resistance of the material. These are known as *equations of equilibrium*.

3. Ensure that the solutions thus achieved are satisfied at the surface of the body - called *boundary conditions*.

While deriving the required stress-strain relations, some assumptions are made as follows:

a) Circular sections remain circular.
b) Plane sections remain plane and do not warp.
c) Straight radial lines remain straight.
d) Twisting moment vector is parallel to the shaft axis.
e) The stresses produced do not exceed the elastic limit.

It will now be shown that the first three assumptions are valid. To do this, different possible modes of deformation will be considered and then it will be logically proven that deformation in only a certain way is possible.

Consider a slice of the shaft of length 'dx', as shown in Fig. 2.1. The faces are plane and normal to the axis of the shaft. Note that, irrespective of the position from which the slice is taken, its shape will be identical to any other slice and the same twisting moment will act on it as is applied at the ends. The pattern of deformation of all these slices will be identical.

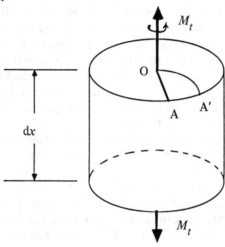

Fig. 2.1 Twisting moment acting on a slice of a shaft

Now suppose that, if one end of the slice bulged out or dished in, symmetry would dictate that the other end should do the same. But then it would be impossible to fit all these deformed disks together to give a continuous shaft. Hence the deformation of the slice cannot leave its plane; i.e. all plane sections must remain plane.

If one radial line OA deforms to OA´. as shown in Fig. 2.2, all the other radial lines will deform in a similar fashion in the same plane. Now, referring to Fig. 2.3, assume that BOA deforms to B´OA´ - deforming in the direction of the twisting moment M_t, and CO_1D deforms to $C´O_1´D´$ in the direction of M_t on the lower face. Now consider the slice immediately below the first one. On the upper face of the lower slice, $C_1O_1D_1$ should deform to $C_1´O_1D´$. But this face is the same as the lower face of the upper slice. This obviously gives rise to the contradiction as the sense of deformation of $C_1O_1D_1$ and CO_1D is different - and this is impossible in a continuous material.   Hence this mode of deformation is ruled out.

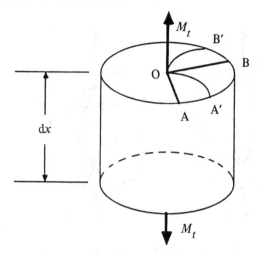

Fig. 2.2 Possible distortion in torsion – Case I

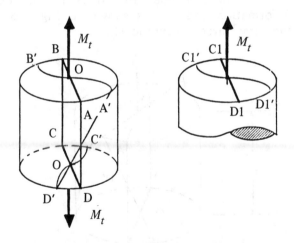

Fig. 2.3 Matching distortion of two adjacent slices

Now assume the curvature of CO_1D is opposite to the one previously considered. This situation is shown in Fig. 2.4. By doing this, adjacent elements can be matched.   Now if this element is turned upside down and placed beside the first figure, it is noticed that the curvature of $B'OA'$ and $C'O_1D'$ are in opposite senses.   Both elements have the same geometry, both are under the same loading and hence both deformations should be identical. This gives another contradiction. Now it can be said that straight diameters do not deform into curves at all and that they remain straight after deformation.

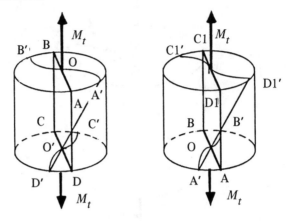

Fig. 2.4 Possible distortion in torsion – Case II

Hence it can be concluded that there is only one possible mode of deformation and that is shown in Fig. 2.5. This has validated the first three assumptions.

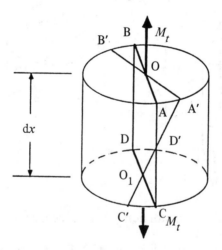

Fig. 2.5 The only possible mode of deformation in torsion

2.1.1 Derivation of Torsion Formula

Fig. 2.6 shows a solid circular shaft. When a torque M_t is applied at its two ends, an outside fibre AB is twisted along the surface to take up a new position that describes a helix as the shaft is twisted through an angle θ.

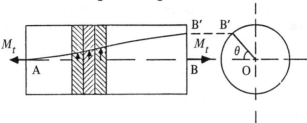

Fig. 2.6 Solid circular shaft

Consider this shaft to be made up of many thin slices each of which is rigid and joined to the adjacent slice with elastic fibres. Slice 2 will rotate past slice 1 until the elastic fibres joining them deform enough to produce a resisting torque equal to the applied torque. When this happens, slice 2 stops moving and slice 3 starts rotating past slice 2 till the joining fibres develop a resisting torque equal to the applied torque. This type of deformation will propagate through the length of the shaft. As the slices become infinitely thin, the original line AB will define the helix AB´ after twisting. It can be seen from Fig. 2.6 that any point C along the radius OB´ also rotates through an angle θ when the shaft twists.

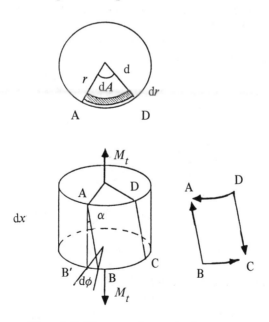

Fig. 2.7 Torque formulation

Fig. 2.7 shows the same situation. The lower face rotates through an angle $d\varphi$ with respect to the upper surface. The lateral surface ABCD takes the new form as shown. The edges AB and CD are deformed through the angle γ. This element is in pure shear. The shear angle γ can be calculated from the triangle AB´B. In this triangle, side BB´ $= r.d\varphi = (d/2).d\varphi$, and the side AB´$=dx$.

Therefore, $\gamma = (d/2).d\varphi/dx$ (2.1)

Here $d\varphi/dx$ represents the angle of twist per unit length. This quantity is constant along the length of the shaft. Let this be denoted by θ. This gives:

$$\gamma = (d/2).\theta \qquad (2.2)$$

The relationship between the shear strain and shear stress is known, namely:

$$\gamma = \tau/G \qquad (2.3)$$

This gives:

$$\tau = G.d.\theta/2 \qquad (2.4)$$

This specifies the stress on the surface of the shaft. Consider an element within the shaft. With the previous assumption that after deformation the diameters remain straight and plane sections remain plane, the shear stress at an internal point is obtained by replacing $d/2$ in the above expression by r. This gives:

$$\tau = G.r.\theta \qquad (2.5)$$

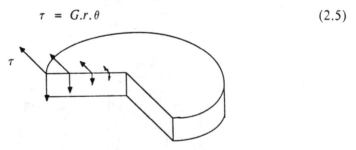

Fig. 2.8 Stress distribution over shaft cross-section

Equation (2.5) shows that the shearing stress varies directly as the radius r. Fig. 2.8 shows the stress distribution over the cross-section. The maximum shear stress occurs at the maximum radius r, i.e. on the surface of the shaft.

2.1.2 *Representation of Pure Shear*

A small element is shown in Fig. 2.9.   The shaft is under a twisting moment only.   This small element will only have the shear stress τ acting along the sides.   Both the other stresses, σ_x and σ_y, are zero.   Referring to the next figure, note that line AB represents this state of pure shear.   If this line is rotated through 90° counterclockwise, a state of stress such that $\tau_{xy} = 0$, $\sigma_x = +\tau$ and $\sigma_y = -\tau$ is obtained.   A´ represents the principal stress σ_1 and B´ represents the principal stress σ_2.   This element is shown in Fig. 2.10.

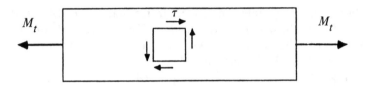

Fig. 2.9a Shaft under torsion

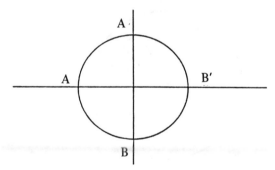

Fig. 2.9b Mohr's circle representation
of a shaft under torsion

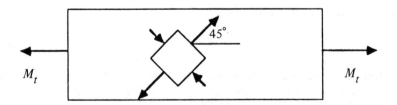

Fig. 2.10 Principal stresses in a shaft under torsion

After observing these figures, it is possible to explain some failures that occur in materials like wood and cast iron. In the case of wood, its longitudinal shear strength is less than its transverse shear strength. Hence failure often manifests itself in the longitudinal direction on the surface in the form of cracks. A material like cast iron is weaker in tension than in compression. It is observed that the maximum tensile stress occurs at 45° to the axis along any point on the surface. This results in a helical crack with the helix angle equal to 45°.

2.1.3 *Equilibrium*

The shear stress τ acts on an infinitesimal area dA producing a force P as shown in Fig. 2.11. This force P produces a moment about the axis O. All such moments add up to resist the externally applied moment M_t.

Using equation (2.5), $M_t = {}_A\!\int \tau.dA.r = {}_A\!\int G.r.\theta.dA.r,$

$$\therefore \quad M_t = G.\theta.{}_A\!\int r^2 dA = G.\theta.I_p \qquad (2.6)$$

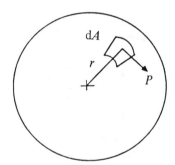

Fig. 2.11 Equilibrium requirements

I_p is the polar moment of inertia of the circular cross-section, and θ is the twist per unit length. If the shaft length is L, then the total deformation is:

$$\varphi = L.\theta = L.M_t/(G.I_p) \qquad (2.7)$$

To find the relation between the shear stress τ and the applied moment M_t, substitute the value of θ given by equation (2.5) into equation (2.6), to give:

$$\tau = M_t r/I_p \qquad (2.8)$$

It can be seen from this that the maximum shear stress occurs at $r = d/2$, thus:

$$\tau_{max} = M_t d/(2.I_p) \qquad (2.9)$$

In the case of the circular shaft under consideration,

$$I_p = \pi.d^4/32 \qquad (2.10)$$

Substituting this value into equations (2.6) and (2.9) gives:

$$M_t = G.\theta.\pi.d^4/32 \qquad (2.11)$$

and

$$\tau_{max} = 16M_t/(\pi.d^3) \qquad (2.12)$$

If the power transmission requirements are known, the diameter of the shaft can be calculated for safe operation. The power is torque × angular velocity. This gives:

$$M_t.2\pi.N/60 = 550{\times}12{\times}(\text{H.P.}) \qquad (2.13)$$

where N is the shaft speed in r.p.m. and M_t is in lb.in. units. From equation (2.12) maximum allowable shaft torque is:

$$M_{tmax} = \pi.d^3.\tau_{max}/16 \qquad (2.12a)$$

Substituting this value of M_t into equation (2.13) gives the value of d (in.) for safe operation as:

$$d = 68.5 \sqrt[3]{\frac{(\text{H.P.})}{N.\tau_{max}}} \qquad (2.13a)$$

Ex 2.1   Determine the minimun diameter(in inches) of a shaft that can be safely used to transmit 50 h.p at 500 r.p.m. The maximum allowable shear stress for the shaft material is 20,000 lb/in^2 (138 MPa).
      Using equation (2.13a),

$$d = 68.5 \sqrt[3]{\frac{50}{500 \times 20,000}}$$

$$d = \frac{68.5}{\sqrt[3]{20,000}} = 1.18 \text{ inches } (30 \text{ mm})$$

2.1.4 Torsion of Hollow Shafts

All the relations derived for the angle of twist and the shear stresses are valid for hollow shafts. The only difference is the change in the limits of integration which change from 0 to r to r_1 to r_2. This gives the new equilibrium equation:

$$M_t = G.\theta.\int_{r_1}^{r_2} r^2 dA = G.\theta.I_p \qquad (2.14)$$

where:

$$I_p = \pi.(r_2^4 - r_1^4)/2 \qquad (2.15)$$

hence:

$$\theta = 2.M_t/[\pi.(r_2^4 - r_1^4).G] \qquad (2.16)$$

and the total twist φ is:

$$\varphi = \theta.L = 2.M_t.L/[\pi.(r_2^4 - r_1^4).G] \qquad (2.17)$$

and:

$$\tau_{max} = M_t.d/[\pi.(r_2^4 - r_1^4).G] \qquad (2.18)$$

2.1.5 Combined Bending and Torsion

This section deals with circular shafts that are acted upon by both torsion and bending. In most practical applications torsion and bending are found to act simultaneously. For example consider a pulley mounted on a shaft. This shaft is acted upon by three loads, as shown in Fig. 2.12, viz.:

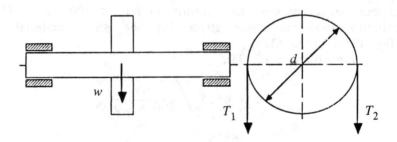

Fig. 2.12 Solid circular shaft with pulley

1.   the weight of the pulley,
2.   the loading due to tensions in the ropes passing over the pulley and,
3.   torsion due to the difference in the tensions in these ropes resulting in a twisting moment.

In the above figure, there is a bending moment in the shaft due to the loads $W+T_2+T_2$ and a twisting moment due to $(T_1-T_2).d$. In addition to this, a circular shaft is often subjected to axial loads and it is necessary to consider the combined stress problem. The resultant stresses are obtained by superposition of the individual effects. The stresses and strains caused by one kind of loading are not affected by the presence of another kind of loading. This superposition is justified owing to the linearity of the equations of equilibrium, compatibility and the stress-strain-temperature relations.

Case 1. Combined stresses due to torsion and tension.

Fig. 2.13 shows a solid uniform homogeneous circular shaft subjected to an axial tensile force P and a twisting moment M_t. The stresses due to individual types of loadings

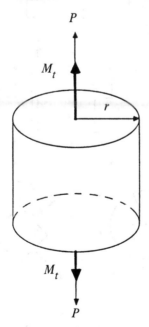

Fig. 2.13 Solid circular shaft under axial force and twist

will be obtained and then added to describe the state of
stress.   The twisting moment gives:

$$\tau_{\theta z} = M_t.r/I_p \qquad (2.19)$$

The tensile force P gives the axial stress in the z direction
as:

$$\sigma_z = P/\pi.r_1{}^2 \qquad (2.20)$$

$\tau_{\theta z}$ varies directly as r and $\tau_{\theta z max}$ occurs at $r = r_1$.   The
most convenient method of describing the combined stress state
is by using the principal stress components.   Fig. 2.14 shows
the individual stress distributions and the stresses acting on an
element on the surface of the cylinder where the shear stress
is a maximum.   The two individual stress states are super-
posed to give the combined stress state and then used to draw
the Mohr's circle to find the principal stresses and directions.

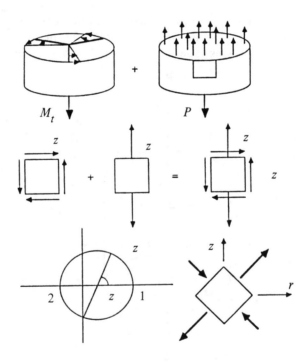

Fig. 2.14 Stresses on a surface element
under combined loading

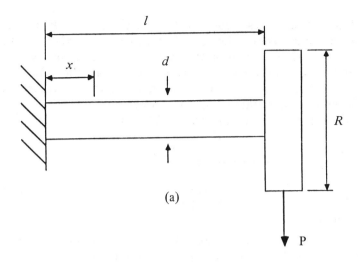

Fig. 2.15a Typical combined loading of a shaft

Case 2. This example considers a circular built-in shaft loaded at the free end at a distance R from the axis as shown in Fig. 2.15a.   This case reduces to one of loading by a torque $M_t = P.R$ and a transverse force P at the free end. The torque is constant along the axis and the bending moment varies along the length of the shaft.   This bending moment is:

$$M_b = -P.(1-x) \qquad (2.21)$$

at any cross-section distant x from the built-in end.   The following stresses need to be determined:

1. Shearing stresses due to the torque M_t,
2. Normal stresses due to the bending moment M_b and
3. Shearing stresses due to the vertical load P.

τ_{max} occurs at the surface of the shaft, and is given by equation (2.12).   The maximum bending moment and hence the maximum bending stress occurs at the built-in end, where $x=0$, at the fibres most distant from the neutral axis of the shaft.   This stress is derived as follows:

$$\sigma_b = \frac{M_b c}{I} = \frac{M_b d/2}{\pi d^4/64}$$

i.e. $\sigma_{max} = 32.M_b/\pi.d^3 \qquad (2.22)$

The maximum shear stress due to the vertical load P occurs on the neutral axis where the bending moment is zero. The maximum combined stress usually occurs at the point where stresses 1 and 2 in the above list have their maxima. In this particular case these occur at the top and bottom surfaces at the built-in end as illustrated in Fig. 2.15b.

Analytically, these are derived as follows:

$$\tau = \frac{M_t r}{I_p} = \frac{M_t \, d/2}{\pi d^4/32}$$

$$= \frac{16 M_t}{\pi \, d^3}$$

The Mohr's circle for this state of stress is shown in Fig. 2.15c. The principal stresses can be determined as:

$$\sigma_{max} = \frac{16}{\pi d^3}\left[M_b + \sqrt{(M_b^2 + M_t^2)}\right]$$

and   (2.23)

$$\sigma_{min} = \frac{16}{\pi d^3}\left[M_b - \sqrt{(M_b^2 + M_t^2)}\right]$$

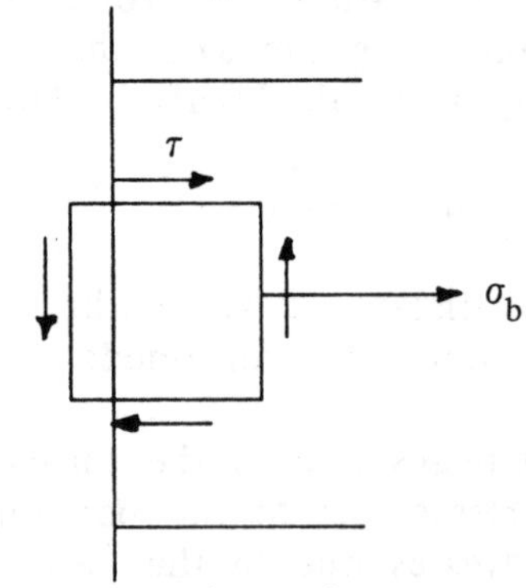

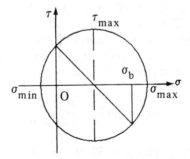

Fig. 2.15b Location of maximum combined stresses

Fig. 2.15c Mohr's circle

The maximum shearing stress in the element is:

$$\tau_{max} = (\sigma_{max} - \sigma_{min})/2 = 16.\sqrt{(M_b{}^2 + M_t{}^2)}/(\pi.d^3) \qquad (2.24)$$

These stresses can of course be found by drawing the Mohr's circle from the known values of σ_x, σ_y and τ_{xy}, exactly as shown for the previous case.

For ductile materials, it is a common practice to use the maximum shear stress to determine a safe diameter for the shaft. From the previous section:

$$\tau_{max} = 16M_{combined}/(\pi.d^3) = 16M_c/(\pi.d^3) =$$
$$16.\sqrt{(M_b{}^2 + M_t{}^2)}/(\pi.d^3) \qquad (2.25)$$

Solving for d gives: $d = \sqrt[3]{(16M_c)/\pi.\tau_{max}} \qquad (2.26)$

where τ_{max} is the maximum working shear stress.

The above example could be extended to hollow circular shafts of outer diameter d_0 and inner diameter d_1. These formulae remain the same except for the value of the section modulus Z. This is given by:

$$Z = \pi d_0{}^3[1-(d_1/d_0)^4]/32 \qquad (2.27)$$

For these sections the ratio d_1/d_0 has to be fixed before finding the value for d_0.

Ex 2.2 A built-in solid shaft of length 150 mm is loaded at the free end at a distance R=100 mm from the axis as shown in Fig 2.15a. If the load is 2 kN and the allowable maximum shear stress is 100 MPa, (a) determine the minimum safe diameter of the shaft. (b) If the shaft is hollow with $d_i/d_o = 3/4$ find the new dimensions of the shaft and estimate the weight saving.

a) Equation (2.26) gives

$$d = \sqrt[3]{(16M_c)/\pi.\tau_{max}}$$

$$M_c = \sqrt{(M_b{}^2 + M_t{}^2)}$$

Maximum $M_b = 2 \times 150 = 300$ kN mm

$$M_t = 2 \times 100 = 200 \text{ kN mm}$$

$$d = \sqrt[3]{\frac{1600 \sqrt{13}}{\pi \times .1}} = 26.38 \text{ mm}$$

Hence, area = 546.7 mm²

b) For the hollow shaft,

$$\tau_{max} = \frac{16 \, d_o}{\pi (d_o^2 - d_i^2)} \cdot \sqrt{M_b^2 + M_t^2}$$

$$0.1 = \frac{16 \, d_o}{\pi d_o^3 (1 - (3/4)^4)} \quad 100\sqrt{13}$$

$$d_o^3 = \frac{1600 \sqrt{13}}{\pi (.1)(1 - .75^4)}$$

$$= 29.95 \text{ mm}$$

Hence, area = 308.2 mm²

So, the percentage weight saving is 44%

2.1.6 Helical Springs

A good application of the torsion formulae presented in this chapter is in the design of a helical spring. Springs are used in machines and devices to exert force, to provide flexibility and to absorb or store energy.

Fig 2.16a shows a circular wire helical compression spring loaded by an axial force, p. D is the mean spring diameter and d is the wire diameter. If the spring is cut at some point as shown in Fig. 2.16b and a portion of it removed as shown in the figure, the force acting at the cut could be represented by a shear force, P, and a torsion, M_t.

The maximum shear stress in the wire may be determined by the superposition of the direct shear stress due to P and the maximum shear stress due to M_t given by equation (2.9). Thus,

$$\tau_{max} = \frac{P}{A} \pm \frac{M_t d}{2I_p} \tag{2.28}$$

where $A = \pi d^2/4$

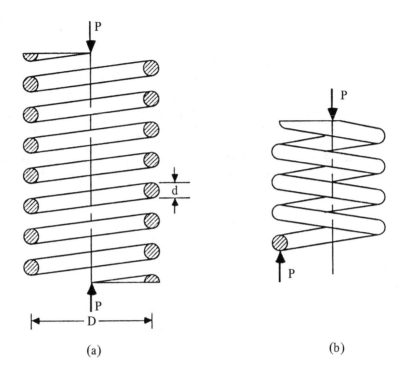

(a)   (b)

Fig. 2.16 Helical springs

$$I_p = \pi d^4/32$$

Using $M_t = PD/2$ we get,

$$\tau_{max} = \frac{4P}{\pi d^2} + \frac{8PD}{\pi d^2} \qquad (2.29)$$

If the spring index is defined as

$$C = D/d$$

then $\tau_{max} = k_s \dfrac{8PD}{\pi d^3}$ \qquad (2.30)

where k_s is the shear stress correction factor and is defined as

$$k_s = \frac{2C + 1}{2C} \tag{2.31}$$

C usually ranges from about 6 to 12 for most springs.

The effect of curvature on the stresses at the inside surface of a helical spring is similar to a stress concentration. The curvature stress is highly localized and it is important only when fatigue is present. For static loads local yielding may releive the high curvature stress.

Wahl and Bergstrasser has suggested factors to correct the stress by replacing K_S by one of the following equations:

$$K_w = \frac{4C - 1}{4C - 4} + \frac{0.615}{C} \tag{2.32}$$

$$K_b = \frac{4C + 2}{4C - 3} \tag{2.33}$$

The first is called the Wahl factor and the second the Bergstrasser factor. They differ by less than one percent and equation (2.33) is preferred because of its simplicity.

The deflection versus force relationship (spring stiffness) is obtained by using Castigliano's Theorem. The total strain energy for a helical spring is composed of a torsional component and a shear component.

The strain energy is given by:

$$U = \frac{P^2 L}{2AG} + \frac{M_t^2 L}{2GI_p} \tag{2.34}$$

where G is the shear modulus of the material.

Substituting

$$M_t = PD/2$$

$$L = \pi D N$$

$$I_p = \pi d^4/32$$

$$A = \pi d^2/4$$

results in

$$U = \frac{2P^2DN}{d^2G} + \frac{4P^2D^3N}{d^4G} \tag{2.35}$$

where N is the number of active coils.
Using Castigliano's theorem,

$$\delta = \frac{\partial u}{\partial p} = \frac{4PDN}{d^2G} + \frac{8PD^3N}{d^4G}$$

$$= \frac{8PD^3N}{d^4G} \left[\frac{1}{2C^2} + 1 \right]$$

$$= \frac{8PD^3N}{d^4G} \tag{2.36}$$

The spring rate is $k = P/\delta$.

$$\text{Hence} \quad k = \frac{d^4G}{8D^3N} \tag{2.37}$$

2.2 TORSION OF NON-CIRCULAR CROSS-SECTIONS

2.2.1 *Introduction*

Elementary torsion theory as described in the previous section deals only with the circular cross-sections. However, there are applications where non-circular and tubular cross-sections are encountered, e.g. prismatic and hollow shafts. These experience deformations that are out of the original plane. Hence the original assumption that plane cross-sections remain plane is no longer valid.

Fig. 2.17 shows a slender prismatic bar under equal and opposite twisting moments applied at the two ends. The centre of twist is chosen as the origin of the coordinate axes x, y and z. At this point the x and y displacements are zero. The z axis passes through the centres of twist of all the planes. When twisting moments are applied, in general the cross-sections deform out of their plane. The warping function is taken to be independent of the axial position.

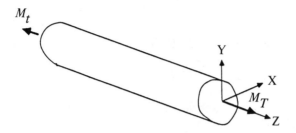

Fig. 2.17 Torsion of a non-circular bar

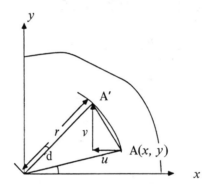

Fig. 2.18 Deflection of a point on a cross-section of a prismatic bar in torsion

This means that this function is identical for all cross-sections. It is assumed that the projection of a warped cross-section rotates as a rigid body. This warping function may be written as follows:

$$w = f(x,y) \tag{2.38}$$

In Fig. 2.18, point A is shown rotating through an angle $d\theta$ into a new position A´. The x coordinate changes from x to x-u and the y coordinate changes from y to y+v as the point A moves owing to torsion. With the assumptions that no rotation occurs at one end and that the angle of rotation $d\theta$ is small, the displacements u and v for the point A may be written as follows

$$u = -AA´\sin\theta = -r\,d\theta\sin\theta = -y\,d\theta \tag{2.39a}$$

and

$$u = - AA'\cos\theta = - rd\theta\cos\theta = - xd\theta \tag{2.39b}$$

On substituting these relations and the defined warping function into the relations for normal and shearing strains, the following relations are obtained:

$$\epsilon_x = \epsilon_y = \epsilon_z = \gamma_{xy} = 0 \tag{2.40a}$$

$$\gamma_{xz} = \partial w/\partial x - y\theta \tag{2.40b}$$

$$\gamma_{xy} = \partial w/\partial y + x\theta \tag{2.40c}$$

The stress-strain relationships, together with the above relations give the following equalities:

$$\sigma_x = \sigma_y = \sigma_z = \tau_{xy} = 0 \tag{2.41a}$$

$$\tau_{xz} = G(\partial w/\partial x - y\theta) \tag{2.41b}$$

$$\tau_{yz} = G(\partial w/\partial y + x\theta) \tag{2.41c}$$

The equations of equilibrium for three-dimensional stress give the following:

$$\partial\tau_{xz}/\partial z = 0 \tag{2.42a}$$

$$\partial\tau_{yz}/\partial z = 0 \tag{2.42b}$$

$$\partial\tau_{xz}/\partial x + \partial\tau_{yz}/\partial y = 0 \tag{2.42c}$$

Differentiating equation (2.41b) with respect to y and equation (2.41c) with respect to x, and then subtracting the second from the first gives the equation of compatibility, namely"

$$\partial\tau_{xz}/\partial y - \partial\tau_{yz}/\partial x = - 2G\theta \tag{2.43}$$

By solving equations (2.42b) and (2.43) together with the boundary conditions, the stress in a bar of non-circular cross-section may be determined.

2.2.2 Saint Venant's Torsion Theory

In the year 1853, Saint Venant presented a solution of the

torsion problem to the French Academy of Sciences. He made the following assumptions for the shaft:

a) The shaft was straight
b) The shaft had a constant cross-section
c) The shaft material was homogeneous
d) The shaft material was isotropic
e) The shaft material was linearly elastic

Equation (2.40a) shows that, since $\tau_{xy} = 0$, a cross-section does not distort in its own plane. The only non-zero stresses are the shear streses τ_{xz} and τ_{yz}. The equations of equilibrium for three dimensions give the following relationship:

$$\partial\sigma/\partial x + \partial\tau_{xy}/\partial y + \partial\tau_{xz}/\partial z = 0 \text{ (Equilibrium equation)}$$

Using equation 2.40b and 2.40c yields

$$\partial^2 w/\partial x^2 + \partial^2 w/\partial y^2 = 0 \qquad\qquad (2.44)$$

alternatively written as:

$$\nabla^2 w = 0 \qquad\qquad (2.44a)$$

This is the Laplace equation. The *warping function w* is known as a *harmonic*. It satisfies the equilibrium and compatibility equations of the elastic theory. For a correct solution, this function also needs to satisfy the boundary conditions. It is often more convenient to use a stress function rather than a warping function.

2.2.3 *Prandtl Stress Function*

In 1903, Ludwig von Prandtl proposed that the two stresses τ_{xz} and τ_{yz} be defined in the form of a stress function φ. The essential difference between Saint Venant's equation and Prandtl's is that the first one is expressed in terms of displacement and the second in terms of force (stress). Stresses are defined in terms of the stress function $\varphi(x,y)$, where:

$$\tau_{xz} = \partial\varphi/\partial y$$
$$\qquad\qquad (2.45)$$
$$\tau_{yz} = -\partial\varphi/\partial x$$

These can be written as:

$$\partial \tau_{xz}/\partial x + \partial \tau_{yz}/\partial y = \partial^2 \varphi/\partial x \partial y - \partial^2 \varphi/\partial z \partial y = 0$$

In this case the equation (2.42) (of equilibrium) are satisfied.

When the above relations are substituted into the equation (2.43) (of compatibility), then the following equation is obtained:

$$\partial^2 \varphi/\partial x^2 + \partial^2 \varphi/\partial y^2 = - 2G\theta$$

alternatively written as:

$$\nabla^2 \varphi = - 2G\theta \qquad (2.46)$$

This shows that the Prandtl stress function has to satisfy Poisson's equation to satisfy compatibility requirements. To obtain the boundary conditions, $\partial \varphi/\partial x . dx + \partial \varphi/\partial y . dy = 0$, or

$$d\varphi = 0 \qquad (2.47)$$

This result shows that the Prandtl stress function φ is constant along the boundary of the cross-section. However, the constant value of φ is arbitrary because stresses depend only on the derivative of φ rather than its magnitude. It is convenient to assume $\varphi = 0$ on the boundary. Except for very special cases, where a mathematical solution can be obtained, numerical methods are recommended.

2.2.4 *Boundary Conditions*

Consider the boundary conditions, first treating the load- free lateral surface. τ_{xz} is a z-directed shearing stress acting on a plane whose normal is parallel to the x axis. τ_{xz} acts on the x-y plane and is x-directed.

Referring to Fig. 2.19 the following relation is obtained.

$$\tau_{xz} . l + \tau_{yz} . m = 0 \qquad (2.48)$$

This shows that the resultant shear stress must be tangent to the boundary. The following relations are also evident from Fig. 2.19:

$$l = dy/ds; \qquad\qquad m = - dx/ds \qquad (2.49)$$

These lead to the following equation, in terms of the Prandtl stress function:

$$\partial\varphi/\partial y . dy/ds + \partial\varphi/\partial x . dx/ds = 0 \qquad (2.50)$$

This states that the directional deviation along a boundary curve is 0. Thus the function φ is an arbitrary constant on the boundary. For solid cross-sections, the value of φ can be set to 0 on the boundaries. For multiply-connected cross-sections, only one of the boundary contours could be assigned the value 0. For a solid cross-section member, the normals are parallel to the z-axis. The summation of forces over the ends of the bar is 0.

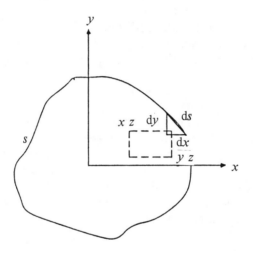

Fig. 2.19 Boundary considerations of a shaft cross-section under torsion

$$\iint T_x \ dxdy = \iint \tau_{xz} \ dxdy = \int \partial\varphi/\partial y \ dxdy =$$

$$\iint dx \ \partial\varphi/\partial y \ dy = \int(\varphi_2 - \varphi_1)dx = 0 \qquad (2.51)$$

It can also be shown that:

$$\iint \tau_{zx} \ dxdy = 0 \qquad (2.52)$$

The end forces provide the twisting moment about the z-axis. This gives:

$$M_t = \iint(x\tau_{yz} - y\tau_{xz}) \ dxdy = \iint x \partial\varphi/\partial x \ dxdy - \iint y \partial\varphi/\partial y \ dxdy$$

$$\int dy \int x \ \partial\varphi/\partial y \ dx - \int dx \int y \ \partial\varphi/\partial y \ dy \qquad (2.53)$$

Since φ is constant on the boundary, the following relation is obtained:

$$M_t = 2 \iint \varphi \ dxdy \qquad (2.54)$$

This represents twice the volume beneath the φ-surface.

Ex 2.3   Fig. 2.20 shows a bar of elliptical cross-section with major axis a and minor axis b. The objective is to determine the maximum shearing stress and the angle of twist per unit length.

In this case the boundary of the elliptical cross-section is known to be given by the equation:

$$x^2/a^2 + y^2/b^2 = 1$$

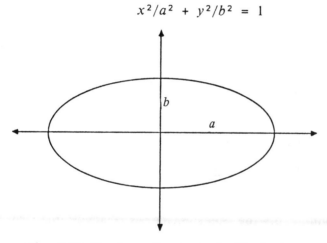

Fig. 2.20 Torsion of a bar of elliptical cross-section

Choose the stress function as follows:

$$\varphi = C.(x^2/a^2 + y^2/b^2 - 1)$$

This choice ensures that the function φ satisfies equations (2.46) and (2.50). This function with equation (2.46) gives:

$$C = a^2.b^2.2.G.\theta/[\,2(a^2+b^2)\,]$$

Substituting this function φ into equation (2.54) gives:

$$M_t = 2.C.[\,(1/a^2)\iint(x^2dxdy + (1/b^2)\iint(y^2dxdy - \iint dxdy\,]$$

$$= [a^2.b^2.2.G.\theta/(a^2+b^2)].[(1/a^2)I_y + (1/b^2)I_x - A]$$

Substituting the expressions for I_x and I_y into the above equation gives:

$$M_t = - \pi a^3.b^3.2.G.\theta/\{2(a^2+b^2)\}$$

This equation can be solved for θ, given the cross-section and the applied external twisting moment.

The shearing stresses can be found using equation (2.55) as:

$$\tau_{xz} = - M_t.y/(2.I_x); \qquad\qquad \tau_{yz} = M_t.x/(2.I_y)$$

Thus, the resultant shearing stress is:

$$\tau_{resultant} = [2.M_t/(\pi ab)]\surd(x^2/a^4+y^2/b^4)$$

The maximum shear stress occurs at the boundary along the minor axis as can be seen from the above expression.

2.2.5 Membrane Analogy

The differential equation describing the deflection of a membrane or soap film subject to pressure is of the same form as the equation for the shear stress in torsion. This leads to an analogy between torsion and membrane problems as illustrated in Fig. 2.21.

The figure shows an edge-supported membrane with its boundary contour of the same shape as that of the bar in torsion. The equation describing the z-deflection of the membrane is derived from equilibrium considerations of an isolated element abcd. The tensile force per unit membrane length is denoted by S. For small deflections, the inclination β can be expressed as approximately equal to $\partial z/\partial x$. As z varies from point to point, the angle of inclination on the other side is given as:

$$\beta + \partial \beta/\partial x.dx \triangleq \partial z/\partial x + \partial^2 z/\partial x^2.dx \qquad (2.55)$$

In the other direction this is given as:

$$\alpha + \partial \alpha/\partial y.dy \triangleq \partial z/\partial y + \partial^2 z/\partial y^2.dy \qquad (2.55b)$$

Neglecting the weight of the membrane and for uniform lateral pressure P, equilibrium considerations give:

$- (S.dy)\partial z/\partial x + S.dy (\partial z/\partial x + \partial^2 z/\partial x^2.dx)$

$-(S.dx)\ \partial z/\partial y + S.dx (\partial z/\partial y+\partial^2 z/\partial y^2.dy) + Pdxdy = 0$ (2.56)

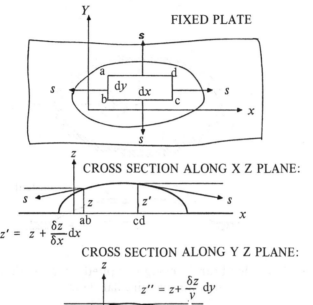

Fig. 2.21 The membrane analogy

This leads to the familiar form of Poisson's equation:

$$\partial^2 z/\partial x^2 + \partial^2 z/\partial y^2 = - p/S$$ (2.57)

This equation has the same form as equation (2.55). The load p/S corresponds to $2G\theta$. Accordingly the Prandtl stress function, φ can be visualized by visualizing the deflected membrane.

Membrane	Torsion
z	φ
$1/S$	G
p	2θ
$\partial z/\partial x,\ \partial z/\partial y$	$\tau_{yz},\ \tau_{xz}$
2.(Volume under membrane)	M_t

The table above lists analogous terms for the torsion and soap bubble problems.

The membrane thus represents the φ-surface. The limitation of this analogy lies in the small deflection of the membrane which is necessary for the analogy to hold.

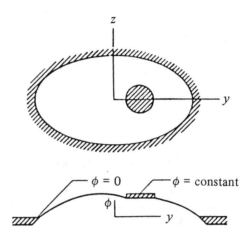

Fig. 2.22 Membrane analogy applied to a section that contains a circular hole

Fig. 2.22 shows a cross-section with a hole. Inside the hole where there is no material, there is no shear stress. Therefore as suggested by Fig. 2.22 the φ surface must have a zero slope over the hole. This situation can be modelled as shown in Fig. 2.22 by having a rigid but weightless plate that 'floats' on the membrane but is constrained to remain horizontal. The volume under the membrane, which is proportional to torque, includes the volume under the floating plate.

2.2.6   Other Sections

The solution of the torsion of rectangular and other sectional bars is very complicated. The angle of twist per unit length, θ, could be related to the torque by an 'effective' polar second moment. Thus

$$\theta = \frac{M_t}{\varphi} \tag{2.58}$$

where φ = the torsional rigidity of the section for a rectangular cross-section.

$$\varphi = \beta ab^3 G \qquad (2.59)$$

β is given in Table 2.1 below for different ratios of a/b. The maximum shear stress could also be evaluated using a parameter α.

Cross section	Shear stress (τ)	Angle of twist per unit length (θ)		
For circular bar $a = b$ (ellipse, $2a$, $2b$)	$\tau_A = \dfrac{2M_t}{\pi ab^2}$	$\dfrac{a^2 + b^2}{\pi a^3 b^3} \cdot \dfrac{M_t}{G}$		
Equilateral triangle	$\tau_A = \dfrac{20M_t}{a^3}$	$\dfrac{46.2}{a^4} \cdot \dfrac{M_t}{G}$		
(rectangle $a \times b$)	$\tau_A = \dfrac{M_t}{\alpha ab^2}$	$\dfrac{1}{\beta ab^3} \cdot \dfrac{M_t}{G}$		
	a/b	β	α	
	1.0	0.141	0.208	
	1.5	0.196	0.231	
	2.0	0.229	0.246	
	2.5	0.249	0.256	
	3.0	0.263	0.267	
	4.0	0.281	0.282	
	5.0	0.291	0.292	
	10.0	0.312	0.312	
	∞	0.333	0.333	
(hollow rectangle)	$\tau_A = \dfrac{M_t}{2abt_1}$ $\tau_B = \dfrac{M_t}{2abt}$	$\theta = \dfrac{at + bt_1}{2t t_1 a^2 b^2} \cdot \dfrac{M_t}{G}$		
For circular tube $a = b$ (elliptic tube $2a$, $2b$)	$\tau_A = \dfrac{M_t}{2\pi abt}$	$\theta = \dfrac{\sqrt{2(a^2 + b^2)}}{4\pi a^2 b^2 t} \cdot \dfrac{M_t}{G}$		

TABLE 2.1 Shear stress and angle of twist for a number of commonly encountered shapes

$$\tau_{max} = \frac{M_t}{Z} \qquad (2.60)$$

where Z can be treated as torsion section modulus. For a rectangle

$$Z = \alpha a b^5 \qquad (2.61)$$

2.2.7 Rectangular Sections

The analysis of the torsion of non-circular cross-sections in detail is beyond the scope of this book. But it is shown experimentally that for a *rectangular shaft* with longer side d and shorter side b, the maximum shear stress occurs at the centre of the longer side and is given by:

$$\tau_{max} = \frac{T}{K_1 d b^2} \qquad (2.62)$$

where K_1 is a constant depending on the ratio d/b and is given in the table below.

d/b	1.0	1.5	1.75	2.0	2.5	3.0	4.0	6.0	8.0	10.0	∞
k_1	0.208	0.231	0.239	0.246	0.258	0.267	0.282	0.299	0.307	0.313	0.333
k_2	0.141	0.196	0.214	0.229	0.249	0.263	0.281	0.299	0.307	0.313	0.333

TABLE 2.2 Values of k_1 and k_2 for rectangular sections

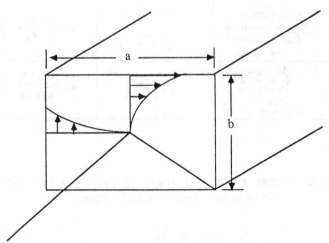

Fig. 2.23 Shear stress distribution

The essential difference between the shear stress distributions in circular and rectangular cross-sections is illustrated in Fig. 2.23, where the shear stress distribution along the major and minor axes of a rectangular section together with a 'radial' line to the corner of the section are shown. The maximum shear stress is found at the centre of the longer side and the stress at the corner is zero.

The angle of twist per unit length is given by:

$$\frac{\theta}{L} = \frac{T}{K_2 db^3 G} \tag{2.63}$$

2.2.8 Torsion of Multiply-Connected Thin-Walled Sections

In Fig. 2.24, a hollow tube of arbitrary cross-section is shown. If a membrane were to span this hollow section, the arc ab would have to be replaced by a plane representing constant φ as the stress in the hollow area must be zero. For bars containing multiply-connected sections, each boundary represents a constant φ value. One boundary may be arbitrarily equated to zero and the others adjusted accordingly.

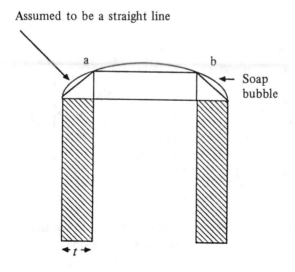

Fig. 2.24 Hollow tube with arbitrary cross-section under torsion

A fixed plate to which the membrane is attached has the same contour as the outer boundary of the tube. The membrane is also attached to a fixed weightless plate having the same shape as the inner boundary of the tube. The inner plate can seek its own position but is not allowed a sideways motion. As the tube is thin-walled, the membrane curvature may be disregarded. Hence the slope is constant over the thickness t and consequently the shearing stress is constant over this thickness. This is given by:

$$\tau = h/t \text{ or } h = \tau.t \tag{2.64}$$

The volume bounded by the mean perimeter is $A.h$. This gives:

$$M_t = 2.A.h \tag{2.65}$$

It can be shown that the angle of twist per unit length is given by:

$$\theta = \{1/(2.G.A)\}.\textstyle\sum\tau ds \tag{2.66}$$

The above three equations can be used to solve problems involving thin hollow non-circular multiply-connected tubes under twisting. An example will be solved to explain this concept.

Ex 2.4

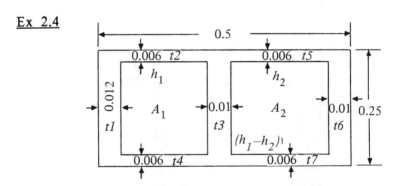

Fig. 2.25 Multiply-connected thin-walled
hollow tube under torsion

Fig. 2.25 shows a multiply-connected rectangular thin tube which is subjected to a twisting moment of $M_t = 56,500$ Nm. Calculate the angle of twist.

$G = 27.6$ GN/m^2 (27.6 × 10^9 N/m^2), $A_1 = A_2 = 0.0625$ m^2.

$$M_t = 2.A_1.h_1 + 2.A_2.h_2$$

this gives the first equation in h:

$$h_1 + h_2 = 452,000 \tag{i}$$

For loop 1, $\sum \tau ds = 2.G.A.\theta$, which gives:

$$0.25\tau_1 + 0.25\tau_2 + 0.25\tau_3 + 0.25\tau_4 = 2.G.A_1.\theta,$$

Substituting for $\tau = h/\text{thickness}$,

$$h_1/0.012 + h_1/0.006 + (h_1 - h_2)/0.01 + h_1/0.006 = 4\times2.G.A_1.\theta,$$

This gives:

$$516.666\, h_1 - 100\, h_2 = 13.8 \times 10^9\, \theta \tag{ii}$$

Similarly for loop 2:

$$0.25\tau_5 + 0.25\tau_6 + 0.25\tau_7 - 0.25\tau_3 = 2.G.A_2.\theta,$$

this gives:

$$h_2/0.006 + h_2/0.01 + h_2/0.006 - (h_1 - h_2)/0.01 = 4\times2.G.A_1.\theta,$$
This leads to the third relation:

$$533.333\, h_2 - 100\, h_1 = 13.8 \times 10^9\, \theta \tag{iii}$$

Solving (i), (ii) and (iii) for h_1 and h_2 gives:

$$h_1 = 229013.333 \text{ and } h_2 = 222986.666$$

With these values, the individual shear stress 'τ' can be calculated from the relation $\tau = h/\text{thickness}$.

From the values of h_1 and h_2 and equation (ii) or (iii), the value of θ can be calculated, and is found to be:

$$\theta = 0.0069583 \text{ rad.}$$

2.3 PLASTIC TORSION - SAND HEAP ANALOGY

If the twist in a bar is gradually increased, at first the material behaves elastically up to a yield point τ_y in shear and then a portion of the bar becomes plastic. It is usual to assume a flat stress-strain diagram (elastic perfectly plastic material) for the plastic region although strain-hardening may also be considered in the analysis. The problem here is to relate the plastic (or partially plastic) torque T_y to the shear stress τ_y, taking into account the geometry of the cross-section.

2.3.1 *Fully Plastic Bar*

Consider the case of the bar having an arbitrary cross-section. Assume that a sufficient twist has been applied and that all material has yielded. With τ_{yz} and τ_{zx} the only non-zero stresses, it can be shown that the principal stresses are:

$$\sigma_1 = -\sigma_2 = \sqrt{(\tau_{yz}^2 + \tau_{zx}^2)}, \text{ and } \sigma_3 = 0 \qquad (2.67)$$

Using Tresca's yield criterion, it can be shown that:

$$\tau_y^2 = \tau_{yz}^2 + \tau_{zx}^2 \qquad (2.68)$$

Now substituting τ_{yz} and τ_{zx} in terms of the Prandtl stress function φ yields:

$$\left[\frac{\partial\varphi}{\partial x}\right]^2 + \left[\frac{\partial\varphi}{\partial y}\right]^2 = \tau_y^2 \qquad (2.69)$$

Using the same boundary conditions as before for φ (i.e. $\varphi=0$ at the boundary) Nadai, in 1923, showed that the *sand heap analogy* can be used to describe equation (2.69), which indicates that the maximum gradient of φ w.r.t. x and y should be equal to the yield shear stress. Thus, if the material is plastic, the maximum slope everywhere in the cross-section is equal. This is analogous to a heap of sand with its surface having the same slope everywhere, as shown in Fig. 2.26.

To use the analogy, sand is poured onto a horizontal flat plate having the same slope as the cross-section of the bar under torsion until the plate is incapable of carrying any more sand. When this point is reached, $\varphi = 0$ on the boundary and the gradient of φ is at its maximum and a constant over the entire cross-section. The torque can be expressed as:

$$T = 2.\int_{\text{area}} \varphi \, dA \qquad (2.70)$$

This also represents the volume of sand on the plate.

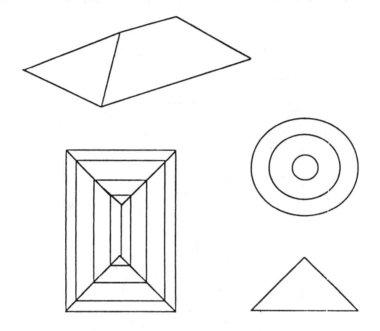

RECTANGULAR CROSS-SECTION CIRCULAR CROSS-SECTION

Fig. 2.26 Sand heap analogy

This torque can readily be calculated for simple geometries of the type shown in Fig. 2.26. For a square cross-section with side 'a'

$$T = 2.\int \varphi \, dA = 2.(1/3)a^2.\tau_y a/2 = \tau_y a^3/3 \qquad (2.71)$$

The sand heap analogy can also be applied to cross-sections with complex geometries, including holes. At the boundary of the cross-section $\varphi=0$, whereas at the boundary of the hole, φ is equal to some constant. In practice this can be achieved by pushing a pipe through the hole and pouring sand over the plate. The pipe is pushed down until any excess sand is spilled through the pipe and the entire pipe under the surface is covered by sand.

2.3.2 *Partially Yielded Bar*

Elastic-plastic torsion where only a portion of the bar has
yielded can be represented by a combination of membrane and
sand heap analogies as shown in Fig. 2.27.   As with the
membrane analogy, a membrane is blown over a hole having
the same cross-section as the bar under torsion.   In addition

MEMBRANE TOUCHING THE ROOF

AT THIS LINE

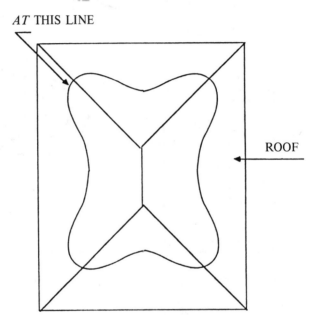

Fig. 2.27 Partially yielded bar combining
membrane and sand heap analogies

to this a roof is built over the top of the membrane, having
the slope of the sand heap analogy.   Individual yielding occurs
where the membrane first touches the roof, under increased
torsion (pressure).   In the case of the rectangular cross-section
shown in Fig. 2.25, yielding will take place at the centre of
the long side of the rectangle.   An increase in the air
pressure (analogous to applying more twist) will further inflate
the membrane and result in a reduction of the elastic portion,
i.e. reduce the free surface of the membrane.

2.4 COMPUTER PROGRAM FOR TORSION

The following section provides a computer listing of a set of FORTRAN programs in TORSION suitable for running on any PC or a larger computer. (The latter may require a few modifications depending on the compiler).

Given the values of torque, the maximum shear stress and the shear modulus of the shaft material, the program will calculate the dimensions of the shaft and the angle of twist per unit length.

The program consists of a main program and a number of subroutines, the functions of which are briefly explained below.

MAIN PROGRAM:
>   Selects the option from the following;
>   1. Circular shaft under pure torsion.
>   2. Non-circular shaft under pure torsion.
>   3. Shafts under combined torsion and
>   bending.

SUBROUTINE ENTER:

>   Asks for the user input of the values for torsion, maximum shear stress and shear Young's Modulus.

SUBROUTINE CIRDIA:

>   This subroutine is called by the main program for circular shaft under pure torsion.
>   It selects the option from:
>   1. Solid shaft
>   2. Hollow shaft
>   and outputs the radius of the shaft and the angle of twist per unit length.

SUBROUTINE NONCIR:

>   This subroutine is called by the main program for non-circular shafts under pure torsion.
>   It selects the option from the following;
>   1. Elliptical cross-section
>   2. Triangular cross-section
>   3. Solid rectangular cross-section

4. Hollow rectangular cross-section
5. Hollow elliptical cross-section
6. Regular Hexagonal cross-section
7. Multiple-connected thin-walled hollow tube.
This subroutine takes the user input and accordingly calls the subroutines ELLIP, EQTRI, RECTA, HORECT, HOELLI, REHEX and MULTI depending on the user input.

SUBROUTINE TORBEN:

This subroutine is called by the main program for shafts under combine torsion and bending. It selects the option from;
1. Combined torsion and bending
2. Built-in shaft.
This subroutine calls TENTOR and DUCDIA according to the user input.

```
         PROGRAM TORSION
c        main   program
c
c        variables
         real   normal(2),shear(2)
         integer   j
c
c     normal: maximum and minimum normal stress, psi(or Pa)
c     shear : maximum and minimum shear stress, psi(or Pa)
c     j      : decided input
c
c     subprograms
      external   cirdia,noncir,torben
c
c     begin( ===========main program=========)
c
10       write(*,*)'select:'
         write(*,*)'1 for circular shaft under pure torsion'
         write(*,*)'2 for noncircular shafts under pure torsion'
         write(*,*)'3 for shafts under combined torsion and
        +bending'
         read(*,*)j
         if (j.eq.1) then
             call   cirdia
         elseif (j.eq.2) then
             call   noncir
         elseif (j.eq.3) then
             call   torben(normal,shear)
         endif
         write(*,100)
100      format(///)
         write(*,*)'enter 1 to continue or any other integer to
        +exit'
         read(*,*)k
         if (k.eq.1) goto 10
         write(*,*)'end of the program'
         stop
         end
c
c
         subroutine enter(mt,ts,g)
         real mt,ts,g
         write(*,*)'enter the value for torsion in lbs.in (or N.m)'
         read(*,*)mt
         write(*,*)'enter the max shear stress in psi (or Pa)'
```

```
        read(*,*)ts
        write(*,*)'enter the shear young's modulus in psi(or Pa)'
        read(*,*)g
        return
        end
c
c

        subroutine   cirdia
        real  mt,ts,g,theta,r,t
        integer  i
        call  enter(mt,ts,g)
        write(*,*)'for solid shaft, enter 1'
        write(*,*)'for hollow shaft, enter 2'
        read(*,*)i
        if (i.eq.1) then
            r=(2.*mt/(ts*3.1416))**(1./3.)
            theta=2.*r**2*mt/(3.1416)
        elseif (i.eq.2) then
            write(*,*)'enter the thickness of the hollow shaft'
            read(*,*)t
            r=(mt/(2.*3.1416*t*ts))**(1./2.)
            theta=mt/(2.*3.1416*t*g*r**3)
        endif
        write(*,*)'radius of the shaft is',r
        write(*,*)'angle of twist per unit length is ', theta
        return
        end
        subroutine   noncir
c
        integer  l
        write(*,*)'select:'
        write(*,*)'1 for elliptical cross section'
        write(*,*)'2 for triangular c.s'
        write(*,*)'3 for solid rectangular c.s'
        write(*,*)'4 for hollow rectangular c.s'
        write(*,*)'5 for hollow elliptical c.s'
        write(*,*)'6 for regular hexagonal c.s'
        write(*,*)'7 for multiply-connected thin walled hollow
        tube'
        read(*,*)l
        if (l.eq.1) call  ellip
        if (l.eq.2) call  eqtri
        if (l.eq.3) call  recta
        if (l.eq.4) call  horect
        if (l.eq.5) call  hoelli
```

```
        if (l.eq.6) call rehex
        if (l.eq.7) call multi
        return
        end
c
c

        subroutine ellip
c
        real mt,ts,g,rat,a,b,theta
        call enter(mt,ts,g)
        write(*,*)'enter the ratio of minor axis length over
        the major axis length'
        read(*,*)rat
        b=(2.*mt/(ts*rat*3.1416))**(1.0/3.0)
        a=rat*b
        theta=mt*(a*a+b*b)/(3.1416*a**3*b**3*g)
        write(*,*)'major axis length is',b,'in (m)'
        write(*,*)'minor axis length is',a,'in (m)'
        write(*,*)'angle of twist per unit length is
     +  ',theta,'degrees'
        return
        end
c
c

        subroutine eqtri
        real mt,ts,g,theta,a
        call enter(mt,ts,g)
        a=(20.*mt/ts)**(1./3.)
        theta=mt*46.2/(a**4*g)
        write(*,*)'side of the triangle =',a,'in (m)'
        write(*,*)'angle of twist per unit length
     +  =',theta,'degrees'
        return
        end


c
c

        subroutine recta
        real mt,ts,g,rat,beta,alpha,a,b,theta
        call enter(mt,ts,g)
        write(*,*)'enter the ratio of length over width '
        write(*,*)'choices:1.0,1.5,2.0,2.5.3.0,4.0,5.0,10.0,11.0'
        read(*,*)rat
        if (rat.eq.1.0) then
            beta=0.141
```

```
      alpha=0.208
elseif (rat.eq.1.5) then
      beta=0.196
      alpha=0.231
elseif (rat.eq.2.0) then
      beta=0.229
      alpha=0.246
elseif (rat.eq.2.5) then
      beta=0.249
      alpha=0.256
elseif (rat.eq.3.0) then
      beta=0.263
      alpha=0.267
elseif (rat.eq.4.0) then
      beta=0.281
      alpha=0.282
elseif (rat.eq.5.0) then
      beta=0.291
      alpha=0.292
elseif (rat.eq.10.0) then
      beta=0.312
      alpha=0.312
elseif (rat.eq.11.0) then
      beta=0.333
      alpha=0.333
  endif
  b=(mt/(ts*rat*alpha))**(1./3.)
  a=rat*b
  theta=mt/(beta*a*b**3*g)
  write(*,*)'large  side  =',a
  write(*,*)'small  side  =',b
  write(*,*)'angle  of  twist  per  unit  length
+  =',theta,'degrees'
  return
  end
c

  subroutine horect
  real  mt,ts,g,rat,tl,t,b,a,theta,max
  call enter(mt,ts,g)
  write(*,*)'enter  the  ratio  of  length  over  the  width'
  read(*,*)rat
  write(*,*)'enter  the  thickness  of  length'
  read(*,*)tl
  write(*,*)' enter  the  thickness  of  the  width'
  read(*,*)t
```

```
      b=(mt/(2*ts*rat*max(t1,t)))**(1.0/2.0)
      a=rat* b
      theta=mt*(a*t+b*t1)/(2*t1*t*a*a*b*b*g)
      write(*,*)'large side =',a
      write(*,*)'small side =',b
      write(*,*)'angle of twist per unit length=',theta,'degrees'
      return
      end
c
c
      subroutine hoelli
c
      real mt,ts,g,rat,b,a,theta,t
c
      call enter(mt,ts,g)
      write(*,*)'enter the ratio of the major axis length
      over the minor axis length'
      read(*,*)rat
      write(*,*)'enter the thickness of the shaft'
      read(*,*)t
      b=(mt/(2*rat*t*ts*3.1416))**(1.0/2.0)
      a=rat*b
      theta=mt*(2*(a*a+b*b))**(1.0/2.0)/(4*a*a*b*b
     +*t*3.1416*g)
      write(*,*)' the major axis is =',a
      write(*,*)' the minor axis is =',b
      write(*,*)' angle of twist per unit length is
     + ',theta,'degrees'
      return
      end
c
c
      subroutine rehex
c
      real mt,ts,g,a,theta,al
      call enter(mt,ts,g)
      a=(5.7*mt/ts)**(1.0/3.0)
      theta=8.8*mt/(a**4*g)
      al=a/(3**(1.0/2.0))
      write(*,*)'the length is ',al
      write(*,*)'angle of twist per unit length is
     + ',theta,'degrees'
      return
      end
c
```

```
c
      subroutine multi
c
      real mt,ts,g,h1,h2,a,s,w,l,theta,max
c
      call enter(mt,ts,g)
      write(*,*)'enter thickness of tube'
      read(*,*)t4
      write(*,*)'enter left thickness'
      read(*,*)t1
      w  e(*,*)'enter right thickness'
      read(*,*)t3
      write(*,*)'enter the centre thickness'
      read(*,*)t2
      h1=ts*max(t4,t1)
      h2=ts*max(t3,t2)
      a=mt/(2*(h1+h2))
      s=a**(1./2.)
      w=s+2.*t4
      l=t1+2.*s+t2+t3
      theta=(h1/t1+2.*h1/t4+(h1-h2)/t2)/(8.*a*g)
      write(*,*)'length of tube is',l
      write(*,*)'width of tube is ',w
      write(*,*)'angle of twist per unit length is
     + ',theta,'degrees'
      return
      end


      subroutine   torben
c
      real normal(2),shear(2)
      real maxnor,minnor,maxshr,minshr
      integer input
c
      write(*,*)'select:'
      write(*,*)'1 for combined torsion and bending'
      write(*,*)'2 for built-in shaft'
      read(*,*)input
      if (input.eq.1) then
           call tentor(maxnor,minnor,maxshr,minshr)
      elseif (input.eq.2) then
           call ducdia(maxnor,minnor,maxshr,minshr)
      endif
      normal(1)=maxnor
```

```
        normal(2)=minnor
        shear(1)=maxshr
        shear(2)=minshr
        write(*,*)'max normal stress is',normal(1),'psi (Pa)'
        write(*,*)'min normal stress is',normal(2),'psi (Pa)'
        write(*,*)'max shear stress is',shear(1),'psi (Pa)'
        write(*,*)'min shear stress is',shear(2),'psi (Pa)'
        return
        end
c
c
c

        subroutine tentor(maxnor,minnor,maxshr,minshr)
c
        real maxnor,minnor,maxshr,minshr
        real mt,f,fstres,mshear,term1,term2,term
        integer sign
        write(*,*)'enter applied twisting moment'
        read(*,*)mt
        write(*,*)'enter diameter'
        read(*,*)d
        write(*,*)'enter magnitude of force'
        read(*,*)f
        write(*,*)'select:'
        write(*,*)'1 for tensile;2 for compressive'
        read(*,*)sign
        fstres=f/(3.1416*d**2/4)
        mshear=(5.0929*mt)/(d**3)
        if (sign.eq.1) then
             term1=fstres/2
        elseif (sign.eq.2) then
             term1=(-1.0*fstres)/2.
        endif
        term=abs(term1)
        term2=sqrt((term)**2+(mshear)**2)
        maxnor=term1+term2
        minnor=term1-term2
        maxshr=term2
        minshr=-1.0*term2
        return
        end
c
c

        subroutine ducdia(maxnor,minnor,maxshr,minshr)
c
```

```
real  maxnor,minnor,maxshr,minshr
real  mt,r,p,x,mb,d
write(*,*)'enter traverse force'
read(*,*)p
write(*,*)'enter distance from neutral axis'
read(*,*)r
write(*,*)'enter distance from built-in edge'
read(*,*)x
write(*,*)'enter the diameter of shaft'
read(*,*)d
mt=p*r
mb=-p*(1.-x)
maxnor=16./(3.1416*(d**3))*(mb+(sqrt(mb**2
+    +mt**2)))
minnor=16./(3.1416*(d**3))*(mb-(sqrt(mb**2
+    +mt**2)))
maxshr=(maxnor-minnor)/2.
minshr=-maxshr
return
end

real function max(x,y)

real  x,y
if (x.gt.y) then
   max=x
else
   max=y
endif
return
end
```

CHAPTER 3

Thick-walled cylinders and discs

3.1 GENERAL CONSIDERATIONS, ELASTIC SOLUTIONS

In this chapter an important class of engineering problems will be considered, often found in practice. Considering **thick-walled cylinders** first of all, it will be assumed that they are generally used as **pressure vessels,** subjected to either internal or external pressure and sometimes an associated end load. All shear stresses will be assumed to be zero, since cylinders are usually long enough to ensure that plane sections will remain plane and that the axial strain will therefore be constant along any radius and also along the length of the cylinder. This will not be the case for **discs,** which may be regarded as very short thick-walled cylinders, so that some shear can take place on axial planes.

The starting point for deriving solutions to these problems is to consider the **equilibrium and compatibility equations** set out on pp. 196 and 202 of the companion volume, for **polar cylindrical coordinates.** In the absence of shear stresses, the equilibrium equations reduce to:

$$\frac{\partial \sigma_r}{\partial r} + \frac{\sigma_r - \sigma_\theta}{r} = 0 \tag{3.1}$$

$$\frac{\partial \sigma_\theta}{\partial \theta} = \frac{\partial \sigma_z}{\partial z} = 0 \tag{3.2}$$

where stresses such as σ_{rr} have been replaced by σ_r, since they are all principal stresses.

Likewise, the compatibility equations reduce to:

$$\epsilon_r = \frac{\partial u_r}{\partial r} \tag{3.3}$$

$$\epsilon_\theta = \frac{u_r}{r} \tag{3.4}$$

$$\epsilon_z = \frac{\partial u_z}{\partial z} \tag{3.5}$$

for small strains, as are found in elastic deformation.

For elastic deformation the stress-strain relations are:

$$\epsilon_r = \frac{\partial u_r}{\partial r} = \frac{1}{E}\left[\sigma_r - \nu(\sigma_\theta + \sigma_z)\right] \qquad (3.6)$$

$$\epsilon_\theta = \frac{u_r}{r} = \frac{1}{E}\left[\sigma_\theta - \nu(\sigma_z + \sigma_r)\right] \qquad (3.7)$$

$$\epsilon_z = \frac{\partial u_z}{\partial z} = \frac{1}{E}\left[\sigma_z - \nu(\sigma_r + \sigma_\theta)\right]. \qquad (3.8)$$

From equation (3.7), $Eu_r = r\left[\sigma_\theta - \nu(\sigma_z + \sigma_r)\right]$ and, differentiating with respect to r,

$$E\frac{\partial u_r}{\partial r} = r\left[\frac{\partial \sigma_\theta}{\partial r} - \nu\left[\frac{\partial \sigma_z}{\partial r} + \frac{\partial \sigma_r}{\partial r}\right]\right] + \sigma_\theta - \nu(\sigma_z + \sigma_r). \qquad (3.9)$$

From equation (3.6),

$$E\frac{\partial u_r}{\partial r} = \sigma_r - \nu(\sigma_\theta + \sigma_z). \qquad (3.10)$$

Subtracting equation (3.10) from equation (3.9),

$$r\left[\frac{\partial \sigma_\theta}{\partial r} - \nu\left[\frac{\partial \sigma_z}{\partial r} + \frac{\partial \sigma_r}{\partial r}\right]\right] + (\sigma_\theta - \sigma_r)(1 + \nu) = 0. \qquad (3.11)$$

Now $\partial \epsilon_z / \partial r = 0$ and, differentiating equation (3.8) with respect to r,

$$\frac{\partial \sigma_z}{\partial r} = \nu\left[\frac{\partial \sigma_r}{\partial r} + \frac{\partial \sigma_\theta}{\partial r}\right]. \qquad (3.12)$$

Substituting equation (3.12) into equation (3.11) gives, after some algebra,

$$\sigma_\theta - \sigma_r + r(1-\nu)\frac{\partial \sigma_\theta}{\partial r} - \nu r \frac{\partial \sigma_r}{\partial r} = 0. \qquad (3.13)$$

From equation (3.1),

$$\sigma_\theta - \sigma_r - r\frac{\partial \sigma_r}{\partial r} = 0 \qquad (3.14)$$

and, subtracting equation (3.14) from equation (3.13) gives:

$$r(1-\nu)\frac{\partial \sigma_\theta}{\partial r} + r\frac{\partial \sigma_r}{\partial r}(1-\nu) = 0, \text{ or } \frac{\partial \sigma_\theta}{\partial r} + \frac{\partial \sigma_r}{\partial r} = 0 \qquad (3.15)$$

Integrating equation (3.15) gives $\sigma_\theta + \sigma_r = 2A$, a constant. (3.16)

Subtracting equation (3.14) from equation (3.16) gives:

$$2\sigma_r + r\frac{\partial \sigma_r}{\partial r} = 2A.$$

The left-hand side of this equation is, in fact,

$$\frac{1}{r}\frac{\partial}{\partial r}(r^2\sigma_r), \text{ thus: } \frac{1}{r}\frac{\partial}{\partial r}(r^2\sigma_r) = 2A,$$

and, integrating, $r^2\sigma_r = Ar^2 + B$. Thus:

$$\sigma_r = A + \frac{B}{r^2} \tag{3.17}$$

and, from equation (3.16),

$$\sigma_\theta = 2A - \sigma_r = A - \frac{B}{r^2}. \tag{3.18}$$

These are the well-known **Lamé equations.**

For a long cylindrical vessel subjected to internal pressure p_1 only, the boundary conditions are $\sigma_r = -p_1$ at $r = r_1$ and $\sigma_r = 0$ at $r = r_2$, where r_1 and r_2 are the internal and external radii. Thus:

$$-p_1 = A + \frac{B}{r_1^2}, \quad \text{and} \quad 0 = A + \frac{B}{r_2^2}.$$

Subtracting,

$$-p_1 = B\left[\frac{1}{r_1^2} - \frac{1}{r_2^2}\right], \quad \text{hence} \quad B = -\frac{p_1}{\dfrac{1}{r_1^2} - \dfrac{1}{r_2^2}}$$

and:

$$A = -\frac{B}{r_2^2} = \frac{p_1}{r_2^2\left[\dfrac{1}{r_1^2} - \dfrac{1}{r_2^2}\right]}.$$

Substituting these values for A and B into equations (3.17) and (3.18) gives the following equations for the radial and circumferential stresses throughout the vessel:

$$\sigma_r = \frac{p_1}{\left[\dfrac{1}{r_1^2} - \dfrac{1}{r_2^2}\right]}\left[\dfrac{1}{r_2^2} - \dfrac{1}{r^2}\right]$$

$$\sigma_\theta = \frac{p_1}{\left[\dfrac{1}{r_1^2} - \dfrac{1}{r_2^2}\right]}\left[\dfrac{1}{r_2^2} + \dfrac{1}{r^2}\right].$$

The axial stress will depend on whether the vessel is **open-ended** or **closed-ended,** as illustrated in Fig. 3.1.

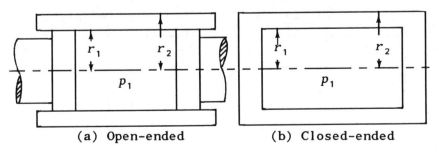

(a) Open-ended   (b) Closed-ended

Fig. 3.1 Open- and closed-ended vessels

In general, vessels have closed ends, as illustrated in Fig. 3.1(b). In both cases, the end load due to the internal pressure, which has to be taken either by the vessel or by plungers of the type shown diagrammatically in Fig. 3.1(a), will be $\pi r_1{}^2 p_1$. In the case of open ends, the axial stress σ_z in the vessel wall itself will obviously be zero. With closed ends, the usual case in practice, the axial stress will be a tensile stress:

$$\sigma_z = \frac{\pi r_1{}^2 p_1}{\pi(r_2{}^2 - r_1{}^2)} = \frac{p_1}{\left[\dfrac{r_2{}^2}{r_1{}^2} - 1\right]}.$$

The diameter ratio of thick-walled cylinders is usually denoted by $K = r_2/r_1$, and the stresses may therefore finally be written more concisely as:

$$\sigma_r = \frac{p_1}{K^2-1}\left[1 - \frac{K^2 r_1{}^2}{r^2}\right] = \frac{p_1}{K^2-1} - \frac{p_1 K^2}{K^2-1}\left[\frac{r_1}{r}\right]^2 \qquad (3.19)$$

$$\sigma_\theta = \frac{p_1}{K^2-1}\left[1 + \frac{K^2 r_1{}^2}{r^2}\right] = \frac{p_1}{K^2-1} + \frac{p_1 K^2}{K^2-1}\left[\frac{r_1}{r}\right]^2 \qquad (3.20)$$

$$\sigma_z = \frac{p_1}{K^2-1}. \qquad (3.21)$$

The distribution of these stresses across the wall of the cylinder is sketched in Fig. 3.2(a).

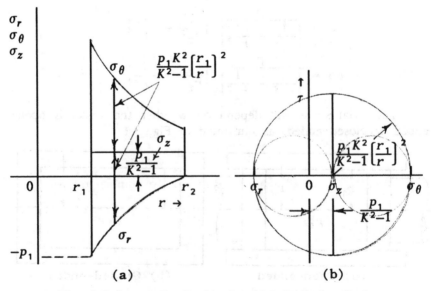

(a)   (b)

Fig. 3.2 Stress distribution in thick-walled cylinder

The state of stress at any point in the closed-ended cylinder is seen to be a hydrostatic tensile stress $\sigma_z = p_1/(K^2-1)$ superimposed on a pure shear stress of $p_1 K^2 (r_1/r)^2/(K^2-1)$ acting on planes at 45° to the r, θ directions as illustrated by the Mohr's circles in Fig. 3.2(b).

If the vessel is subjected to an external pressure p_2 only, the boundary conditions will be $\sigma_r = 0$ at $r = r_1$ and $\sigma_r = -p_2$ at $r = r_2$, where r_1 and r_2 are the internal and external radii. Thus:

$$0 = A + \frac{B}{r_1^2}, \text{ and } -p_2 = A + \frac{B}{r_2^2}.$$

Subtracting,

$$+p_2 = B\left[\frac{1}{r_1^2} - \frac{1}{r_2^2}\right], \text{ hence } B = \frac{p_2}{\dfrac{1}{r_1^2} - \dfrac{1}{r_2^2}}$$

and:

$$A = -\frac{B}{r_1^2} = -\frac{p_2}{r_1^2\left[\dfrac{1}{r_1^2} - \dfrac{1}{r_2^2}\right]}.$$

Substituting these values for A and B into equations (3.17) and (3.18) gives the following equations for the radial and circumferential stresses throughout the vessel:

$$\sigma_r = -\frac{p_2}{\left[\dfrac{1}{r_1^2} - \dfrac{1}{r_2^2}\right]} \left[\frac{1}{r_1^2} - \frac{1}{r^2}\right]$$

$$\sigma_\theta = -\frac{p_2}{\left[\dfrac{1}{r_1^2} - \dfrac{1}{r_2^2}\right]} \left[\frac{1}{r_1^2} + \frac{1}{r^2}\right].$$

Since $1/r_1^2 \geqslant 1/r^2 \geqslant 1/r_2^2$ both σ_r and σ_θ are negative (compressive) in this case. If the pressure p_2 is also applied to the outside ends of the cylinder illustrated in Fig. 3.1(b) the axial stress in the tubular wall will be a compressive stress of:

$$\sigma_z = -\frac{\pi r_2^2 p_2}{\pi(r_2^2 - r_1^2)} = -\frac{p_2}{1 - \dfrac{r_1^2}{r_2^2}}.$$

With $r_2/r_1 = K$, these stresses can be summarized as:

$$\sigma_r = -\frac{p_2 K^2}{K^2-1} + \frac{p_2 K^2}{K^2-1}\left[\frac{r_1}{r}\right]^2 \tag{3.22}$$

$$\sigma_\theta = -\frac{p_2 K^2}{K^2-1} - \frac{p_2 K^2}{K^2-1}\left[\frac{r_1}{r}\right]^2 \tag{3.23}$$

$$\sigma_z = -\frac{p_2 K^2}{K^2-1}. \tag{3.24}$$

The distribution of these stresses across the wall of the cylinder is sketched in Fig. 3.3(a), their Mohr's circle representation being shown in Fig. 3.3(b).

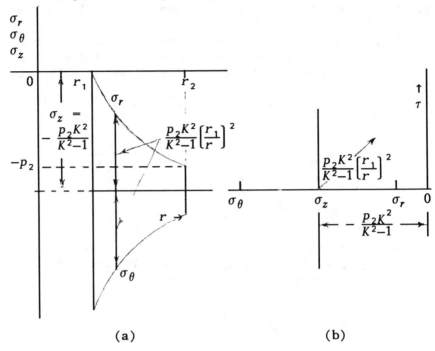

(a)   (b)

Fig. 3.3 Stress distribution in thick–walled cylinder subjected to external pressure

The general case of a thick–walled cylinder subjected to both internal pressure p_1 and external pressure p_2 is also of interest, particularly for **compound pressure vessels** made up of several thick–walled cylinders. The distribution of the stresses can be found easily by simply summing the previous two cases, as follows:

$$\sigma_r = \frac{p_1 - p_2 K^2}{K^2 - 1} + \frac{(p_2 - p_1) K^2}{K^2 - 1} \left[\frac{r_1}{r}\right]^2 \qquad (3.25)$$

$$\sigma_\theta = \frac{p_1 - p_2 K^2}{K^2 - 1} - \frac{(p_2 - p_1) K^2}{K^2 - 1} \left[\frac{r_1}{r}\right]^2 \qquad (3.26)$$

$$\sigma_z = \frac{p_1 - p_2 K^2}{K^2 - 1}. \qquad (3.27)$$

As will be discussed later, a compound vessel comprising two cylinders can be made by heating the larger cylinder and cooling the smaller cylinder until the smaller one can be inserted into the larger. The amount of interference between the outer diameter of the smaller cylinder and the inner diameter of the larger cylinder is determined by the temperature difference which can be practically achieved without damaging the cylinder. Thus, it is of interest to determine the inward radial displacement of the outer diameter of the inner cylinder when subjected to an external pressure and the outward radial displacement of the inner diameter of the outer cylinder when subjected to the same (internal) pressure. The sum of these radial displacements can then be equated to the total radial interference to determine the interfacial pressure which can be achieved.

Therefore expressions are required for the internal radial displacement u_{r_1} due to an internal pressure p_1 and the external radial displacement u_{r_2} due to an external pressure p_2. The axial stress σ_z induced by compounding is small and is therefore generally neglected.

Using equation (3.7), the following two expressions are obtained, neglecting σ_z,

$$u_{r\,1} = \frac{r_1}{E}\left[\sigma_{\theta\,1} - \nu\sigma_{r\,1}\right], \qquad u_{r\,2} = \frac{r_2}{E}\left[\sigma_{\theta\,2} - \nu\sigma_{r\,2}\right],$$

Substituting the values of $\sigma_{\theta\,1}$ and σ_{r_1} from equations (3.19) and (3.20) for an internal pressure p_1 and putting $r = r_1$ gives, after some algebra:

$$u_{r\,1} = \frac{r_1 p_1}{E}\left[\frac{(1-\nu)+K^2(1+\nu)}{K^2-1}\right]. \qquad (3.28)$$

Likewise, substituting the values of $\sigma_{\theta\,2}$ and $\sigma_{r\,2}$ from equations (3.22) and (3.23) for an external pressure p_2 and putting $r = r_2$ gives:

$$u_{r\,2} = -\frac{r_2 p_2}{E}\left[\frac{(1+\nu)+K^2(1-\nu)}{K^2-1}\right]. \qquad (3.29)$$

For the compound cylinder problem, of course, the two cylinders will probably have different K-ratios and possibly also different values of Young's Modulus E, as discussed later.

3.2 ELASTIC–PLASTIC SOLUTIONS, AUTOFRETTAGE

For the open–ended case $\sigma_z = 0$ and it can be seen from Fig. 3.2 that there is a state of pure stress everywhere. Fig. 3.2(a) shows that the maximum stresses occur at the bore and that there is a poor utilization of material in the outer layers of the vessel. As the internal pressure is increased, yielding will first occur at the bore. From Vol. 1 p.178, the Maxwell–von Mises criterion may conveniently be written in the form $\bar{\sigma} = (1/\sqrt{2})[\,(\sigma_1-\sigma_2)^2+(\sigma_2-\sigma_3)^2+(\sigma_3-\sigma_1)^2\,]^{\frac{1}{2}}$ in terms of the principal stresses, which are in this case σ_r, σ_θ, σ_z, i.e.

$$\bar{\sigma} = (1/\sqrt{2})[\,(\sigma_r-\sigma_\theta)^2+(\sigma_\theta-\sigma_z)^2+(\sigma_z-\sigma_r)^2\,]^{\frac{1}{2}}. \qquad (3.30)$$

At the bore, $r = r_1$, and for the closed end case, from equations (3.19), (3.20) and (3.21):

$$\sigma_r = \frac{p_1(1-K^2)}{K^2-1}, \quad \sigma_\theta = \frac{p_1(1+K^2)}{K^2-1}, \quad \sigma_z = \frac{p_1}{K^2-1}.$$

Hence, after substituting these values into the Maxwell criterion:

$$\bar{\sigma} = \sqrt{3}\,\frac{p_1 K^2}{K^2-1}. \qquad (3.31)$$

Thus, yield will occur when $p_1 = p_y$, where:

$$p_y = \frac{\bar{\sigma}}{\sqrt{3}}\left[1 - \frac{1}{K^2}\right]. \qquad (3.32)$$

For the Tresca criterion, from Vol. 1 p.181, $\bar{\sigma} = \sigma_1-\sigma_3$, but with $\sigma_1 = \sigma_r$ and $\sigma_3 = \sigma_\theta$ in this case, i.e.

$$\bar{\sigma} = \sigma_r - \sigma_\theta. \qquad (3.33)$$

Substituting σ_r and σ_θ at the bore into equation (3.33) gives:

$$p_y = \frac{\bar{\sigma}}{2}\left[1 - \frac{1}{K^2}\right]. \qquad (3.34)$$

For open ends, $\sigma_z = 0$ and it is found that, for the Maxwell–von Mises criterion:

$$p_y = \frac{\bar{\sigma}}{\sqrt{3}}\left[1 - \frac{1}{K^2}\right]\frac{1}{\sqrt{1 + \dfrac{1}{3K^4}}}. \qquad (3.35)$$

Consideration of equations (3.19) to (3.21) reveals that the axial strain in an elastic vessel with closed ends under internal pressure p_1 would be $\epsilon_z = (1/E)[\sigma_z-\nu(\sigma_r+\sigma_\theta)] = (1-2\nu)p_1 \div \{E(K^2-1)\}$, which is very small and would be zero if $\nu = \frac{1}{2}$, as is the case for an incompressible material. The case of zero axial or longitudinal strain is often referred to as a condition of "plane strain", and will often be referred to in the general discussion of plasticity theory which follows later. For the present case (a cylindrical tube with closed ends under internal pressure) plane strain conditions would require $\epsilon_z = 0$ and therefore $\sigma_z = \nu(\sigma_r+\sigma_\theta) = 2\nu p_1/(K^2-1)$. Since the Lamé equations require σ_r

and σ_θ to have the values given by equations (3.19) and (3.20), substitution of these values into the criterion of von Mises [equation (3.30)] gives, after some algebra,:

$$p_y = \frac{\bar{\sigma}}{\sqrt{3}} \left[1 - \frac{1}{K^2} \right] \frac{1}{\sqrt{1 + \frac{(1-2\nu)^2}{3K^4}}}.$$ (3.36)

For open ends and the Tresca criterion the result is the same as for the closed-ended case, namely:

$$p_y = \frac{\bar{\sigma}}{2} \left[1 - \frac{1}{K^2} \right].$$ (3.34)

To achieve a better utilization of the material in such a thick cylinder, the internal pressure can be increased above p_y so that the inner layers of the vessel are subjected to plastic flow. As long as the external layers of the vessel remain elastic there will be no excessive deformation of the vessel. On release of the internal pressure the outer layers of the vessel will try to relax to their original dimensions but will be unable to so because of the plastically deformed inner layers which will not return to their original dimensions. Thus the vessel will have residual compressive stresses in its inner layers and residual tensile stresses in its outer layers.

Subsequent application of internal pressure can reach higher levels than p_y before yielding again occurs, due to the residual compressive stresses in the inner layers. This strengthening of pressure vessels is known as **autofrettage** and is often applied to gun barrels. A similar effect can be achieved by shrinking one cylinder onto another, usually by expanding the outer cylinder by heating, so that the compound vessel has its inner cylinder in compression and the outer cylinder in tension.

Considering autofrettage in the first instance, the simplest criterion of plastic flow which can be used for this case is that of Tresca, which may conveniently be written for this problem, in which σ_z is the intermediate principal stress, as:

$$\sigma_\theta - \sigma_r = Y.$$ (3.37)

The equation of equilibrium may be written as:

$$\frac{d\sigma_r}{dr} = \frac{\sigma_\theta - \sigma_r}{r}$$ (3.38)

so that, for the inner plastic layers of the vessel, by substituting equation (3.37) into equation (3.38):

$$\frac{d\sigma_r}{dr} = \frac{Y}{r}. \tag{3.39}$$

Hence $\qquad \sigma_r = Y\ln r + \text{Constant}. \tag{3.40}$

At the plastic–elastic interface $(r = c)$, the radial stress to cause yielding of the outer layers will be, from equation (3.34), with $\sigma_r = -p_y$, $\bar{\sigma} = Y$ and $K = r_2/c$:

$$\sigma_r = -\frac{Y}{2}\left[1 - \frac{c^2}{r_2^2}\right].$$

Substituting this value into equation (3.40), with $r = c$, gives the value of the constant as:

$$\sigma_r - Y\ln c = -\frac{Y}{2}\left[1 - \frac{c^2}{r_2^2} + 2\ln c\right].$$

Thus, in the plastic zone, from equation (3.40):

$$\sigma_r = \frac{Y}{2}\left[\frac{c^2}{r_2^2} -1-2\ln c+2\ln r\right] = \frac{Y}{2}\left[\frac{c^2}{r_2^2} -1+2\ln \frac{r}{c}\right] \tag{3.41}$$

and $\sigma_\theta = \sigma_r + Y$, from equation (3.36),

$$= \frac{Y}{2}\left[\frac{c^2}{r_2^2} +1+2\ln \frac{r}{c}\right]. \tag{3.42}$$

The axial stress σ_z can only be determined by applying the complete equations of plasticity theory and will therefore be ignored for the present discussion.

The pressure at the bore to cause plastic flow to occur at radius c will be, from equation (3.41) with $r = r_1$ and $\sigma_r = -p$,

$$p = -\frac{Y}{2}\left[\frac{c^2}{r_2^2} -1+2\ln \frac{r_1}{c}\right] = \frac{Y}{2}\left[1 - \frac{c^2}{r_2^2} +2\ln \frac{c}{r_1}\right]. \tag{3.43}$$

The stresses in the elastic region can be obtained from equations (3.19) to (3.21) by substituting c for r_1, r_2/c for K and $(Y/2)(1-c^2/r_2^2)$ for p_1. Hence:

$$\sigma_r = \frac{Yc^2}{2r_2^2}\left[1 - \frac{r_2^2}{r^2}\right], \qquad \text{for } c \leqslant r \leqslant r_2 \tag{3.44}$$

$$\sigma_\theta = \frac{Yc^2}{2r_2^2}\left[1 + \frac{r_2^2}{r^2}\right], \qquad \text{for } c \leqslant r \leqslant r_2. \tag{3.45}$$

The distributions of the stresses σ_r and σ_θ in a thick–walled cylinder which has been expanded until about half the cylinder wall is plastic are sketched in Fig.3.4.

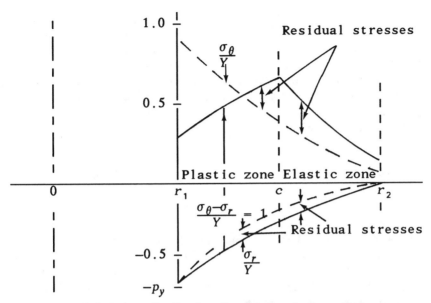

Fig. 3.4 Stress distribution in elastic–plastic cylinder

The full lines are the stresses induced by the pressure p_y and the dotted lines show the 'fictitious' elastic stresses which would be associated with p_y if the material did not yield. When the pressure p_y is released the stresses generally reduce according to this fictitious elastic stress distribution, although sometimes yielding can occur in the reverse direction at the bore. The final residual state of stress in the cylinder is obtained from the difference between these sets of curves, as sketched in Fig. 3.5.

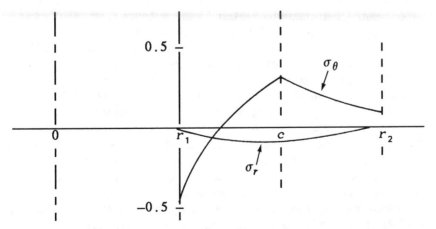

Fig. 3.5 Residual stress distribution in autofrettaged cylinder

It can be seen from this final state of stress that the inner layers of the autofrettaged cylinder are left in compression due to the residual circumferential tensile stresses which remain in the outer layers. Subsequent applications of internal pressure can be made to a higher level, since the residual compressive stresses have to be overcome in the inner layer. The maximum elastic pressure which can be applied to an autofrettaged thick–walled cylinder can be twice that for a non–autofrettaged cylinder.

3.3 HIGH–PRESSURE CONTAINERS, MANNING'S METHOD

In practice, the material of the thick–walled cylinder (usually a strong steel) work hardens and the simple Tresca criterion is an over–simplication of the actual situation. To obtain a more realistic estimate of the actual stress distribution in the **elastic–plastic cylinder** a practical method was developed by Manning which will now be described. A complete discussion of **Manning's method** has been given in the book by Ford with Alexander, so what follows is a summary of the method.

The previous section was devoted to a description of the simplest complete elastic–plastic solution, achieved by assuming the Tresca criterion, zero work–hardening, small deformations and negligible axial stresses. In practice, one of the main problems of interest is that of designing high–pressure vessels, generally made of high–strength steels. For such materials the von Mises criterion of yielding and its associated work–hardening laws must be employed if a realistic solution is required, and large bore deformations taken into account, so that small–deformation "elastic type" theory must be discarded. The complexity of obtaining exact elastic–plastic solutions to such problems, even with the simplification of axial symmetry, has been pointed out by many experts in plasticity (notably by Hill in his definitive book on the mathematical theory, for example).

Manning's great contribution was to provide a simplified model which would take **work–hardening** into account, using the yield criterion of von Mises. Because the strains involved in the plastic region are large he assumed that the elastic strains in the plastic region could be neglected and that plane strain conditions prevailed. The same result is achieved by assuming that $\nu = \frac{1}{2}$, in which case it can be seen by comparing equations (3.34) and (3.36) (with $\nu = \frac{1}{2}$), that the plane strain case gives the same initial yield pressure as the closed end case. Manning actually chose to use a yield stress–strain curve expressed in terms of shear stress and shear strain, readily determined from a **torsion test,** by using a method described by Nadai. This was because of the fact, which has already been pointed out, that the **Lamé equations** for a thick–walled tube subjected to either internal or external pressure represent a state of pure shear stress superimposed on a hydrostatic stress. However, in this discussion, the method follows the one often used in the solution of problems involving plasticity theory, i.e. by using the **yield criterion of von Mises** together with the **Lévy–Mises flow rule** employing **total strain–increments.** This method preceded the more precise **Prandtl–Reuss equations** based on **incremental strain–increments,** as discussed in Vol. 1. The method of solution is essentially Manning's and begins with the assumption of plane strain conditions, viz. $\delta\epsilon_z=0$.

Referring to the basic equations of the plasticity theory set out in Vol. 1, at the top of p.184, the axial plastic strain increment is $d\epsilon^P_z = {}^2/_3\ d\lambda[\ \sigma_z-\tfrac{1}{2}(\sigma_r+\sigma_\theta)\ ]$, which is added to the associated <u>elastic</u> strain increment $d\epsilon^e_z = {}^1/_E\ d\lambda[\ \sigma_z-\nu(\sigma_r+\sigma_\theta)\ ]$ to produce the total strain increment $d\epsilon_z=d\epsilon^P_z+d\epsilon^e_z$, where single suffixes have been used everywhere since all stresses and strain increments are principal components. It is perhaps worthwhile repeating here that there is a complete analogy between the elastic and plastic strain increments, with Poisson's Ratio ν being replaced by the fraction $\tfrac{1}{2}$ required to give zero volume change of the plastic strain increments, and ${}^1/_E$ being replaced by the incremental quantity ${}^2/_3d\lambda$ (to give a consistent mathematical relationship between incremental quantities). It is easy to prove (as shown in Vol. 1, p. 184), that ${}^2/_3\ d\lambda = d\overline{\epsilon}^P/\overline{\sigma}$, a quantity which is completely analogous to ${}^1/_E$ (dimensions strain/stress).

Thus Manning's assumptions of plane strain and neglect of the elastic strains means that he assumed $d\epsilon^P_z=0$, leading to the equation $\sigma_z=\tfrac{1}{2}(\sigma_r+\sigma_\theta)$. Substituting this value of σ_z into the yield criterion of von Mises [equation (3.30)] gives simply:

$$\sigma_\theta - \sigma_r = \frac{2}{\sqrt{3}}\ \overline{\sigma}. \tag{3.46}$$

The equation of equilibrium [equation (3.1)] then becomes:

$$\frac{d\sigma_r}{dr} = \frac{2}{\sqrt{3}}\ \frac{\overline{\sigma}}{r} \tag{3.47}$$

[cf. equation (3.38) for the Tresca yield criterion]. Integrating equation (3.47) gives, for internally applied pressure p_1:

$$\int_{r_1}^{c} d\sigma_r = 0-(-p_1) = p_1 = \frac{2}{\sqrt{3}}\int_{r_1}^{c}\frac{\overline{\sigma}}{r}\ dr \tag{3.48}$$

since the largest deformation and hence the plastic zone will be restricted to the internal layers of the vessel, extending from $r=r_1$ to $r=c$, the radius of the plastic–elastic interface.

Similarly, the equivalent plastic strain increment $d\bar{\epsilon}^p$ (see Vol. 1, p.180) can be written as:

$$d\bar{\epsilon}^p = [\, ^2/_3(d\epsilon_r^{p2}+d\epsilon_\theta^{p2}+d\epsilon_z^{p2})\,]^{\frac{1}{2}} \tag{3.49}$$

and, since the volumetric plastic strain $(d\epsilon_r^p+d\epsilon_\theta^p+d\epsilon_z^p)$ is zero, and $d\epsilon_z^p$ is zero, (neglecting elastic strains), then $d\epsilon_r^p=-d\epsilon_\theta^p$. Substituting these values into equation (3.49) gives:

$$d\bar{\epsilon}^p = \frac{2}{\sqrt{3}}d\epsilon_\theta^p = \frac{2}{\sqrt{3}}\frac{dr}{r}. \tag{3.50}$$

Integrating this equation and neglecting elastic strains gives:

$$\bar{\epsilon}^p = \frac{2}{\sqrt{3}}\epsilon_\theta^p = \frac{2}{\sqrt{3}}\ln\frac{r+u}{r} \tag{3.51}$$

where r is any radius in the plastic zone and u is the radial displacement at that radius.

Manning's method of solution was to assume a bore strain large enough to cover the range of problems of particular interest and find the strain in a number of concentric annuli. If the initial internal radius r_{10} becomes $r_1=r_{10}+u_{r1}$ on application of the internal pressure p_1, then the bore strain is u_{r1}/r_{10}. If any generic annulus at radius r has the initial thickness δr, its thickness will change to $\delta(r+u)$ at radius $(r+u)$ which can be found by assuming constant volume (because $\nu=\frac{1}{2}$ everywhere). For example, the initial cross–sectional area between r_1 and r is $\pi(r^2-r_1^2)$, which must equal its final area $\pi[\,(r+u)^2-(r_1+u_{r1})^2\,]$, since $\epsilon_z=0$ everywhere. Therefore $r^2-r_1^2 = (r+u)^2-(r_1+u_{r1})^2$, hence:

$$(r+u)^2 = r^2+2r_1u_{r1}+u_{r1}^2 = r^2+C^2, \tag{3.52}$$

where $C^2 = 2r_1u_{r1}+u_{r1}^2$, a constant for any chosen bore strain u_{r1}/r_{10}, using Manning's notation.

The flow stress $\bar{\sigma}$ at any radius r for the bore strain u_{r1}/r_{10} can now be found from equation (3.51) for the equivalent strain $\bar{\epsilon}^p$ above, which can be rewritten following Manning, by using equation (3.52), as:

$$\bar{\epsilon}^p = \frac{1}{\sqrt{3}}\ln\left[1+\frac{C^2}{r^2}\right]. \tag{3.53}$$

All that is required is the stress–strain curve for the

material of the pressure vessel, either in the form of a set of values of $\bar{\sigma}$ versus $\bar{\epsilon}^p$ as used by Manning, to create a table of values, or by using an analytical expression for the stress–strain curve. An example of the former is given on p.469 of the book by Ford and Alexander. An example of the latter is given below, achieved by using a curve–fitting computer program called Swift and described on pp. 136–140 of the book by Alexander, Brewer and Rowe. For convenience the program is repeated here as Program 3.1, followed by Solution 3.1 which relates to the material used in the tabular procedure described in the book by Ford and Alexander, viz. a Ni–Cr–Mo steel to EN 25T. The stress–strain values for this steel have been taken from various values entered in Table C of that book.

PROGRAM 3.1 Curve–fitting Swift's equation and others covered by the generic equation $\bar{\sigma} = \sigma_0 + A(B+\bar{\epsilon})^n$.

```
  1:*          PROGRAM SWIFT
  2:           DIMENSION E(20),S(20),X(20),Y(20),XY(20),XX(20),B(20),R(20),
  3:        1  SS(20),S0(20),R1(20)
  4:           WRITE (*,1)
  5: 1         FORMAT(1X' SWIFT FITS GIVEN S/E DATA POINTS TO VARIANTS OF',/
  6:        1  ' EQUATION S = S0+A(B+E)**N   BY A LEAST SQUARE FIT OF M',/1X,
  7:        2  ' LOG-LOG VALUES FOR A NUMBER (J) OF ASSUMED VALUES OF B',/1X
  8:        3  '   OR ALTERNATIVELY A NUMBER (K) OF ASSUMED VALUES OF S0 ',/
  9:        4  ' WHICH ARE OPTIMIZED BY DIRECT SEARCH, CHOOSING SOLUTIONS',/
 10:        5  ' TO MINIMIZE R = SUMMATION OF (S(GIVEN)-S(CALCULATED))**2',/
 11:        6  ' TRY NOT TO USE ANY STRAIN VALUE LESS THAN E=0.01 ',/)
 12:           WRITE(*,2)
 13: 2         FORMAT( 1X' TYPE IN M ')
 14:           READ(*,*) M
 15:           DO 4 I = 1,M
 16:           WRITE(*,3) I,I
 17: 3         FORMAT( 1X' TYPE IN E(',I2,'),S(',I2,')')
 18:           READ(*,*) E(I),S(I)
 19: 4         CONTINUE
 20:           N = 1
 21:           B(1) = .001
 22:           F = 2.
 23:           IF (E(1).EQ.0.) E(1) =.1E-01
 24:           IF (S(1).EQ.0.) S(1) =.1E-01
 25:           WRITE (*,59)
 26: 59        FORMAT(/2X,' FITTING EQUATION S=A(B+E)**N GIVES:- ',/)
 27: 12        DO 6 J = 1,9
 28:           CALL SUM(E,S,B,A ,EPO,XN,M,J)
 29:           R(J) = 0.
 30:           DO 5 I = 1, M
 31:           SS(I) = A*(B(J)+E(I))**XN
 32: 5         R(J) = R(J)+(S(I)-SS(I))**2
 33: 6         B(J+1) = F*B(J)
 34:           DO 7 J = 1,8
 35:           IF (R(J+1)-R(J)) 7,8,8
 36: 7         CONTINUE
 37: 8         WRITE(*,9) R(J), B(J)
 38: 9         FORMAT( 1X,'LATEST MINIMUM VALUE OF R IS ',E12.6,' FOR B = ',
 39:        1  E12.6,)
 40:           IF (ABS((R(J+1)-R(J))/R(J+1))-.5E-6)11,11,10
 41: 10        IF(N.EQ.20)GO TO 19
 42:           B(1) = B(J)/F
 43:           F = F**.25
 44:           N = N+1
 45:           GO TO 12
 46: 11        CALL SUM(E,S,B,A ,EPO,XN,M,J)
 47:           WRITE(*,13)
 48: 13        FORMAT(/ 2X,' (R(J+1)-R(J))/R(J+1) LESS THAN .5E-6 NEAR',/1X,
 49:        1  ' MINIMUM R(J), SO THE BEST SOLUTION IS AS FOLLOWS ',/)
```

```
50:            WRITE(*,14)
51: 14         FORMAT( 6X,' N',10X,' A',10X,'EP0', 9X,'B',11X,'R',)
52:            WRITE(*,15) XN, A,EP0,B(J),R(J)
53: 15         FORMAT(6E12.6,I2)
54:            WRITE(*,16)
55: 16         FORMAT(/11X,'STRAIN',10X,'STRESS(GIVEN)',5X,'STRESS(CALCULATED)
56:         1  ,)
57:            DO 18 I = 1,M
58:            SS(I) = A*(B(J)+E(I))**XN
59:            WRITE (*,17) E(I),S(I),SS(I)
60: 17         FORMAT(3E20.7)
61: 18         CONTINUE
62:            N = 0
63:            NEND = 0
64:            S0(1) = 0.
65:            SR = S(1)
66:            WRITE (*,60)
67: 60         FORMAT(/2X,' FITTING EQUATION S=S0+AE**N GIVES:- ',/)
68: 42         DO 36 K = 1,9
69:            CALL SUM1(E,S,S0,A,XN,M,K)
70:            IF(NEND.EQ.1) GO TO 23
71:            R1(K) = 0.
72:            DO 35 I = 1,M
73:            SS(I) = S0(K)+A*E(I)**XN
74: 35         R1(K) = R1(K)+(S(I)-SS(I))**2
75: 36         S0(K+1) = S0(K)+SR*.125
76:            DO 37 K = 1,8
77:            IF (R1(K+1)-R1(K)) 37,38,38
78: 37         CONTINUE
79: 38         WRITE(*,39) R1(K), S0(K)
80: 39         FORMAT(1X,' LATEST MINIMUM VALUE OF R IS ',E12.6,' FOR S0 = '
81:         1  E12.6,)
82:            IF (K.EQ.1) GO TO 22
83:            IF (ABS((R1(K+1)-R1(K))/R1(K+1))-.5E-6) 41,41,40
84: 40         IF (N.EQ.20) GO TO 19
85:            S0(1) = S0(K)-SR*.125
86:            SR = SR*.25
87:            N = N+1
88:            GO TO 42
89: 22         NEND = 1
90: 41         CALL SUM1(E,S,S0,A,XN,M,K)
91: 23         WRITE(*,43)
92: 43         FORMAT(/2X,'(R(K+1)-R(K)/R(K+1)LESS THAN .5E-6 NEAR',/1X,
93:         1  ' MINIMUM R(K), SO THE BEST SOLUTION IS AS FOLLOWS ',/)
94:            WRITE(*,44)
95: 44         FORMAT( 6X, 'N',10X,'S0',10X,'A',11X,'R',6X,/)
96:            WRITE(*,45) XN,S0(K),A,R1(K)
97: 45         FORMAT(4E12.6,I2)
98:            WRITE(*,46)
99: 46         FORMAT(/11X,'STRAIN',10X,'STRESS(GIVEN)'5X,'STRESS(CALCULATED)
100:        1  ,)
101:           DO 48 I = 1,M
102:           SS(I) = S0(K)+A*E(I)**XN
103:           WRITE(*,47) E(I),S(I),SS(I)
104: 47        FORMAT(3E20.7)
105: 48        CONTINUE
106:           WRITE (*,61)
107: 61        FORMAT(/2X,' FITTING EQUATION S=AE**N  GIVES:- ',/)
108:           SUM X = 0.
109:           SUM Y = 0.
110:           SUMXY = 0.
111:           SUMXX = 0.
112:           DO 49 I = 1,M
113:           X(I) = ALOG (E(I))
114:           Y(I) = ALOG (S(I))
115:           XY(I) = X(I)*Y(I)
116:           XX(I) = X(I)**2
117:           SUMX = SUMX  +  X(I)
118:           SUMY = SUMY  +  Y(I)
119:           SUMXY = SUMXY + XY(I)
120: 49        SUMXX = SUMXX + XX(I)
121:           XM = M
122:           XN = (XM*SUMXY-SUMX*SUMY)/(XM*SUMXX-SUMX**2)
123:           A = EXP((SUMY-XN*SUMX)/XM)
124:           R2 = 0.
125:           DO 50 I = 1,M
126:           SS(I) = A*E(I)**XN
127: 50        R2 = R2+(S(I)-SS(I))**2
128:           WRITE (*,51) XN,A,R2
129: 51        FORMAT(1X,' N = ',E12.6,' A = ',E12.6,' R  = ',E12.6,/1X,
130:        1  ' WITH THE FOLLOWING SOLUTION:- ',/)
131:           WRITE(*,46)
```

```
132:           DO 52 I = 1,M
133:           SS(I) = A*E(I)**XN
134:           WRITE (*,47) E(I),S(I),SS(I)
135: 52        CONTINUE
136:           IF (R(J).LE.R1(K).AND.R(J).LE.R2) GO TO 53
137:           IF (R2.LE.R1(K)  .AND.R2.LE.R(J)) GO TO 57
138:           IF (R1(K).LE.R(J).AND.R1(K).LT.R2)GO TO 55
139: 53        WRITE(*,54)
140:           GO TO 21
141: 55        WRITE(*,56)
142:           GO TO 21
143: 57        WRITE(*,58)
144: 54        FORMAT(//9X,29HS = A(B+E)**N GIVES BEST FIT./)
145: 56        FORMAT(//9X,29HS = S0+AE**N  GIVES BEST FIT./)
146: 58        FORMAT(//9X,29HS = AE**N    GIVES BEST FIT./)
147:           GO TO 21
148: 19        WRITE(*,20)
149: 20        FORMAT(/1X,' NUMBER OF ITERATIONS = 20',/)
150: 21        STOP
151:           END
152:
153:           SUBROUTINE SUM (E,S,B, A,EP0,XN,M,J)
154:           DIMENSION E(20),S(20),X(20),Y(20),XY(20),XX(20),B(20),R(20),
155:         1 SS(20),S0(20),R1(20)
156:           SUMX  = 0.
157:           SUMY  = 0.
158:           SUMXY = 0.
159:           SUMXX = 0.
160:           DO 1 I = 1,M
161:           X(I) = ALOG(B(J)+E(I))
162:           Y(I) = ALOG(S(I))
163:           XY(I) = X(I)*Y(I)
164:           XX(I) = X(I)**2
165:           SUMX  = SUMX  +  X(I)
166:           SUMY  = SUMY  +  Y(I)
167:           SUMXY = SUMXY + XY(I)
168: 1         SUMXX= SUMXX + XX(I)
169:           XM = M
170:           XN = (XM*SUMXY-SUMX*SUMY)/(XM*SUMXX-SUMX**2)
171:           A  = EXP((SUMY-XN*SUMX)/XM)
172:           EP0 = XN*A*(B(J))**(XN-1.)
173:           RETURN
174:           END
175:
176:           SUBROUTINE SUM1(E,S,S0,A,XN,M,K)
177:           DIMENSION E(20),S(20),X(20),Y(20),XY(20),XX(20),B(20),R(20),
178:         1 SS(20),S0(20),R1(20)
179:           SUMX  = 0.
180:           SUMY  = 0.
181:           SUMXY = 0.
182:           SUMXX = 0.
183:           DO 4 I = 1,M
184:           X(I) = ALOG(E(I))
185:           IF (S(I)-S0(K)) 1,1,2
186: 1         Y(I) = ALOG(ABS(S(I)-S0(K))+.1E-6)
187:           GO TO 3
188: 2         Y(I) = ALOG(S(I)-S0(K))
189: 3         XY(I) = X(I)*Y(I)
190:           XX(I) = X(I)**2
191:           SUMX  = SUMX  +  X(I)
192:           SUMY  = SUMY  +  Y(I)
193:           SUMXY = SUMXY + XY(I)
194: 4         SUMXX= SUMXX + XX(I)
195:           XM = M
196:           XN = (XM*SUMXY-SUMX*SUMY)/(XM*SUMXX-SUMX**2)
197:           A  = EXP((SUMY-XN*SUMX)/XM)
198:           RETURN
199:           END
```

SOLUTION 3.1 Nickel–chrome–molybdenum steel to EN 25T.

```
SWIFT FITS GIVEN S/E DATA POINTS TO VARIANTS OF
EQUATION S = S0+A(B+E)**N  BY A LEAST SQUARE FIT OF M
LOG-LOG VALUES FOR A NUMBER (J) OF ASSUMED VALUES OF B
  OR ALTERNATIVELY A NUMBER (K) OF ASSUMED VALUES OF S0
WHICH ARE OPTIMIZED BY DIRECT SEARCH, CHOOSING SOLUTIONS
TO MINIMIZE R = SUMMATION OF (S(GIVEN)-S(CALCULATED))**2
TRY NOT TO USE ANY STRAIN VALUE LESS THAN E=0.01

TYPE IN M
 11
TYPE IN E( 1),S( 1)
 .0547   785.
TYPE IN E( 2),S( 2)
 .0633   795.7
TYPE IN E( 3),S( 3)
 .0690   803.9
TYPE IN E( 4),S( 4)
 .0777   813.9
TYPE IN E( 5),S( 5)
 .0867   823.6
TYPE IN E( 6),S( 6)
 .0973   833.2
TYPE IN E( 7),S( 7)
 .11     841.
TYPE IN E( 8),S( 8)
 .1254   849.4
TYPE IN E( 9),S( 9)
 .1434   858.8
TYPE IN E(10),S(10)
 .1664   869.8
TYPE IN E(11),S(11)
 .1951   883.6

     FITTING EQUATION S=A(B+E)**N GIVES:-

LATEST MINIMUM VALUE OF R IS   .378850E+02 FOR B =  .100000E-02
LATEST MINIMUM VALUE OF R IS   .370106E+02 FOR B =  .500000E-03
LATEST MINIMUM VALUE OF R IS   .368631E+02 FOR B =  .420448E-03
LATEST MINIMUM VALUE OF R IS   .368346E+02 FOR B =  .402623E-03
LATEST MINIMUM VALUE OF R IS   .368251E+02 FOR B =  .399365E-03
LATEST MINIMUM VALUE OF R IS   .368372E+02 FOR B =  .398286E-03
LATEST MINIMUM VALUE OF R IS   .368240E+02 FOR B =  .398151E-03
LATEST MINIMUM VALUE OF R IS   .368247E+02 FOR B =  .398083E-03
LATEST MINIMUM VALUE OF R IS   .368245E+02 FOR B =  .398067E-03

     (R(J+1)-R(J))/R(J+1) LESS THAN .5E-6 NEAR
  MINIMUM R(J), SO THE BEST SOLUTION IS AS FOLLOWS

        N           A          EP0          B           R
  .920804E-01 .102804E+04 .115650E+06 .398067E-03 .368245E+02

           STRAIN            STRESS(GIVEN)       STRESS(CALCULATED)
         .5470000E-01         .7850000E+03         .7872156E+03
         .6330000E-01         .7957000E+03         .7977996E+03
         .6900000E-01         .8039000E+03         .8041205E+03
         .7770000E-01         .8139000E+03         .8129133E+03
         .8670000E-01         .8236000E+03         .8211186E+03
         .9730000E-01         .8332000E+03         .8298481E+03
         .1100000E+00         .8410000E+03         .8392393E+03
         .1254000E+00         .8494000E+03         .8493916E+03
         .1434000E+00         .8588000E+03         .8599156E+03
         .1664000E+00         .8698000E+03         .8717448E+03
         .1951000E+00         .8836000E+03         .8845827E+03
```

```
FITTING EQUATION S=S0+AE**N GIVES:-

LATEST MINIMUM VALUE OF R IS  .361386E+02 FOR S0 =  .000000E+00

(R(K+1)-R(K)/R(K+1) LESS THAN .5E-6 NEAR
MINIMUM R(K), SO THE BEST SOLUTION IS AS FOLLOWS

    N          S0            A            R

.917169E-01 .000000E+00 .102758E+04 .361386E+02

              STRAIN         STRESS(GIVEN)      STRESS(CALCULATED)
          .5470000E-01       .7850000E+03        .7871724E+03
          .6330000E-01       .7957000E+03        .7977856E+03
          .6900000E-01       .8039000E+03        .8041194E+03
          .7770000E-01       .8139000E+03        .8129252E+03
          .8670000E-01       .8236000E+03        .8211379E+03
          .9730000E-01       .8332000E+03        .8298709E+03
          .1100000E+00       .8410000E+03        .8392614E+03
          .1254000E+00       .8494000E+03        .8494080E+03
          .1434000E+00       .8588000E+03        .8599219E+03
          .1664000E+00       .8698000E+03        .8717346E+03
          .1951000E+00       .8836000E+03        .8845499E+03

FITTING EQUATION S=AE**N  GIVES:-

N =  .917169E-01 A =  .102758E+04 R  =  .361386E+02
WITH THE FOLLOWING SOLUTION:-

              STRAIN         STRESS(GIVEN)      STRESS(CALCULATED)
          .5470000E-01       .7850000E+03        .7871724E+03
          .6330000E-01       .7957000E+03        .7977856E+03
          .6900000E-01       .8039000E+03        .8041194E+03
          .7770000E-01       .8139000E+03        .8129252E+03
          .8670000E-01       .8236000E+03        .8211379E+03
          .9730000E-01       .8332000E+03        .8298709E+03
          .1100000E+00       .8410000E+03        .8392614E+03
          .1254000E+00       .8494000E+03        .8494080E+03
          .1434000E+00       .8588000E+03        .8599219E+03
          .1664000E+00       .8698000E+03        .8717346E+03
          .1951000E+00       .8836000E+03        .8845499E+03


     S = AE**N     GIVES BEST FIT.

Stop - Program terminated.
```

Although Swift's equation actually does not give absolutely the best fit to the stress–strain curve of this particular specialized steel, it will be used in the computer programs which follow for Manning's method, since Swift's equation is generally superior to the others given in Program 3.1, in its ability to fit the stress–strain curves of most metals.

Program 3.2 gives a computer program called "Manning" which utilizes Manning's basic method of constructing a table for a thick–walled tube of unit bore radius, for any value of K–ratio up to a value of 9, which is large enough to cover the strong materials of interest in this particular problem of high–pressure vessel technology. The variables which have to be inserted in order to obtain a solution are the variables involved in Swift's equation together with the yield–point strain, the bore displacement and the K–ratio (both of which latter variables should be made large enough to cover the range of interest for the particular application involved).

PROGRAM 3.2 Manning's tabular solution for expanding or compressing a tube of unit bore radius.

```
 1:*        PROGRAM MANNING
 2:         DIMENSION R(83),RU(83),DRU(83),RUR(83),ARUR(83),
 3:       1 EQ(83),S(83),DP(83),P(83),UR(83),RUM(83)
 4:         WRITE(*,1)
 5: 1       FORMAT(1X' "MANNING" PRINTS A MANNING TABLE FOR ANY K-',/1X,
 6:       & ' RATIO UP TO A VALUE OF 9 FOR EXPANDING OR COMPRESSING A',/1
 7:       & ' TUBE OF UNIT BORE RADIUS. NOTATION-R=RADIUS,U=RADIAL',/1X,
 8:       & ' DISPLACEMENT,UR=U/R,UA=BORE DISPLACEMENT,DR=DELTA R',/1X
 9:       & ' RU=R+U,DRU=DELTA(R+U),RUR=(R+U)/R,S=YIELD STRESS,',/1X,
10:       & ' ARUR=(2/SQ RT(3))LN(R+U)/R,EQ=EQUIVALENT STRAIN,',/1X,
11:       & ' P=PRESSURE,DP=DELTA P,E=YOUNGS MODULUS,EO=INITIAL',/1X,
12:       & ' EQUIVALENT STRAIN,XN=EXPONENT OF EMPIRICAL STRESS-',/1X,
13:       & ' STRAIN RELATION S=A*(B+EQ(I))**XN.MAKE K-RATIO NEAREST',/1X
14:       & ' POSSIBLE MULTIPLE OF 0.1,THEN PUT N=INTEGER(K-.9)*10.',/1X,
15:       & ' EACH DIMENSION STORE SHOULD BE AT LEAST N+2.',/1X,
16:       & ' FOR COMPRESSION MAKE UA NEGATIVE:-P IS THEN EXTERNAL.',/)
17:         WRITE(*,2)
18: 2       FORMAT(/1X,'  TYPE IN E,A,B,XN,UA,N',/)
19:         READ(*,*) E,A,B,XN,UA,N
20:         WRITE(*,3)
21: 3       FORMAT(/10X,1HE,10X,1HA,10X,1HB,10X,2HXN,8X,2HUA,7X,1HN/)
22:         WRITE(*,4) E,A,B,XN,UA,N
23: 4       FORMAT(4X,5E11.5,I4)
24:         WRITE(*,5)
25: 5       FORMAT(/2X,1HI,7X,1HR,11X,2HEQ,11X,1HS,9X,2HDP,11X,1HP,11X,
26:       & 2HUR/)
27:         EO=A*B**XN/E
28:         R(1)=1.
29:         P(1)=0.
30:         T=2./SQ RT(3.)
31:         DR=.1
32:         C2=UA*(2.+UA)
33:         DO 7 I=1,N+1
34:         RU(I)=SQ RT(R(I)**2+C2)
35:         RUR(I)=RU(I)/R(I)
36:         ARUR(I)=ABS(T*ALOG(RUR(I)))
37:         UR(I)=RUR(I)-1.
38: 7       R(I+1)=R(I)+DR
39:         DO 10 I=1,N
40:         EQ(I)=(ARUR(I)+ARUR(I+1))*.5
41:         S(I)=A*(B+EQ(I))**XN
42:         DRU(I)=RU(I+1)-RU(I)
43:         RUM(I)=(RU(I+1)+RU(I))*.5
44:         DP(I)=T*S(I)*DRU(I)/RUM(I)
45:         P(I+1)=P(I)+DP(I)
46:         WRITE(*,8) I,R(I),P(I),UR(I)
47: 8       FORMAT(1X,I2,E12.5,36X,2E12.5)
48:         WRITE(*,9) I,EQ(I),S(I),DP(I)
49: 9       FORMAT(3X,I2,11X,3E12.5)
50: 10      CONTINUE
51:         WRITE(*,12) EO
52: 12      FORMAT(//20X,5HEO = ,E12.5//)
53:         STOP
54:         END
```

SOLUTION 3.2 Manning table for Nickel–chrome–molybdenum steel to EN 25T and a bore strain of 20%.

```
"MANNING" PRINTS A MANNING TABLE FOR ANY K-
RATIO UP TO A VALUE OF 9 FOR EXPANDING OR COMPRESSING A
TUBE OF UNIT BORE RADIUS. NOTATION-R=RADIUS,U=RADIAL
DISPLACEMENT,UR=U/R,UA=BORE DISPLACEMENT,DR=DELTA R
RU=R+U,DRU=DELTA(R+U),RUR=(R+U)/R,S=YIELD STRESS,
ARUR=(2/SQ RT(3))LN(R+U)/R,EQ=EQUIVALENT STRAIN,
P=PRESSURE,DP=DELTA P,E=YOUNGS MODULUS,EO=INITIAL
EQUIVALENT STRAIN,XN=EXPONENT OF EMPIRICAL STRESS-
STRAIN RELATION S=A*(B+EQ(I))**XN.MAKE K-RATIO NEAREST
POSSIBLE MULTIPLE OF 0.1,THEN PUT N=INTEGER(K-.9)*10.
EACH DIMENSION STORE SHOULD BE AT LEAST N+2.
FOR COMPRESSION MAKE UA NEGATIVE:-P IS THEN EXTERNAL.
```

```
TYPE IN E,A,B,XN,UA,N

206786.    1028.04    .000398067    .0920804    .2    81
```

	E	A	B	XN	UA	N
	.20679E+06	.10280E+04	.39807E-03	.92080E-01	.20000E+00	81

I	R	EQ	S	DP	P	UR
1	.10000E+01				.00000E+00	.20000E+00
1		.19480E+00	.88446E+03	.69488E+02		
2	.11000E+01				.69488E+02	.16775E+00
2		.16650E+00	.87179E+03	.65660E+02		
3	.12000E+01				.13515E+03	.14261E+00
3		.14377E+00	.86012E+03	.61979E+02		
4	.13000E+01				.19713E+03	.12266E+00
4		.12526E+00	.84930E+03	.58504E+02		
5	.14000E+01				.25563E+03	.10657E+00
5		.11002E+00	.83926E+03	.55258E+02		
6	.15000E+01				.31089E+03	.93415E-01
6		.97346E-01	.82988E+03	.52247E+02		
7	.16000E+01				.36314E+03	.82532E-01
7		.86695E-01	.82111E+03	.49463E+02		
8	.17000E+01				.41260E+03	.73429E-01
8		.77670E-01	.81288E+03	.46894E+02		
9	.18000E+01				.45949E+03	.65740E-01
9		.69960E-01	.80514E+03	.44526E+02		
10	.19000E+01				.50402E+03	.59190E-01
10		.63326E-01	.79783E+03	.42341E+02		
11	.20000E+01				.54636E+03	.53565E-01
11		.57580E-01	.79092E+03	.40325E+02		
12	.21000E+01				.58668E+03	.48701E-01
12		.52572E-01	.78437E+03	.38463E+02		
13	.22000E+01				.62515E+03	.44466E-01
13		.48182E-01	.77814E+03	.36740E+02		
14	.23000E+01				.66189E+03	.40757E-01
14		.44314E-01	.77222E+03	.35143E+02		
15	.24000E+01				.69703E+03	.37492E-01
15		.40889E-01	.76657E+03	.33661E+02		
16	.25000E+01				.73069E+03	.34601E-01
16		.37843E-01	.76118E+03	.32283E+02		
17	.26000E+01				.76297E+03	.32031E-01
17		.35121E-01	.75602E+03	.31000E+02		
18	.27000E+01				.79397E+03	.29736E-01
18		.32681E-01	.75109E+03	.29803E+02		
19	.28000E+01				.82378E+03	.27678E-01
19		.30484E-01	.74635E+03	.28685E+02		
20	.29000E+01				.85246E+03	.25826E-01
20		.28500E-01	.74180E+03	.27638E+02		
21	.30000E+01				.88010E+03	.24153E-01
21		.26702E-01	.73743E+03	.26657E+02		
22	.31000E+01				.90676E+03	.22637E-01
22		.25068E-01	.73321E+03	.25736E+02		
23	.32000E+01				.93249E+03	.21258E-01
23		.23579E-01	.72916E+03	.24870E+02		
24	.33000E+01				.95736E+03	.20002E-01
24		.22218E-01	.72524E+03	.24055E+02		

25	.34000E+01				.98142E+03	.18853E-01
25		.20971E-01	.72146E+03	.23286E+02		
26	.35000E+01				.10047E+04	.17801E-01
26		.19825E-01	.71781E+03	.22560E+02		
27	.36000E+01				.10273E+04	.16834E-01
27		.18770E-01	.71428E+03	.21874E+02		
28	.37000E+01				.10491E+04	.15943E-01
28		.17797E-01	.71086E+03	.21225E+02		
29	.38000E+01				.10704E+04	.15121E-01
29		.16897E-01	.70755E+03	.20609E+02		
30	.39000E+01				.10910E+04	.14361E-01
30		.16064E-01	.70434E+03	.20025E+02		
31	.40000E+01				.11110E+04	.13657E-01
31		.15290E-01	.70123E+03	.19470E+02		
32	.41000E+01				.11305E+04	.13003E-01
32		.14571E-01	.69820E+03	.18943E+02		
33	.42000E+01				.11494E+04	.12395E-01
33		.13901E-01	.69527E+03	.18441E+02		
34	.43000E+01				.11679E+04	.11828E-01
34		.13276E-01	.69241E+03	.17962E+02		
35	.44000E+01				.11858E+04	.11300E-01
35		.12693E-01	.68964E+03	.17506E+02		
36	.45000E+01				.12033E+04	.10806E-01
36		.12146E-01	.68693E+03	.17070E+02		
37	.46000E+01				.12204E+04	.10344E-01
37		.11635E-01	.68430E+03	.16654E+02		
38	.47000E+01				.12370E+04	.99102E-02
38		.11154E-01	.68174E+03	.16256E+02		
39	.48000E+01				.12533E+04	.95035E-02
39		.10703E-01	.67925E+03	.15875E+02		
40	.49000E+01				.12692E+04	.91212E-02
40		.10279E-01	.67681E+03	.15510E+02		
41	.50000E+01				.12847E+04	.87616E-02
41		.98791E-02	.67444E+03	.15160E+02		
42	.51000E+01				.12998E+04	.84229E-02
42		.95021E-02	.67212E+03	.14824E+02		
43	.52000E+01				.13147E+04	.81033E-02
43		.91463E-02	.66986E+03	.14502E+02		
44	.53000E+01				.13292E+04	.78015E-02
44		.88101E-02	.66765E+03	.14192E+02		
45	.54000E+01				.13434E+04	.75164E-02
45		.84920E-02	.66549E+03	.13894E+02		
46	.55000E+01				.13573E+04	.72465E-02
46		.81908E-02	.66339E+03	.13608E+02		
47	.56000E+01				.13709E+04	.69909E-02
47		.79054E-02	.66132E+03	.13332E+02		
48	.57000E+01				.13842E+04	.67486E-02
48		.76345E-02	.65931E+03	.13066E+02		
49	.58000E+01				.13973E+04	.65186E-02
49		.73773E-02	.65734E+03	.12810E+02		
50	.59000E+01				.14101E+04	.63002E-02
50		.71329E-02	.65540E+03	.12563E+02		
51	.60000E+01				.14226E+04	.60925E-02
51		.69004E-02	.65351E+03	.12325E+02		
52	.61000E+01				.14350E+04	.58950E-02
52		.66790E-02	.65166E+03	.12095E+02		
53	.62000E+01				.14471E+04	.57069E-02
53		.64682E-02	.64985E+03	.11872E+02		
54	.63000E+01				.14589E+04	.55277E-02
54		.62671E-02	.64808E+03	.11658E+02		
55	.64000E+01				.14706E+04	.53568E-02
55		.60753E-02	.64634E+03	.11450E+02		
56	.65000E+01				.14820E+04	.51936E-02
56		.58920E-02	.64463E+03	.11249E+02		
57	.66000E+01				.14933E+04	.50378E-02
57		.57170E-02	.64296E+03	.11054E+02		
58	.67000E+01				.15043E+04	.48889E-02
58		.55496E-02	.64131E+03	.10866E+02		
59	.68000E+01				.15152E+04	.47464E-02
59		.53895E-02	.63971E+03	.10683E+02		
60	.69000E+01				.15259E+04	.46103E-02
60		.52363E-02	.63813E+03	.10506E+02		
61	.70000E+01				.15364E+04	.44798E-02
61		.50894E-02	.63658E+03	.10335E+02		
62	.71000E+01				.15467E+04	.43547E-02
62		.49485E-02	.63505E+03	.10168E+02		
63	.72000E+01				.15569E+04	.42348E-02
63		.48135E-02	.63356E+03	.10007E+02		
64	.73000E+01				.15669E+04	.41199E-02
64		.46840E-02	.63209E+03	.98501E+01		

```
65  .74000E+01                                         .15768E+04  .40095E-02
    65          .45596E-02  .63065E+03  .96978E+01
66  .75000E+01                                         .15864E+04  .39035E-02
    66          .44401E-02  .62924E+03  .95499E+01
67  .76000E+01                                         .15960E+04  .38017E-02
    67          .43251E-02  .62785E+03  .94061E+01
68  .77000E+01                                         .16054E+04  .37037E-02
    68          .42146E-02  .62648E+03  .92662E+01
69  .78000E+01                                         .16147E+04  .36095E-02
    69          .41083E-02  .62513E+03  .91302E+01
70  .79000E+01                                         .16238E+04  .35189E-02
    70          .40060E-02  .62381E+03  .89980E+01
71  .80000E+01                                         .16328E+04  .34317E-02
    71          .39074E-02  .62252E+03  .88693E+01
72  .81000E+01                                         .16417E+04  .33476E-02
    72          .38124E-02  .62124E+03  .87438E+01
73  .82000E+01                                         .16504E+04  .32666E-02
    73          .37208E-02  .61998E+03  .86217E+01
74  .83000E+01                                         .16590E+04  .31884E-02
    74          .36324E-02  .61874E+03  .85028E+01
75  .84000E+01                                         .16675E+04  .31130E-02
    75          .35472E-02  .61753E+03  .83870E+01
76  .85000E+01                                         .16759E+04  .30404E-02
    76          .34650E-02  .61633E+03  .82739E+01
77  .86000E+01                                         .16842E+04  .29701E-02
    77          .33856E-02  .61515E+03  .81639E+01
78  .87000E+01                                         .16924E+04  .29025E-02
    78          .33089E-02  .61399E+03  .80564E+01
79  .88000E+01                                         .17004E+04  .28369E-02
    79          .32347E-02  .61285E+03  .79515E+01
80  .89000E+01                                         .17084E+04  .27736E-02
    80          .31631E-02  .61173E+03  .78493E+01
81  .90000E+01                                         .17162E+04  .27125E-02
    81          .30938E-02  .61062E+03  .77494E+01

              EO =    .24178E-02

Stop - Program terminated.
```

Solution 3.2 which follows Program 3.2 prints out a Manning table of exactly the same format as that shown on p.469 of the book by Ford and Alexander and it is interesting to observe the close correspondence of the values obtained for the internal pressure p_1 (P in the program) and u/r at any radius. The detailed method of calculation is set out in that book and will not be repeated in its entirety here. For an annulus between $r = 1.4$ and $r = 1.5$, for example, the mean $\bar{\epsilon}^p$ = 0.11002, the mean $\bar{\sigma}$ is 839.26 MPa and a pressure of 55.258 MPa would be needed to cause this amount of strain (corresponding to a bore strain of 0.10657 at the inner radius of the annulus, namely 1.4) if this annulus were regarded as existing on its own.

The whole cylinder can be considered as a large number of these concentric tubes, since σ_r, σ_θ, and σ_z are principal stresses and no shear stresses are transmitted between these tubes. The internal pressure for any wall ratio can therefore be found simply by summing the pressure increments for the annuli between. Thus a cylinder of wall ratio 2.8 would require an internal pressure of 823.78 MPa to expand the bore by 20%, whilst a cylinder of half that K–ratio, namely 1.4, would require an internal pressure of only 255.63 MPa to achieve the same bore strain, i.e. $< \frac{1}{3} \times 823.78$ MPa.

Values of pressure required for other values of bore strain and K–ratio can also be found from Manning's table, by interpolation. However, it is more convenient to use Program 3.2 for various values of bore strain and combine the results for a tube of a given K–ratio to give a set of results for internal pressure versus bore strain. This is the information normally required in any given situation and allows the maximum (burst) pressure to be determined. Program 3.3 called Manning1 allows this to be achieved.

PROGRAM 3.3 For determining internal or external pressure versus bore displacement, using Manning's method.

```
 1:*           PROGRAM MANNING1
 2:            DIMENSION R(83),RU(83),DRU(83),RUR(83),ARUR(83),EQ(83),S(83),
 3:         1  DP(83),P(83),UR(83),RUM(83)
 4:            WRITE(*,1)
 5: 1          FORMAT(1X' "MANNING1" TABULATES P VERSUS UA FOR ANY K-',/1X,
 6:         &  ' RATIO UP TO A VALUE OF 9 FOR EXPANDING OR COMPRESSING A',/1
 7:         &  ' TUBE OF UNIT BORE RADIUS.NOTATION-R=RADIUS,U=RADIAL',/1X,
 8:         &  ' DISPLACEMENT,UR=U/R,UA=BORE DISPLACEMENT,DR=DELTA R',/1X,
 9:         &  ' RU=R+U,DRU=DELTA(R+U),RUR=(R+U)/R,S=YIELD STRESS,',/1X,
10:         &  ' ARUR=(2/SQ RT(3))LN(R+U)/R,EQ=EQUIVALENT STRAIN,',/1X,
11:         &  ' P=PRESSURE,DP=DELTA P,E=YOUNGS MODULUS,EO=INITIAL',/1X,
12:         &  ' EQUIVALENT STRAIN,XN=EXPONENT OF EMPIRICAL STRESS-',/1X,
13:         &  ' STRAIN RELATION S=A*(B+EQ(I))**XN.MAKE K-RATIO NEAREST',/1X
14:         &  ' POSSIBLE MULTIPLE OF .1, THEN PUT N=INTEGER(K-.9)*10.',/1X,
15:         &  ' EACH DIMENSION STORE SHOULD BE AT LEAST N+2.',/1X,
16:         &  ' FOR COMPRESSION MAKE UA NEGATIVE:-P IS THEN EXTERNAL.',/)
17:            WRITE(*,2)
18: 2          FORMAT(1X,' TYPE IN E,A,B,XN,UA,N'/)
19:            READ(*,*) E,A,B,XN,UA,N
20:            WRITE(*,3)
21: 3          FORMAT(/,10X,1HE,10X,1HA,10X,1HB,10X,2HXN,8X,2HUA,7X,1HN/)
22:            WRITE(*,4) E,A,B,XN,UA,N
23: 4          FORMAT(4X,5E11.5,I4)
24:            WRITE(*,5)
25: 5          FORMAT(/20X,2HUA,20X,1HP/)
26:            EO=A*B**XN/E
27:            R(1)=1.
28:            P(1)=0.
29:            T=2./SQ RT(3.)
30:            DR=.1
31: 6          C2=UA*(2.+UA)
32:            DO 7 I=1,N+1
33:            RU(I)=SQ RT (R(I)**2+C2)
34:            RUR(I)=RU(I)/R(I)
35:            ARUR(I)=ABS(T*ALOG(RUR(I)))
36:            UR(I)=RUR(I)-1.
37: 7          R(I+1)=R(I)+DR
38:            DO 10 I=1,N
39:            EQ(I)=(ARUR(I)+ARUR(I+1))*.5
40:            S(I)=A*(B+EQ(I))**XN
41:            DRU(I)=RU(I+1)-RU(I)
42:            RUM(I)=(RU(I+1)+RU(I))*.5
43:            DP(I)=T*S(I)*DRU(I)/RUM(I)
44:            P(I+1)=P(I)+DP(I)
45: 10         CONTINUE
46:            WRITE(*,11) UR(1),P(N)
47: 11         FORMAT(7X,2(8X,E12.5))
48:            IF(UA.LT.0.0) GOTO 13
49:            UA=UA+.02
50:            GOTO 14
51: 13         UA=UA-.02
52: 14         IF(ABS(UA).LT.0.31) GOTO 6
53:            WRITE(*,12) EO
54: 12         FORMAT(/20X,4HEO =,E12.5//)
55:            STOP
56:            END
```

SOLUTION 3.3a Listing internal pressure versus outward bore displacement, using Manning's method, for Nickel–chrome–molybdenum steel to EN 25T.

```
"MANNING1" TABULATES P VERSUS UA FOR ANY K-
RATIO UP TO A VALUE OF 9 FOR EXPANDING OR COMPRESSING A
TUBE OF UNIT BORE RADIUS.NOTATION-R=RADIUS,U=RADIAL
DISPLACEMENT,UR=U/R,UA=BORE DISPLACEMENT,DR=DELTA R
RU=R+U,DRU=DELTA(R+U),RUR=(R+U)/R,S=YIELD STRESS,
ARUR=(2/SQ RT(3))LN(R+U)/R,EQ=EQUIVALENT STRAIN,
P=PRESSURE,DP=DELTA P,E=YOUNGS MODULUS,EO=INITIAL
EQUIVALENT STRAIN,XN=EXPONENT OF EMPIRICAL STRESS-
STRAIN RELATION S=A*(B+EQ(I))**XN.MAKE K-RATIO NEAREST
POSSIBLE MULTIPLE OF .1, THEN PUT N=INTEGER(K-.9)*10.
EACH DIMENSION STORE SHOULD BE AT LEAST N+2.
FOR COMPRESSION MAKE UA NEGATIVE:-P IS THEN EXTERNAL.

TYPE IN E,A,B,XN,UA,N

  206786.    1028.04    .000398067    .0920804    .02    81

      E          A          B          XN         UA        N

 .20679E+06 .10280E+04 .39807E-03 .92080E-01 .20000E-01   81

            UA                       P

         .20000E-01               .15324E+04
         .40000E-01               .16035E+04
         .60000E-01               .16435E+04
         .80000E-01               .16691E+04
         .10000E+00               .16864E+04
         .12000E+00               .16984E+04
         .14000E+00               .17064E+04
         .16000E+00               .17117E+04
         .18000E+00               .17148E+04
         .20000E+00               .17162E+04
         .22000E+00               .17163E+04
         .24000E+00               .17152E+04
         .26000E+00               .17132E+04
         .28000E+00               .17105E+04
         .30000E+00               .17071E+04

            EO = .24178E-02
```

Stop - Program terminated.

Solution 3.3a lists internal pressure versus (outward) bore displacement up to 30% for a high–pressure vessel of K–ratio 9 made from the high–strength steel under discussion. The maximum pressure is seen to occur at a bore strain somewhere between 20% and 22%, actually at a value of about 21.2%, as obtained by interpolation and shown in Fig. 3.6.

Program 3.3 also allows determination of the external pressure required to cause inward bore displacement and Solution 3.3b lists external pressure versus (inward) bore displacement up to 30% for a vessel of the same K–ratio and steel. As would be expected, the external pressure increases continuously, never reaching a peak value, as also indicated on Fig. 3.6.

SOLUTION 3.3b Listing external pressure versus inward bore displacement, using Manning's method, for Nickel-chrome-molybdenum steel to EN 25T.

```
TYPE IN E,A,B,XN,UA,N

   206786.    1028.04    .000398067    .0920804    -.02    81

          E          A          B          XN          UA        N

   .20679E+06 .10280E+04 .39807E-03 .92080E-01-.20000E-01   81
```

UA	P
-.20000E-01	.15621E+04
-.40000E-01	.16662E+04
-.60000E-01	.17408E+04
-.80000E-01	.18023E+04
-.10000E+00	.18565E+04
-.12000E+00	.19060E+04
-.14000E+00	.19525E+04
-.16000E+00	.19970E+04
-.18000E+00	.20400E+04
-.20000E+00	.20821E+04
-.22000E+00	.21235E+04
-.24000E+00	.21646E+04
-.26000E+00	.22056E+04
-.28000E+00	.22466E+04
-.30000E+00	.22879E+04

```
          EO =   .24178E-02
```

Stop - Program terminated.

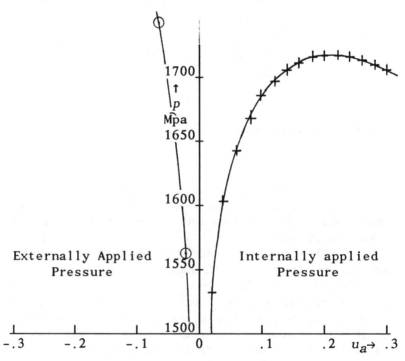

Fig. 3.6 Pressure vs. bore displacement for both internal and external applied pressure

An alternative way of presenting the results of applying Manning's method to such problems is to combine computer programs Manning and Manning1 to give Program Manning2, which displays Manning tables for increasing values of bore displacement. This tabulation takes up a lot of space but does display the maximum amount of information. Program Manning2 is tabulated below, as Program 3.4.

PROGRAM 3.4 For displaying Manning tables and values of internal or external pressure versus bore displacement.

```
 1:*          PROGRAM MANNING2
 2:           DIMENSION R(83),RU(83),DRU(83),RUR(83),ARUR(83),
 3:        1  EQ(83),S(83),DP(83),P(83),UR(83),RUM(83)
 4:           WRITE(*,1)
 5: 1         FORMAT(1X' "MANNING2" DISPLAYS MANNING TABLES AND VALUES',/1X
 6:        &  ' OF P VS.UA FOR ANY K-VALUE UP TO 9, FOR EXPANDING OR',/1X,
 7:        &  ' COMPRESSING A TUBE OF UNIT BORE RADIUS.NOTATION-R=',/1X,
 8:        &  ' RADIUS,U=RADIAL DISPLACEMENT,UR=U/R,UA=BORE DIS-',/1X,
 9:        &  ' PLACEMENT,DR=DELTA R,RU=R+U,DRU=DELTA(R+U),RUR=',/1X,
10:        &  ' (R+U)/R,S=YIELD STRESS,ARUR=(2/SQ RT(3))LN(R+U)/R,EQ=',/1X,
11:        &  ' EQUIVALENT STRAIN,P=PRESSURE,DP=DELTA P,E=YOUNGS',/1X,
12:        &  ' MODULUS,EO=INITIAL EQUIVALENT STRAIN,XN=EXPONENT OF',/1X,
13:        &  ' SWIFTS EQUATION S=A*(B+(EQ))**XN.MAKE K-RATIO NEAREST',/1X,
14:        &  ' POSSIBLE MULTIPLE OF .1,THEN PUT N=INTEGER(K-.9)*10.',/1X,
15:        &  ' EACH DIMENSION STORE SHOULD BE AT LEAST N+2.',/1X,
16:        &  ' FOR COMPRESSION MAKE UA NEGATIVE:-P IS THEN EXTERNAL.',/)
17:           WRITE(*,2)
18: 2         FORMAT(1X,'  TYPE IN E,A,B,XN,UA,N'/)
19:           READ(*,*) E,A,B,XN,UA,N
20:           WRITE(*,3)
21: 3         FORMAT(/10X,1HE,10X,1HA,10X,1HB,10X,2HXN,9X,2HUA,7X,1HN/)
22:           WRITE(*,4) E,A,B,XN,UA,N
23: 4         FORMAT(4X,5E11.5,I4)
24:           WRITE(*,5)
25: 5         FORMAT(/2X,1HI,7X,1HR,11X,2HEQ,11X,1HS,9X,2HDP,11X,1HP,11X,
26:        &  2HUR/)
27:           E0=A*B**XN/E
28:           R(1)=1.
29:           P(1)=0.
30:           T=2./SQRT(3.)
31:           DR=.1
32: 6         C2=UA*(2.+UA)
33:           DO 7 I=1,N+1
34:           RU(I)=SQRT (R(I)**2+C2)
35:           RUR(I)=RU(I)/R(I)
36:           ARUR(I)=ABS(T*ALOG(RUR(I)))
37:           UR(I)=RUR(I)-1.
38: 7         R(I+1)=R(I)+DR
39:           DO 10 I=1,N
40:           EQ(I)=(ARUR(I)+ARUR(I+1))*.5
41:           S(I)=A*(B+EQ(I))**XN
42:           DRU(I)=RU(I+1)-RU(I)
43:           RUM(I)=(RU(I+1)+RU(I))*.5
44:           DP(I)=T*S(I)*DRU(I)/RUM(I)
45:           P(I+1)=P(I)+DP(I)
46:           WRITE(*,8) I,R(I),P(I),UR(I)
47: 8         FORMAT(1X,I2,E12.5,36X,2E12.5)
48:           WRITE(*,9) I,EQ(I),S(I),DP(I)
49: 9         FORMAT(3X,I2,11X,3E12.5)
50: 10        CONTINUE
51:           WRITE(*,11) UR(1),P(N)
52: 11        FORMAT(/10X,4HUA =,E12.5,5X,3HP =,E12.5/)
53:           IF(UA.LT.0.0) GOTO 13
54:           UA=UA+.02
55:           GOTO 14
56: 13        UA=UA-.02
57: 14        IF(ABS(UA).LT.0.31) GOTO 6
58:           WRITE(*,12) E0
59: 12        FORMAT(//20X,4HEO =,E12.5//)
60:           STOP
61:           END
```

Solution 3.4 was obtained from Program 3.4 and displays Manning tables and values of internal pressure versus outward bore displacement for a vessel having $K = 1.5$ (chosen to conserve space; similarly, intermediate values between $u_a = 0.1$ and $u_a = 0.28$ have been omitted).

SOLUTION 3.4 Displays Manning tables and values of internal pressure versus outward bore displacement.

```
"MANNING2" DISPLAYS MANNING TABLES AND VALUES
OF P VS.UA FOR ANY K-VALUE UP TO 9, FOR EXPANDING OR
COMPRESSING A TUBE OF UNIT BORE RADIUS.NOTATION-R=
RADIUS,U=RADIAL DISPLACEMENT,UR=U/R,UA=BORE DIS-
PLACEMENT,DR=DELTA R,RU=R+U,DRU=DELTA(R+U),RUR=
(R+U)/R,S=YIELD STRESS,ARUR=(2/SQ RT(3))LN(R+U)/R,EQ=
EQUIVALENT STRAIN,P=PRESSURE,DP=DELTA P,E=YOUNGS
MODULUS,EO=INITIAL EQUIVALENT STRAIN,XN=EXPONENT OF
SWIFTS EQUATION S=A*(B+(EQ))**XN.MAKE K-RATIO NEAREST
POSSIBLE MULTIPLE OF .1,THEN PUT N=INTEGER(K-.9)*10.
EACH DIMENSION STORE SHOULD BE AT LEAST N+2.
FOR COMPRESSION MAKE UA NEGATIVE:-P IS THEN EXTERNAL.

TYPE IN E,A,B,XN,UA,N

 206786.   1028.04   .000398067   .0920804   .02   6
```

	E	A	B	XN	UA	N
	.20679E+06	.10280E+04	.39807E-03	.92080E-01	.20000E-01	6

I	R	EQ	S	DP	P	UR
1	.10000E+01				.00000E+00	.20000E-01
1		.20914E-01	.72129E+03	.76511E+02		
2	.11000E+01				.76511E+02	.16557E-01
2		.17468E-01	.70967E+03	.69141E+02		
3	.12000E+01				.14565E+03	.13931E-01
3		.14807E-01	.69921E+03	.62960E+02		
4	.13000E+01				.20861E+03	.11882E-01
4		.12709E-01	.68972E+03	.57713E+02		
5	.14000E+01				.26633E+03	.10254E-01
5		.11027E-01	.68105E+03	.53211E+02		
6	.15000E+01				.31954E+03	.89378E-02
6		.96574E-02	.67309E+03	.49313E+02		

        UA = .20000E-01   P = .31954E+03

I	R	EQ	S	DP	P	UR
1	.10000E+01				.00000E+00	.40000E-01
1		.41483E-01	.76758E+03	.78584E+02		
2	.11000E+01				.78584E+02	.33169E-01
2		.34751E-01	.75530E+03	.71423E+02		
3	.12000E+01				.15001E+03	.27943E-01
3		.29524E-01	.74418E+03	.65328E+02		
4	.13000E+01				.21533E+03	.23857E-01
4		.25387E-01	.73405E+03	.60092E+02		
5	.14000E+01				.27543E+03	.20604E-01
5		.22059E-01	.72477E+03	.55558E+02		
6	.15000E+01				.33099E+03	.17972E-01
6		.19342E-01	.71622E+03	.51601E+02		

        UA = .40000E-01   P = .33099E+03

I	R	EQ	S	DP	P	UR
1	.10000E+01				.00000E+00	.60000E-01
1		.61719E-01	.79595E+03	.78692E+02		
2	.11000E+01				.78692E+02	.49833E-01
2		.51849E-01	.78337E+03	.71924E+02		
3	.12000E+01				.15062E+03	.42033E-01
3		.44148E-01	.77195E+03	.66075E+02		
4	.13000E+01				.21669E+03	.35923E-01
4		.38029E-01	.76152E+03	.60994E+02		
5	.14000E+01				.27769E+03	.31049E-01
5		.33091E-01	.75194E+03	.56552E+02		
6	.15000E+01				.33424E+03	.27099E-01
6		.29049E-01	.74309E+03	.52646E+02		

        UA = .60000E-01   P = .33424E+03

```
1  .10000E+01                                               .00000E+00  .80000E-01
   1              .81629E-01  .81660E+03  .78006E+02
2  .11000E+01                                               .78006E+02  .66546E-01
   2              .68763E-01  .80387E+03  .71681E+02
3  .12000E+01                                               .14969E+03  .56199E-01
   3              .58677E-01  .79228E+03  .66135E+02
4  .13000E+01                                               .21582E+03  .48075E-01
   4              .50633E-01  .78168E+03  .61259E+02
5  .14000E+01                                               .27708E+03  .41584E-01
   5              .44120E-01  .77191E+03  .56958E+02
6  .15000E+01                                               .33404E+03  .36318E-01
   6              .38776E-01  .76287E+03  .53147E+02

        UA =  .80000E-01     P =  .33404E+03

1  .10000E+01                                               .00000E+00  .10000E+00
   1              .10123E+00  .83286E+03  .76913E+02
2  .11000E+01                                               .76913E+02  .83307E-01
   2              .85496E-01  .82007E+03  .71043E+02
3  .12000E+01                                               .14796E+03  .70436E-01
   3              .73109E-01  .80839E+03  .65818E+02
4  .13000E+01                                               .21377E+03  .60311E-01
   4              .63193E-01  .79768E+03  .61171E+02
5  .14000E+01                                               .27494E+03  .52209E-01
   5              .55141E-01  .78779E+03  .57033E+02
6  .15000E+01                                               .33198E+03  .45626E-01
   6              .48518E-01  .77864E+03  .53339E+02

        UA =  .10000E+00     P =  .33198E+03
```

↓   ↓   ↓   ↓   ↓
↓   ↓   ↓   ↓   ↓
↓   ↓   ↓   ↓   ↓
↓   ↓   ↓   ↓   ↓
↓   ↓   ↓   ↓   ↓

```
1  .10000E+01                                               .00000E+00  .28000E+00
   1              .26484E+00  .90978E+03  .63328E+02
2  .11000E+01                                               .63328E+02  .23596E+00
   2              .22824E+00  .89743E+03  .60748E+02
3  .12000E+01                                               .12408E+03  .20139E+00
   3              .19844E+00  .88596E+03  .58083E+02
4  .13000E+01                                               .18216E+03  .17378E+00
   4              .17390E+00  .87528E+03  .55430E+02
5  .14000E+01                                               .23759E+03  .15140E+00
   5              .15350E+00  .86530E+03  .52847E+02
6  .15000E+01                                               .29044E+03  .13302E+00
   6              .13637E+00  .85596E+03  .50371E+02

        UA =  .28000E+00     P =  .29044E+03

1  .10000E+01                                               .00000E+00  .30000E+00
   1              .28174E+00  .91497E+03  .61855E+02
2  .11000E+01                                               .61855E+02  .25310E+00
   2              .24327E+00  .90271E+03  .59538E+02
3  .12000E+01                                               .12139E+03  .21621E+00
   3              .21184E+00  .89130E+03  .57094E+02
4  .13000E+01                                               .17849E+03  .18671E+00
   4              .18590E+00  .88067E+03  .54625E+02
5  .14000E+01                                               .23311E+03  .16277E+00
   5              .16428E+00  .87072E+03  .52195E+02
6  .15000E+01                                               .28531E+03  .14310E+00
   6              .14611E+00  .86139E+03  .49844E+02

        UA =  .30000E+00     P =  .28531E+03


              E0 =  .24178E-02


Stop - Program terminated.
```

3.4 STRESSES IN ROTATING DISCS, THERMAL STRESSES

The familiar practical example of a rotating disc is the **simple flywheel,** in which the disc has a central hole thus forming an annulus, and radial and axial thicknesses which are constant both circumferentially and radially and are also small in comparison with the outer radius. A particularly simple solution can then be obtained in which the stresses σ_r and σ_θ at any radius may be assumed not to vary axially.

As discussed on p.33 of Vol. 1, an element of such a rim subtending an angle $\delta\theta$, having a cross–sectional area a, mean radius r, density ρ and rotational speed ω, must suffer the **centripetal force** F = mass × **centripetal acceleration,** i.e. $F = \rho r a \delta\theta.\omega^2 r$. If the circumferential stress induced in the ends of the element is σ_θ, then $\rho r a \delta\theta.\omega^2 r = 2\sigma_\theta a \delta\theta/2$, from which:

$$\sigma_\theta = \rho\omega^2 r^2. \tag{3.54}$$

For this simple case of a rim of small dimensions, the radial stress σ_r is assumed to be zero through the thickness of the disc (more accurately the **rim**), and the circumferential stress σ_θ constant everywhere. As an example, consider a steel flywheel of this simple type having a mean diameter of 0.5 m, rotating at a speed of 7000 rpm (typical maximum for a modern car engine), ρ = 7.85 Mg/m^3, r = 0.25 m and ω = 2π×7000/60 = 733 rad/sec. From equation (3.54), σ_θ = 7850×(733×0.25)^2/10^6 = 263.6 MPa ($\stackrel{\triangle}{=}$ 17 T/in^2). Such a stress would certainly require a high–strength steel. Note that doubling the radius would increase the stress by a factor of 4.

Considering now the case of a solid thin disc, the basic equations of equilibrium for a cylindrical body have already been derived [equations (3.1) to (3.18)], and these now need to be modified to include the centripetal body forces.

On p.195 of the companion volume Fig. 5.2 shows in detail all the stresses which could act on the faces of a small element in cylindrical polar coordinates. For a rotating disc free of any imposed external forces there would be zero shear stresses on all the faces of such an element. On p.196 of that volume are shown the equations derived from consideration of the equilibrium of the forces acting on the basic element. In the absence of shear stresses on the radial, circumferential and axial faces of the element, the radial equilibrium equation becomes (where stresses such as σ_{rr} have been replaced by σ_r, since they are all principal stresses):

$$\left[\sigma_r + \frac{\partial\sigma_r}{\partial r}\delta r\right](r+\delta r)\delta\theta\delta z - \left[\sigma_\theta + \frac{\partial\sigma_\theta}{\partial\theta}\delta\theta\right]\delta r\delta z\frac{\delta\theta}{2}$$
$$- \sigma_r r\delta\theta\delta z - \sigma_\theta\delta r\delta z\delta\theta/2 = 0.$$

Neglecting second order terms this becomes:

$$\sigma_r \,\delta r \delta\theta\delta z + r\frac{\partial\sigma_r}{\partial r}\delta r\delta\theta\delta z - \sigma_\theta\delta r\delta\theta\delta z = 0. \qquad (3.55)$$

If centrifugal force is to be included, the radially outward body force $\rho r\delta\theta\delta r\delta z.\omega^2 r$ must be added to the left-hand side of equation (3.55) to give:

$$\sigma_r \,\delta r\delta\theta\delta z + r\frac{\partial\sigma_r}{\partial r}\delta r\delta\theta\delta z - \sigma_\theta\delta r\delta\theta\delta z + \rho r\delta\theta\delta r\delta z.\omega^2 r = 0. \qquad (3.56)$$

Dividing this equation by $r\delta\theta\delta r\delta z$

$$\frac{\partial\sigma_r}{\partial r} + \frac{\sigma_r - \sigma_\theta}{r} + \rho\omega^2 r = 0 \qquad (3.57)$$

thus adding the term $\rho\omega^2 r$ to the left-hand side of equation (3.1). For a thin disc, the axial stress σ_z may be assumed to be zero so that the stress-strain relations for the radial and circumferential directions [equations (3.6) and (3.7)] become:

$$\epsilon_r = \frac{\partial u_r}{\partial r} = \frac{1}{E}\Big[\sigma_r - \nu\sigma_\theta\Big] \qquad (3.58)$$

$$\epsilon_\theta = \frac{u_r}{r} = \frac{1}{E}\Big[\sigma_\theta - \nu\sigma_r\Big]. \qquad (3.59)$$

Differentiating equation (3.59) with respect to r gives:

$$E\frac{\partial u_r}{\partial r} = r\Big[\frac{\partial\sigma_\theta}{\partial r} - \nu\frac{\partial\sigma_r}{\partial r}\Big] + \sigma_\theta - \nu\sigma_r.$$

Equating this to equation (3.58) gives

$$\sigma_r - \nu\sigma_\theta = r\Big[\frac{\partial\sigma_\theta}{\partial r} - \nu\frac{\partial\sigma_r}{\partial r}\Big] + \sigma_\theta - \nu\sigma_r$$

i.e.

$$(\sigma_\theta - \sigma_r)(1+\nu) + r\Big[\frac{\partial\sigma_\theta}{\partial r} - \nu\frac{\partial\sigma_r}{\partial r}\Big] = 0$$

from which:

$$\frac{\sigma_r - \sigma_\theta}{r} = \frac{1}{1+\nu}\frac{\partial}{\partial r}\Big[\sigma_\theta - \nu\sigma_r\Big]. \qquad (3.60)$$

Substituting equation (3.60) into equation (3.57) gives

$$\frac{\partial\sigma_r}{\partial r} + \frac{1}{1+\nu}\frac{\partial}{\partial r}\Big[\sigma_\theta - \nu\sigma_r\Big] + \rho\omega^2 r = 0$$

from which:

$$\frac{\partial}{\partial r}\Big[\sigma_r + \sigma_\theta\Big] = -(1+\nu)\rho\omega^2 r.$$

Integrating, $\sigma_r + \sigma_\theta = 2A - (1+\nu)\rho\omega^2 r^2/2$, $\qquad (3.61)$
where A is an arbitrary constant of integration.

From equation (3.57),

$$\sigma_r - \sigma_\theta = -r\frac{\partial \sigma_r}{\partial r} - \rho\omega^2 r^2. \tag{3.62}$$

Adding equations (3.61) and (3.62) yields

$$2\sigma_r + r\frac{\partial \sigma_r}{\partial r} = 2A - \frac{3+\nu}{2}\rho\omega^2 r^2.$$

Now

thus

$$\frac{1}{r}\frac{\partial}{\partial r}(\sigma_r r^2) = \frac{1}{r}\left[\sigma_r . 2r + r^2\frac{\partial \sigma_r}{\partial r}\right] = 2\sigma_r + r\frac{\partial \sigma_r}{\partial r}$$

$$\frac{\partial}{\partial r}(\sigma_r r^2) = 2A - \frac{3+\nu}{2}\rho\omega^2 r^3$$

and integrating,

$$\sigma_r = A + \frac{B}{r^2} - \frac{3+\nu}{8}\rho\omega^2 r^2, \tag{3.63}$$

where B is the second constant of integration.

Substituting equation (3.63) into equation (3.61) gives, eventually:

$$\sigma_\theta = A - \frac{B}{r^2} - \frac{1+3\nu}{8}\rho\omega^2 r^2. \tag{3.64}$$

Comparing equations (3.63) and (3.64) with equations (3.17) and (3.18) shows that these are simply the Lamé equations with additional body force terms. To determine the constants A and B, the boundary conditions for an unloaded disc are $\sigma_r = 0$ at the internal radius of the disc ($r = r_1$), and at the external radius ($r = r_2$). From equation (3.63)

$$0 = A + \frac{B}{r_1^2} - \frac{3+\nu}{8}\rho\omega^2 r_1^2 \tag{3.65}$$

$$0 = A + \frac{B}{r_2^2} - \frac{3+\nu}{8}\rho\omega^2 r_2^2. \tag{3.66}$$

Subtracting equation (3.66) from (3.65)

$$0 = B\left[\frac{1}{r_1^2} - \frac{1}{r_2^2}\right] + \frac{3+\nu}{8}\rho\omega^2(r_2^2 - r_1^2),$$

from which

$$B = -\frac{3+\nu}{8}\rho\omega^2 r_1^2 r_2^2. \tag{3.67}$$

Substituting equation (3.67) into equation (3.65)

$$A = \frac{3+\nu}{8}\rho\omega^2(r_1^2 + r_2^2). \tag{3.68}$$

Substituting equations (3.67) and (3.68) into equation (3.63)

$$\sigma_r = \frac{3+\nu}{8}\rho\omega^2\left[r_1^2 + r_2^2 - \frac{r_1^2 r_2^2}{r^2} - r^2\right]. \tag{3.69}$$

Substituting equations (3.67) and (3.68) into equation (3.64)

$$\sigma_\theta = \frac{\rho\omega^2}{8}\left[(3+\nu)\left(r_1{}^2+r_2{}^2 + \frac{r_1{}^2r_2{}^2}{r^2}\right) - (1+3\nu)r^2\right]. \quad (3.70)$$

At the inside radius r_1 of the disc, from equations (3.69) and (3.70)

$$\sigma_{r\,1} = \frac{3+\nu}{8}\rho\omega^2\left[r_1{}^2+r_2{}^2 - r_2{}^2 - r_1{}^2\right] = 0 \qquad (3.71)$$

$$\sigma_{\theta\,1} = \frac{\rho\omega^2}{4}\left[(1-\nu)r_1{}^2+(3+\nu)r_2{}^2\right]. \qquad (3.72)$$

At the outside radius r_2 of the disc, from equations (3.69) and (3.70)

$$\sigma_{r\,2} = \frac{3+\nu}{8}\rho\omega^2\left[r_1{}^2+r_2{}^2 - r_1{}^2 - r_2{}^2\right] = 0 \qquad (3.73)$$

$$\sigma_{\theta\,2} = \frac{\rho\omega^2}{4}\left[(3+\nu)r_1{}^2+(1-\nu)r_2{}^2\right]. \qquad (3.74)$$

The radius at which $\sigma_{r\text{max}} = \hat{\sigma}_r$ occurs can be found by differentiating equation (3.69) and equating the result to zero, to give

$$\frac{d\sigma_r}{dr} = \frac{3+\nu}{8}\rho\omega^2\left[\frac{2r_1{}^2r_2{}^2}{r^3} - 2r\right] = 0, \text{ i.e. when } r = \sqrt{r_1 r_2}.$$

At that radius σ_r is found from equation (3.69) to be

$$\hat{\sigma}_r = \frac{3+\nu}{8}\rho\omega^2(r_2 - r_1)^2. \qquad (3.75)$$

The radius for $\hat{\sigma}_\theta$ *cannot* be found by differentiating equation (3.70) and equating the result to zero, because a *real* solution does not exist. In fact, $\hat{\sigma}_\theta$ always occurs at the inside radius of the disc, its value being given by equation (3.72) as

$$\hat{\sigma}_\theta = \sigma_{\theta\,1} = \frac{\rho\omega^2}{4}\left[(1-\nu)r_1{}^2+(3+\nu)r_2{}^2\right]. \qquad (3.72a)$$

To obtain the solution for a rotating thin <u>solid</u> disc, with r_1 = 0, it might be thought that the above equations could be used, simply by setting r_1 equal to zero. However, the frequent presence of the term $r_1{}^2r_2{}^2/r^2$ in those equations in which it is desired to put $r_1 = r = 0$ precludes this, since the term then equals 0/0, which is indeterminate. In fact, it is necessary to return to the basic equations (3.63) and (3.64) and start again. To avoid infinities in both these equations when $r = 0$ the integration constant B must be made zero.

Considering this case of the thin solid disc, putting $B = 0$ in equation (3.63) gives the result

$$\sigma_r = A - \frac{3+\nu}{8} \rho\omega^2 r^2 \qquad (3.76)$$

with the boundary condition that $\sigma_r = 0$ at the outside radius r_2 of the solid disc (in which $r_1 = 0$).

Thus

$$A = \frac{3+\nu}{8} \rho\omega^2 r_2^2$$

and

$$\sigma_r = \frac{3+\nu}{8}\rho\omega^2 \left[r_2^2 - r^2 \right] \qquad (3.77)$$

$$\sigma_\theta = \frac{\rho\omega^2}{8} \left[(3+\nu) r_2^2 - (1+3\nu) r^2 \right]. \qquad (3.78)$$

At the centre of the solid disc, $r = 0$ and

$$\hat{\sigma}_r = \hat{\sigma}_\theta = \frac{3+\nu}{8}\rho\omega^2 r_2^2. \qquad (3.79)$$

At the outside of the solid disc, $\sigma_{r2} = 0$ and, from equation (3.78)

$$\sigma_{\theta 2} = \frac{\rho\omega^2}{8} \left[(3+\nu) r_2^2 - (1+3\nu) r_2^2 \right] = \frac{2(1-\nu)}{8}\rho\omega^2 r_2^2. \qquad (3.80)$$

Compare these peak values with those for the thin disk with a minute central hole. From equations (3.75) and (3.72a) when $r_1 \to 0$

$$\hat{\sigma}_r = \frac{3+\nu}{8}\rho\omega^2 r_2^2 \qquad (3.75a)$$

$$\hat{\sigma}_\theta = \sigma_{\theta 1} = \frac{3+\nu}{4}\rho\omega^2 r_2^2. \qquad (3.72b)$$

Thus the maximum stress in such a disk is double the stress at the centre of a solid disk. This is a manifestation of the *stress concentration factor*, of value 2 in this case.

It is also of interest to compare the actual values given by the above equations with those obtained previously for a thin rim. Assuming an external radius of a thin steel disc of 0.25 m in which there is a very small central hole, the maximum stress is the circumferential stress at the edge of the central hole, given by equation (3.72b). Its value, with $\rho = 7850$ kg/m^3, $\omega = 7000$ rpm (733 rad/sec), $\nu = 0.3$ and $r_2 = 0.25$ is $\hat{\sigma}_\theta = (3+\nu)\times \rho\omega^2 r_2^2/4 = 3.3\times7850\times(733\times0.25)^2/(4\times10^6) = 217.5$ MPa $\triangleq 14$ T/in^2, which is less than that found in the thin rim (17 T/in^2). As a check on equation (3.54) for $\hat{\sigma}_\theta$ in a thin rim, the equations for a thin disk with a <u>large</u> central hole may be used, by putting $r_1 = r_2 = r$ in equation (3.72a), to give $\hat{\sigma}_\theta = \rho\omega^2 r^2$ as for equation (3.54). The lower value of $\hat{\sigma}_\theta$ in the disc with the small hole is due to the support of the material inside the outer rim.

PROGRAM 3.5 Stresses in rotating thin discs

```
 1:*       PROGRAM DISCS35
 2:        WRITE(*,11)
 3: 11     FORMAT(1X' DISCS35 CALCULATES STRESSES IN ROTATING',/1X,
 4:       1        ' THIN DISCS FROM ASSUMED INPUT:- DENSITY RHO,',/1X,
 5:       2        ' ROTATIONAL SPEED OMEGA=O, POISSONS RATIO XNU,',/1X,
 6:       3        ' GENERIC RADIUS=R, INSIDE RADIUS=R1, OUTSIDE ',/1X,
 7:       4        ' RADIUS=R2, RADIAL STRESS=SR, CIRCUMFERENTIAL',/1X,
 8:       5        ' STRESS=SC, FLOW STRESS=SF. ',/)
 9:        WRITE(*,12)
10: 12     FORMAT(1X' TYPE IN RHO,O,XNU,R1,R2')
11:        READ(*,*) RHO,O,XNU,R1,R2
12:        WRITE(*,13) RHO,O,XNU,R1,R2
13: 13     FORMAT(' RHO=',E10.5,' O=',E10.5,' XNU=',E10.5,' R1=',E10.5,
14:       1        ' R2=',E10.5,/)
15:        D = (3.+XNU)*RHO*O**2/8.
16:        E = RHO*O**2/8.
17:        R = R1
18:        IF(R1.EQ.0.) GO TO 18
19: 14     F = R1**2+R2**2-(R1*R2/R)**2-R**2
20:        G = (3.+XNU)*(R1**2+R2**2+(R1*R2/R)**2)
21:        H = (1+3.*XNU)*R**2
22:        SR = D*F*1E-6
23:        SC = E*(G-H)*1E-6
24: 15     SF = SQRT(SR**2-SR*SC+SC**2)
25:        WRITE(*,16) R,SR,SC,SF
26: 16     FORMAT(' R=',E10.5,' SR=',E10.5,' SC=',E10.5,' SF=',E10.5,)
27:        IF(R1.EQ.0.) GO TO 17
28:        R = R+(R2-R1)/4.
29:        IF(R.GT.R2) GO TO 20
30:        GO TO 14
31: 17     R = R+R2/5.
32:        IF(R.GT.R2) GO TO 20
33:        GO TO 19
34: 18     R = 0.
35: 19     SR = D*(R2**2-R**2)*1E-6
36:        SC = E*((3.+XNU)*R2**2-(1.+3.*XNU)*R**2)*1E-6
37:        GO TO 15
38: 20     STOP
39:        END
```

SOLUTION 3.5 Stresses in rotating thin discs
(from PROGRAM 3.5 in S.I. units to 5 figures)

```
DISCS35 CALCULATES STRESSES IN ROTATING
THIN DISCS FROM ASSUMED INPUT:- DENSITY RHO,
ROTATIONAL SPEED OMEGA=O, POISSONS RATIO XNU,
GENERIC RADIUS=R, INSIDE RADIUS=R1, OUTSIDE
RADIUS=R2, RADIAL STRESS=SR, CIRCUMFERENTIAL
STRESS=SC, FLOW STRESS=SF.

TYPE IN RHO,O,XNU,R1,R2
 7850. 733. .3 .0 .25
RHO=.78500E+04 O=.73300E+03 XNU=.30000E+00 R1=.00000E+00 R2=.25000E+00

R=.00000E+00 SR=.10874E+03 SC=.10874E+03 SF=.10874E+03
R=.50000E-01 SR=.10439E+03 SC=.10623E+03 SF=.10532E+03
R=.10000E+00 SR=.91340E+02 SC=.98721E+02 SF=.95245E+02
R=.15000E+00 SR=.69592E+02 SC=.86200E+02 SF=.79213E+02
R=.20000E+00 SR=.39146E+02 SC=.68670E+02 SF=.59664E+02
R=.25000E+00 SR=.00000E+00 SC=.46131E+02 SF=.46131E+02

TYPE IN RHO,O,XNU,R1,R2
 7850. 733. .3 .05 .25
RHO=.78500E+04 O=.73300E+03 XNU=.30000E+00 R1=.50000E-01 R2=.25000E+00

R=.50000E-01 SR=.00000E+00 SC=.21932E+03 SF=.21932E+03
R=.10000E+00 SR=.68505E+02 SC=.13026E+03 SF=.11285E+03
R=.15000E+00 SR=.61860E+02 SC=.10263E+03 SF=.89504E+02
R=.20000E+00 SR=.36699E+02 SC=.79815E+02 SF=.69197E+02
R=.25000E+00 SR=.00000E+00 SC=.54830E+02 SF=.54830E+02
Stop - Program terminated.
```

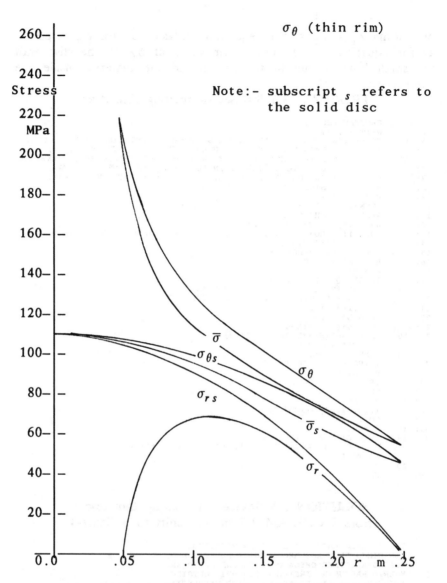

Fig. 3.7 Stress distributions in rotating thin disks

Typical distributions of the stresses in both solid and hollow thin disks are shown in Fig. 3.7. In practice, thin disks such as those used to form flywheels or gas turbine disks are not solid but have central holes in them so that they can be secured to shafts. Thus it is their higher stress distribution which is of more importance in practical engineering manufacture.

As far as failure of the disc is concerned, it may well be that the maximum values of stresses such as $\hat{\sigma}_r$ and $\hat{\sigma}_\theta$ are important, but for ductile materials it is often the effective or equivalent stress $\bar{\sigma}$ which is of more importance. As shown later in Chapter 5, the value of $\bar{\sigma}$ in an axially symmetrical system of stress with $\sigma_z = 0$ is given by the equation

$$\bar{\sigma} = \sqrt{\sigma_r{}^2 - \sigma_r \sigma_\theta + \sigma_\theta{}^2}. \tag{3.81}$$

For a thin disc with a central hole it is possible to obtain an equation for $\bar{\sigma}$ by substituting the expressions for σ_r and σ_θ given by equations (3.69) and (3.70) into equation (3.81). The resulting equation is too cumbersome to be useful, however, so it is necessary to calculate σ_r and σ_θ separately and substitute their values into equation (3.81), to determine $\bar{\sigma}$ for any radius r, as also shown in Fig. 3.7. The data for Fig. 3.7 was obtained from Program and Solution 3.5, for a solid disc and a disc with a central hole.

Probably the most important example of a disc in modern engineering is that of the gas turbine. Such discs have to carry turbine blades at fairly large radii and rotate at high speeds.

They also have to be thin, to enable several discs to be carried on a single shaft. Therefore, their thickness must be allowed to vary radially, to minimize the amount of disc material and its weight. Materials often used for the discs are nickel-based alloys of high strength and heat resistance.

Considering this problem more closely, neglect any axial (σ_z) stresses through the small thickness of such a disc, since they would be very small. To make full use of the material by having the highest possible stresses it seems sensible to put $\sigma_r = \sigma_\theta$ $(= \bar{\sigma}$, say) everywhere. That would also necessitate having $\sigma_{r1} = \sigma_{\theta 1} = \sigma_{r2} = \sigma_{\theta 2} = \bar{\sigma}$ at the inner and outer radii. That is not unreasonable, since the disc has to be secured to the shaft, and the blades are fixed to the outer edge of the disc.

If the radial thickness is to vary with radius, it is necessary to set up a new equilibrium equation, based on the element illustrated in Fig. 3.8. Resolving forces radially and neglecting second order quantities:

$$2\sigma_\theta t \delta r . \sin \delta\theta/2 + \sigma_r tr\delta\theta = \sigma_r(t+\delta t)(r+\delta r)\delta\theta + \rho\omega^2 r . r\delta\theta . t\delta r.$$

If $\sigma_r = \sigma_\theta = \bar{\sigma}$ say, then, in the limit

$$\frac{dt}{dr} = -\frac{\rho\omega^2 rt}{\bar{\sigma}}, \qquad \therefore \int_{t_1}^{t}\frac{dt}{t} = -\frac{\rho\omega^2}{\bar{\sigma}}\int_{r_1}^{r}rdr. \tag{3.82}$$

Integrating,
$$\ln\frac{t}{t_1} = -\frac{\rho\omega^2}{2\bar{\sigma}}(r^2 - r_1{}^2),$$

or
$$\frac{t}{t_1} = \exp\left[-\frac{\rho\omega^2}{2\bar{\sigma}}(r^2 - r_1{}^2)\right]. \tag{3.83}$$

PROGRAM 3.6 Profiles of uniform strength discs

```
 1:*      PROGRAM DISCS36
 2:       WRITE(*,11)
 3: 11    FORMAT(1X' DISCS36 CALCULATES PROFILES OF UNIFORM ',/1X,
 4:       1         ' STRENGTH DISCS FROM ASSUMED INPUT:- DENSITY ',/1X,
 5:       2         ' RHO, FLOW STRESS=SF, INSIDE RADIUS=R1,',/1X,
 6:       3         ' THICKNESS AT R1=T1, GENERIC RADIUS=R, GENERIC',/1X,
 7:       4         ' THICKNESS=T, ROTATIONAL SPEED OMEGA=O. ',/)
 8:       WRITE(*,12)
 9: 12    FORMAT(1X' TYPE IN RHO,SF,R1,T1,O')
10:       READ(*,*) RHO,SF,R1,T1,O
11:       WRITE(*,13) RHO,SF,R1,T1
12: 13    FORMAT(' RHO=',E10.5,' SF=',E10.5,' R1=',E10.5,' T1=',E10.5,/)
13: 14    WRITE(*,15) O
14: 15    FORMAT(/8X' O=',E10.5,)
15:       R = R1
16: 16    T = T1*EXP(-(RHO*O**2*(R**2-R1**2)/(2.*SF)))
17:       WRITE(*,17) R,T
18: 17    FORMAT(' R=',E10.5,' T=',E10.5,)
19:       R = R+.05
20:       IF(R.GT..3) GO TO 18
21:       GO TO 16
22: 18    O = O+261.8
23:       IF(O.GT.3142) GO TO 19
24:       GO TO 14
25: 19    STOP
26:       END
```

SOLUTION 3.6 Profiles of uniform strength discs
(from Program 3.6 in S.I. units to 5 figures)

```
DISCS36 CALCULATES PROFILES OF UNIFORM
STRENGTH DISCS FROM ASSUMED INPUT:- DENSITY
RHO, FLOW STRESS=SF, INSIDE RADIUS=R1,
THICKNESS AT R1=T1, GENERIC RADIUS=R, GENERIC
THICKNESS=T, ROTATIONAL SPEED OMEGA=O.

TYPE IN RHO,SF,R1,T1,O
 8900. 300E6 .05 .268 1047.
RHO=.89000E+04 SF=.30000E+09 R1=.50000E-01 T1=.26800E+00


        O=.10470E+04
R=.50000E-01 T=.26800E+00
R=.10000E+00 T=.23723E+00
R=.15000E+00 T=.19360E+00
R=.20000E+00 T=.14565E+00              O=.23560E+04
R=.25000E+00 T=.10102E+00       R=.50000E-01 T=.26800E+00
R=.30000E+00 T=.64599E-01       R=.10000E+00 T=.14453E+00
                                R=.15000E+00 T=.51639E-01
        O=.13088E+04            R=.20000E+00 T=.12224E-01
R=.50000E-01 T=.26800E+00       R=.25000E+00 T=.19171E-02
R=.10000E+00 T=.22150E+00       R=.30000E+00 T=.19921E-03
R=.15000E+00 T=.16123E+00
R=.20000E+00 T=.10335E+00
R=.25000E+00 T=.58350E-01              O=.26178E+04
R=.30000E+00 T=.29012E-01       R=.50000E-01 T=.26800E+00
                                R=.10000E+00 T=.12504E+00
        O=.15706E+04            R=.15000E+00 T=.35092E-01
R=.50000E-01 T=.26800E+00       R=.20000E+00 T=.59244E-02
R=.10000E+00 T=.20368E+00       R=.25000E+00 T=.60165E-03
R=.15000E+00 T=.12892E+00       R=.30000E+00 T=.36755E-04
R=.20000E+00 T=.67954E-01
R=.25000E+00 T=.29831E-01
R=.30000E+00 T=.10906E-01              O=.28796E+04
                                R=.50000E-01 T=.26800E+00
        O=.18324E+04            R=.10000E+00 T=.10654E+00
R=.50000E-01 T=.26800E+00       R=.15000E+00 T=.22897E-01
R=.10000E+00 T=.18446E+00       R=.20000E+00 T=.26605E-02
R=.15000E+00 T=.98975E-01       R=.25000E+00 T=.16713E-03
R=.20000E+00 T=.41400E-01       R=.30000E+00 T=.56762E-05
R=.25000E+00 T=.13499E-01
R=.30000E+00 T=.34315E-02
                                       O=.31414E+04
        O=.20942E+04            R=.50000E-01 T=.26800E+00
R=.50000E-01 T=.26800E+00       R=.10000E+00 T=.89401E-01
R=.10000E+00 T=.16453E+00       R=.15000E+00 T=.14344E-01
R=.15000E+00 T=.72959E-01       R=.20000E+00 T=.11070E-02
R=.20000E+00 T=.23370E-01       R=.25000E+00 T=.41094E-04
R=.25000E+00 T=.54072E-02       R=.30000E+00 T=.73373E-06
R=.30000E+00 T=.90370E-03       Stop - Program terminated.
```

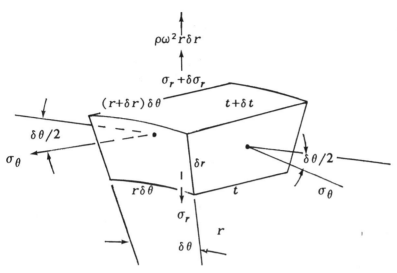

Fig. 3.8 Element of disc of variable thickness

PROGRAM 3.7 Outer radii for discs with equal outer thickness, at various speeds

```
 1:*      PROGRAM DISCS37
 2:       WRITE(*,11)
 3: 11    FORMAT(1X' DISCS37 CALCULATES OUTER RADII RX OF UNIFORM ',/1X
 4:       1       ' STRENGTH DISCS HAVING THE SAME OUTER THICKNESS ',/1
 5:       2       ' TX FOR VARIOUS ROTATIONAL SPEEDS OMEGA=O. INPUT',/1
 6:       3       ' DATA:- DENSITY RHO, FLOW STRESS=SF, INSIDE',/1X,
 7:       4       ' RADIUS=R1, THICKNESS AT R1=T1. ',/)
 8:       WRITE(*,12)
 9: 12    FORMAT(1X' TYPE IN RHO,O,SF,R1,T1,TX')
10:       READ(*,*) RHO,O,SF,R1,T1,TX
11:       WRITE(*,13) RHO,SF,R1,T1,TX
12: 13    FORMAT(' RHO=',E10.5,' SF=',E10.5,' R1=',E10.5,' T1=',E10.5,
13:       1       ' TX=',E10.5,/)
14: 14    RX = SQRT(R1**2-2.*SF*ALOG(TX/T1)/(RHO*O**2))
15:       WRITE(*,15) O,RX
16: 15    FORMAT(' O=',E10.5,' RX=',E10.5,)
17:       O = O+261.8
18:       IF(O.GT.3142.) GO TO 16
19:       GO TO 14
20: 16    STOP
21:       END
```

SOLUTION 3.7 Outer radii for discs with equal outer thickness, at various speeds

```
DISCS37 CALCULATES OUTER RADII RX OF UNIFORM
STRENGTH DISCS HAVING THE SAME OUTER THICKNESS
TX FOR VARIOUS ROTATIONAL SPEEDS OMEGA=O. INPUT
DATA:- DENSITY RHO, FLOW STRESS=SF, INSIDE
RADIUS=R1, THICKNESS AT R1=T1.

TYPE IN RHO,O,SF,R1,T1,TX
  8900. 1047. 300E6   .05 .268 .02899
RHO=.89000E+04 SF=.30000E+09 R1=.50000E-01 T1=.26800E+00 TX=.28990E-01

O=.10470E+04 RX=.37320E+00
O=.13088E+04 RX=.30005E+00
O=.15706E+04 RX=.25156E+00
O=.18324E+04 RX=.21715E+00
O=.20942E+04 RX=.19154E+00
O=.23560E+04 RX=.17179E+00
O=.26178E+04 RX=.15614E+00
O=.28796E+04 RX=.14346E+00
O=.31414E+04 RX=.13302E+00
Stop - Program terminated.
```

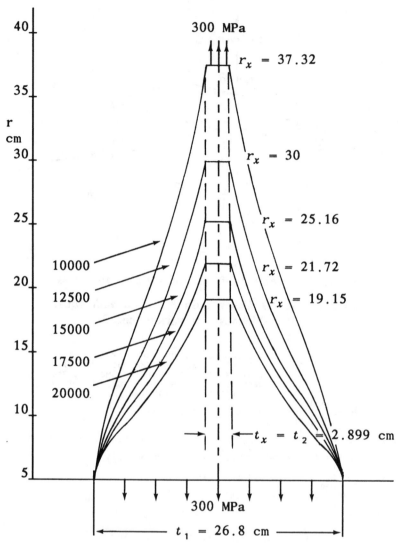

Fig. 3.9 Disc profiles for various rotational speeds

A typical example of a gas turbine disc alloy is a nickel alloy having a working stress of $\bar{\sigma} = 300$ MPa and a density of 8900 kg/m³. Determine the profile of such a disc rotating at speeds up to 20000 rpm, if it has an outer radius of 30 cm and a thickness of 26.8 cm at its inner radius of 5 cm. For σ_r and σ_θ to be everywhere equal to 300 MPa, the required profile can be determined from equation (3.83), using Program and Solution 3.6, from which the profile has been sketched in Fig. 3.9.

Fig. 3.8 was based on the inner thickness, this being common to all the discs, and reveals that the outer thickness has to be made infinitesimally small for the higher rotational speeds, because of the very high centrifugal forces induced in the outer layers of the discs. In practice, the outer thickness t_2 (= t_x, say) has to be large enough to accommodate the customary fir tree root of the gas turbine blade, which would probably not be less than the 2.899 cm ($\stackrel{\triangle}{=}$ $1\frac{1}{8}$ in) found for the disc rotating at 12500 rpm. Thus, as can be seen from Fig. 3.9, discs of uniform strength rotating at higher speeds would require their outer radii to be reduced from 30 cm to 25.16 cm at 15000 rpm, 21.72 cm at 17500 rpm, and 19.15 cm at 20000 rpm, as calculated from equation (3.83) with $t = t_x$ and $r = r_x$. Program and Solution 3.7 tabulate these values.

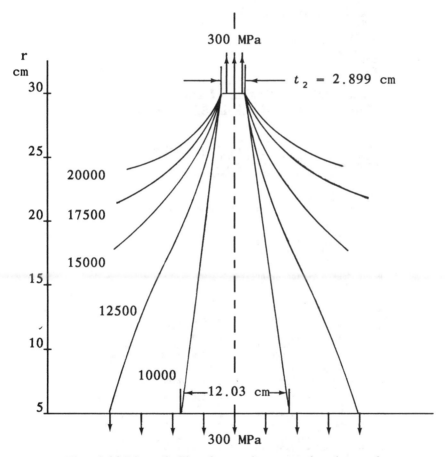

Fig. 3.10 Disc profiles for various rotational speeds

From the need to design for manufacture, it seems more sensible to base the calculations on the outer thickness t_2. If that is done, the required shape of disc can be found from

PROGRAM 3.8 Profiles of uniform strength discs

```
1:*     PROGRAM DISCS38
2:      WRITE(*,11)
3: 11   FORMAT(1X' DISCS38 CALCULATES PROFILES OF UNIFORM ',/1X,
4:      1        ' STRENGTH DISCS FOR VARIOUS ROTATIONAL SPEEDS',/1X,
5:      2        ' OMEGA=O. INPUT DATA:-DENSITY=RHO, FLOW ',/1X,
6:      3        ' STRESS=SF, OUTER THICKNESS=T2 AT RADIUS R2,',/1X,
7:      4        ' GENERIC RADIUS=R, GENERIC THICKNESS=T. ',/)
8:      WRITE(*,12)
9: 12   FORMAT(1X' TYPE IN RHO,O,SF,T2,R2')
10:     READ(*,*) RHO,O,SF,T2,R2
11:     WRITE(*,13) RHO,SF,T2,R2
12: 13  FORMAT(' RHO=',E10.5,' SF=',E10.5,' T2=',E10.5,' R2=',E10.5,/)
13: 14  WRITE(*,15) O
14: 15  FORMAT(/8X' O=',E10.5,)
15:     R = R2
16: 16  T = T2*EXP((RHO*O**2*(R2**2-R**2)/(2.*SF)))
17:     WRITE(*,17) R,T
18: 17  FORMAT(' R=',E10.5,' T=',E10.5,)
19:     R = R-.05
20:     IF(R.LT..05) GO TO 18
21:     GO TO 16
22: 18  O = O+261.8
23:     IF(O.GT.3142.) GO TO 19
24:     GO TO 14
25: 19  STOP
26:     END
```

SOLUTION 3.8 Profiles of uniform strength discs (from PROGRAM 3.8 in S.I. units to 5 figures)

```
DISCS38 CALCULATES PROFILES OF UNIFORM
STRENGTH DISCS FOR VARIOUS ROTATIONAL SPEEDS
OMEGA=O. INPUT DATA:-DENSITY=RHO, FLOW
STRESS=SF, OUTER THICKNESS=T2 AT RADIUS R2,
GENERIC RADIUS=R, GENERIC THICKNESS=T.

TYPE IN RHO,O,SF,T2,R2
  8900. 1047. 300E6 .02899 .3
RHO=.89000E+04 SF=.30000E+09 T2=.28990E-01 R2=.30000E+00


         O=.10470E+04
R=.30000E+00 T=.28990E-01
R=.25000E+00 T=.45337E-01
R=.20000E+00 T=.65364E-01
R=.15000E+00 T=.86880E-01
R=.10000E+00 T=.10646E+00
R=.50000E-01 T=.12027E+00


         O=.13088E+04                      O=.23560E+04
R=.30000E+00 T=.28990E-01        R=.30000E+00 T=.28990E-01
R=.25000E+00 T=.58305E-01        R=.25000E+00 T=.27899E+00
R=.20000E+00 T=.10327E+00        R=.20000E+00 T=.17789E+01
R=.15000E+00 T=.16110E+00        R=.15000E+00 T=.75148E+01
R=.10000E+00 T=.22133E+00        R=.10000E+00 T=.21033E+02
R=.50000E-01 T=.26780E+00        R=.50000E-01 T=.39001E+02


         O=.15706E+04                      O=.26178E+04
R=.30000E+00 T=.28990E-01        R=.30000E+00 T=.28990E-01
R=.25000E+00 T=.79296E-01        R=.25000E+00 T=.47454E+00
R=.20000E+00 T=.18064E+00        R=.20000E+00 T=.46727E+01
R=.15000E+00 T=.34269E+00        R=.15000E+00 T=.27678E+02
R=.10000E+00 T=.54142E+00        R=.10000E+00 T=.98620E+02
R=.50000E-01 T=.71239E+00        R=.50000E-01 T=.21138E+03


         O=.18324E+04                      O=.28796E+04
R=.30000E+00 T=.28990E-01        R=.30000E+00 T=.28990E-01
R=.25000E+00 T=.11405E+00        R=.25000E+00 T=.85358E+00
R=.20000E+00 T=.34976E+00        R=.20000E+00 T=.13588E+02
R=.15000E+00 T=.83618E+00        R=.15000E+00 T=.11694E+03
R=.10000E+00 T=.15584E+01        R=.10000E+00 T=.54411E+03
R=.50000E-01 T=.22641E+01        R=.50000E-01 T=.13687E+04


         O=.20942E+04                      O=.31414E+04
R=.30000E+00 T=.28990E-01        R=.30000E+00 T=.28990E-01
R=.25000E+00 T=.17346E+00        R=.25000E+00 T=.16237E+01
R=.20000E+00 T=.74969E+00        R=.20000E+00 T=.43740E+02
R=.15000E+00 T=.23405E+01        R=.15000E+00 T=.56676E+03
R=.10000E+00 T=.52779E+01        R=.10000E+00 T=.35323E+04
R=.50000E-01 T=.85972E+01        R=.50000E-01 T=.10589E+05
                                 Stop - Program terminated.
```

the following equation (3.84), derived from equation (3.82):

$$\frac{t}{t_2} = \exp\left[\frac{\rho\omega^2}{2\bar{\sigma}}(r_2{}^2 - r^2)\right].\qquad(3.84)$$

It is a simple matter to modify Program 3.6 to achieve this end, as shown in Program and Solution 3.8, the resultant profiles being shown in Fig. 3.10.

It can be seen from Fig. 3.10 that a disc of this particular material and internal radius rotating at a speed of 10000 rpm has a profile which can actually be approximated by a straight line. Such a profile would be easier to manufacture and to check by quality control than the precise exponential form.

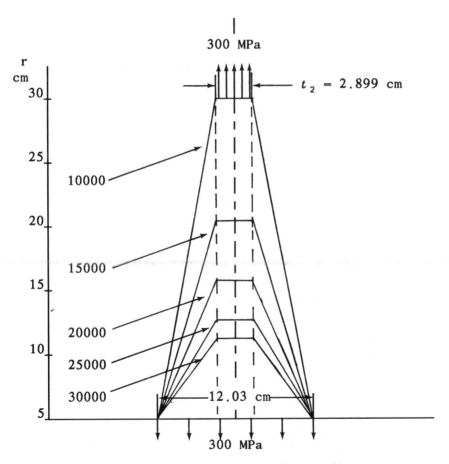

Fig. 3.11 Discs with straight-line profiles

A rotational speed of 10000 rpm is quite low for any gas turbine and, as has been shown, higher speeds of rotation would require the outer radius to be reduced. The detailed design of turbines depends on many factors, often conflicting, but the rotor design is certainly one of the most important, from the point of view of safety. Assuming that the inside and outside thicknesses t_1 and t_2 and the inner diameter r_1 are fixed at the values just found for a speed of 10000 rpm, viz: $r_1 = 0.05$, $t_1 = 0.1203$, $t_2 = 0.02899$ m, then r_2 can be expressed as a function of ω, using equation (3.84), from which

$$\frac{t_1}{t_2} = \exp\left[\frac{\rho\omega^2}{2\bar\sigma}(r_2{}^2 - r_1{}^2)\right],$$

hence

$$r_2 = \sqrt{\left[r_1{}^2 + \frac{2\bar\sigma}{\rho\omega^2}\ln\frac{t_1}{t_2}\right]}. \qquad (3.85)$$

A FORTRAN program for equation (3.85) is listed as Program 3.9, giving Solution 3.9 with the above values of the parameters. The approximately straight profiles obtained are shown to scale in Fig. 3.11.

PROGRAM 3.9 Comparison of uniform strength discs

```
1:*        PROGRAM DISCS39
2:         WRITE(*,11)
3: 11      FORMAT(1X' DISCS39 COMPARES STRAIGHT-LINE PROFILES OF',/1X,
4:         1      ' UNIFORM STRENGTH DISCS FOR VARIOUS ROTATIONAL',/1X,
5:         2      ' SPEEDS OMEGA=O. INPUT DATA:-DENSITY=RHO, FLOW',/1X,
6:         3      ' STRESS=SF, INNER THICKNESS=T1 AT RADIUS=R1',/1X,
7:         4      ' OUTER THICKNESS=T2 AT RADIUS R2.',/)
8:         WRITE(*,12)
9: 12      FORMAT(1X' TYPE IN RHO,O,SF,T1,R1,T2')
10:        READ(*,*) RHO,O,SF,T1,R1,T2
11:        WRITE(*,13) RHO,SF,T1,R1,T2
12: 13     FORMAT(' RHO=',E10.5,' SF=',E10.5,' T1=',E10.5,' R1=',E10.5,
13:        1      ' T2=',E10.5,/)
14: 14     R2 = SQRT(R1**2+2.*SF*ALOG(T1/T2)/(RHO*O**2))
15:        WRITE(*,15) O,R2
16: 15     FORMAT(' O=',E10.5,' R2=',E10.5,)
17:        O = O+261.8
18:        IF(O.GT.3142.) GO TO 16
19:        GO TO 14
20: 16     STOP
21:        END
```

SOLUTION 3.9 Comparison of uniform strength discs (from PROGRAM 3.9 in S.I. units to 5 figures)

```
DISCS39 COMPARES STRAIGHT-LINE PROFILES OF
UNIFORM STRENGTH DISCS FOR VARIOUS ROTATIONAL
SPEEDS OMEGA=O. INPUT DATA:-DENSITY=RHO, FLOW
STRESS=SF, INNER THICKNESS=T1 AT RADIUS=R1
OUTER THICKNESS=T2 AT RADIUS R2.

TYPE IN RHO,O,SF,T1,R1,T2
  8900. 1047. 300E6 .1203 .05 .02899
RHO=.89000E+04 SF=.30000E+09 T1=.12030E+00 R1=.50000E-01 T2=.28990E-01

O=.10470E+04  R2=.30003E+00
O=.13088E+04  R2=.24188E+00
O=.15706E+04  R2=.20345E+00
O=.18324E+04  R2=.17627E+00
O=.20942E+04  R2=.15612E+00
O=.23560E+04  R2=.14065E+00
O=.26178E+04  R2=.12845E+00
O=.28796E+04  R2=.11861E+00
O=.31414E+04  R2=.11055E+00
Stop - Program terminated.
```

It should be noted that, in all these calculations of the profiles of uniform strength discs, the stress has been referred to as the 'flow' stress $\bar{\sigma}$, whereas a value of $\bar{\sigma} = 300$ MPa has been used as a safe 'working' stress for a typical nickel alloy. Care must be taken to select a working stress which will give an adequate length of life for the turbine rotor or disc at its maximum operating temperature, bearing in mind that some flow or creep of the disc material will inevitably occur at the high temperatures involved.

The high temperatures encountered in gas turbines require consideration of the thermal stresses induced, particularly in the rotors or discs. Consider equations (3.58) and (3.59) which apply to a thin disc of uniform thickness, the simplest case. If T is the temperature rise above the unstressed state, an additional strain of αT (where $\alpha =$ linear coefficient of thermal expansion) will be induced, so that equations (3.58) and (3.59) will be modified to become

$$\epsilon_r = \frac{\partial u_r}{\partial r} = \frac{1}{E}\left[\sigma_r - \nu\sigma_\theta\right] + \alpha T \qquad (3.86)$$

$$\epsilon_\theta = \frac{u_r}{r} = \frac{1}{E}\left[\sigma_\theta - \nu\sigma_r\right] + \alpha T. \qquad (3.87)$$

Differentiating equation (3.87) with respect to r and equating to equation (3.86) gives

$$(\sigma_\theta - \sigma_r)(1+\nu) + r\left[\frac{\partial\sigma_\theta}{\partial r} - \nu\frac{\partial\sigma_r}{\partial r}\right] + E\alpha r\frac{\partial T}{\partial r} = 0. \qquad (3.88)$$

The equilibrium equation [equation (3.57)] remains unchanged, and may be written as

$$\sigma_\theta - \sigma_r = r\frac{\partial\sigma_r}{\partial r} + \rho\omega^2 r^2 = 0. \qquad (3.89)$$

Substituting equation (3.89) into equation (3.88) gives

$$\frac{\partial\sigma_\theta}{\partial r} + \frac{\partial\sigma_r}{\partial r} + (1+\nu)\rho\omega^2 r + E\alpha\frac{\partial T}{\partial r} = 0$$

which may be integrated to give

$$\sigma_\theta + \sigma_r = -(1+\nu)\frac{\rho\omega^2 r^2}{2} - E\alpha T + 2A. \qquad (3.90)$$

Subtracting equation (3.89) from equation (3.90) gives

$$2\sigma_r + r\frac{\partial\sigma_r}{\partial r} = \frac{3+\nu}{2}\rho\omega^2 r^2 - E\alpha T + 2A = \frac{1}{r}\frac{\partial}{\partial r}(r^2\sigma_r)$$

which may be integrated to give

$$r^2 \sigma_r = \frac{3+\nu}{8} \rho \omega^2 r^4 - E\alpha \int T r \, dr + A r^2 - B$$

i.e.

$$\sigma_r = A - \frac{B}{r^2} + \frac{3+\nu}{8} \rho \omega^2 r^2 - \frac{E\alpha}{r^2} \int T r \, dr. \qquad (3.91)$$

From equation (3.90)

$$\sigma_\theta = A + \frac{B}{r^2} - \frac{1+3\nu}{8} \rho \omega^2 r^2 - E\alpha T + \frac{E\alpha}{r^2} \int T r \, dr. \qquad (3.92)$$

Equations (3.91) and (3.92) enable the stresses in a rotating thin disc of uniform thickness and known distribution of temperature T to be found. For discs of non–uniform thickness the problem of including thermal stresses is more complicated and requires numerical methods of solution, such as the finite element method.

Chapter 4

STRESS CONCENTRATIONS AND FRACTURE MECHANICS

4.1 STRESS CONCENTRATIONS

4.1.1 *Introduction*

When a part fails, it is likely that the failure will occur at some irregularity in the part such as the shoulder in a shaft. Other likely places for failure are at a weld or at the point of application of a concentrated load. Static stress analysis predicts the stress under load of continuous materials with uniform cross-sectional areas, but non-uniform or discontinuous areas cause a discontinuity in the stress distribution. These irregularities are called *stress raisers* and the irregularities in stress they cause are called *stress concentrations*. The average stress over an area with a stress raiser is still the stress predicted by statics, but a concentration of stress exists at the discontinuity. (See Fig. 4.1.)

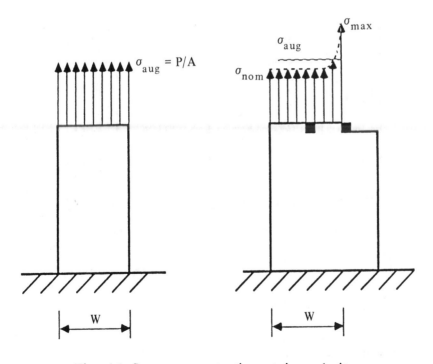

Fig. 4.1 Stress concentration at irregularity

As can be seen from Fig. 4.1, the average stress on the cross-section is the same as if there were no concentration present, but the stress raiser redistributes the amount of stress felt by infinitesimal elements to the member.   If the maximum stress at the discontinuity is greater than the yield strength of the material, failure could occur.   As can also be seen from Fig. 4.1, the stress concentration is only important very near the discontinuity.   This demonstrates Saint Venant's principle of rapid dissipation of localized streses.

A simple model of stress concentration can perhaps give the reader a better feel of why stress concentrations occur.   If stress can be thought of as a series of force lines flowing through a material, a uniform cross-section results in a uniform flow as in Fig. 4.2(b).   However, if a notch in the material exists, the flow of stress cannot pass through the area where there is no material.   The flow lines must bend to get around the notch, which results in a localized buildup of stress as shown in Fig. 4.2(a).

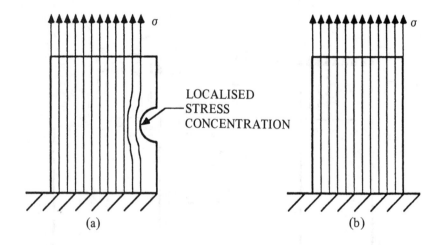

Fig. 4.2 Flow model of stress concentration

4.1.2   Stress Concentration Factors

Now that it has been shown that stress concentrations exist, there must be a way of accounting for them in practical problems.   It is usually not of any practical importance to know the stress distribution, but it is important to know the maximum stress that occurs at the discontinuity. These values of maximum stress are found through the use of stress concentration factors. These factors are defined as

$$K_t = \frac{\sigma_{max}}{\sigma_{nom}} \quad \text{for normal stress} \quad (4.1)$$

$$K_{ts} = \frac{\tau_{max}}{\tau_{nom}} \quad \text{for shear stress} \quad (4.2)$$

The subscript 't' is to indicate that these factors are theoretical. Theoretical stress concentration factors are dependent only on the geometry of the part; material properties have no importance to the factor whatsoever. Once a stress calculation is made, it can be used for any part with the same geometry. There is a vast wealth of stress concentration factors that have been calculated over the years. These factors are usually recorded in graphical form. Figs. 4.4 to 4.11 show graphs for various different stress concentration factors. These are reproduced from the *Machine Design* journal Feb. 1951. In addition to this, R.E. Peterson's book *Stress Concentration Factors* is a good source for factors relating to almost any geometry for shafts and plates.

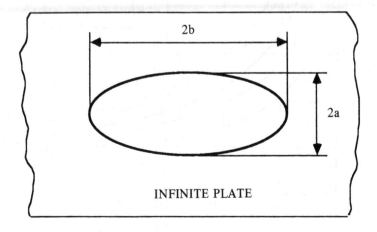

Fig. 4.3 K for elliptical hole in infinite plate

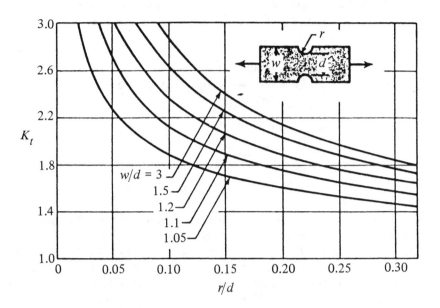

Fig. 4.4 Notched rectangular bar in tension or simple
compression

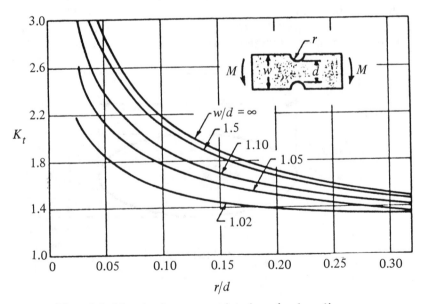

Fig. 4.5 Notched rectangular bar in bending

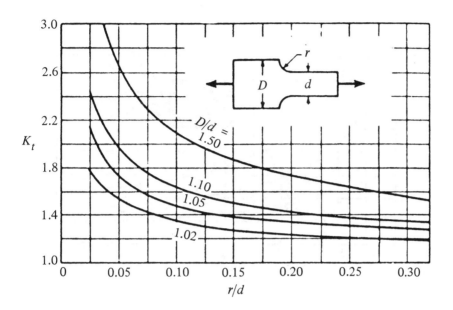

Fig. 4.6 Rectangular filleted bar in tension or simple
compression

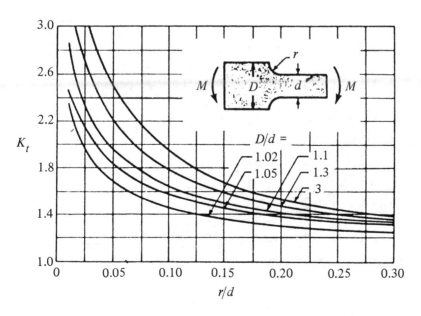

Fig. 4.7 Rectangular filleted bar in bending

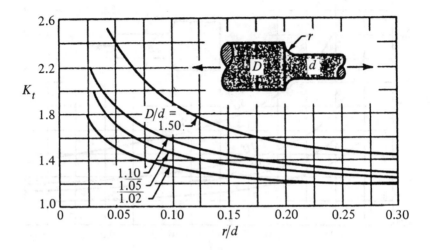

Fig. 4.8 Round shaft with shoulder fillet in tension

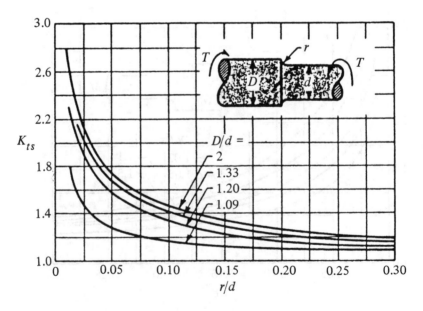

Fig. 4.9 Round shaft with shoulder fillet in torsion

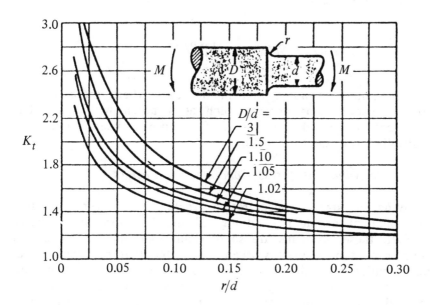

Fig. 4.10 Round shaft with shoulder fillet in bending

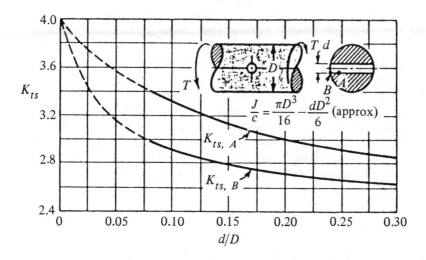

Fig. 4.11 Round shaft in torsion with transverse hole

To find the maximum stress it is only necessary to select a factor from the correct graph and multiply the calculated stress by the factor.

$$\sigma_{\max} = K_t \sigma_{\text{nom}} = K_t(P/A) \tag{4.3}$$

It should be noted that in the case where there are two stress concentrations present, such as a notch within a notch, the factors should be found for both concentrations and multiplied for the total factor.

4.1.3 Derivation of Stress Concentration Factors

Deriving a stress concentration factor is a very complex mathematical problem involving advanced theory of elasticity. In fact, these problems are so difficult that only a few of the simpler ones have been worked out. One that has been solved mathematically is the case of an elliptical hole in an infinite plate as shown in Fig. 4.3. However, the most common method for solving for the maximum stress is by using physical or computer generated models to find the stress distributions for certain geometries.

The photoelastic method is an accurate way to find the stress distribution by constructing a model from photoelastic material such as celluloid or bakelite and passing polarized light through it. The stresses in the model disturb the polarized light, causing light fringes. The stress distribution can be determined by noting the number and the pattern of fringes that occur in the model. There are other methods available to model stress concentrations physically, but most are not as accurate or easy as the photoelastic method. Currently, finite element modelling on a computer is used most often to determine stress concentrations where ordinary strength of materials methods cannot be used. The finite element method is very convenient and accurate and it does not involve building any physical models. Chapter 6 deals with application of finite element method and stress analysis problems.

4.1.4 Summary of How to Apply Stress Concentration Factors

1. Find irregularities in the part and measure their geometry.

2. Use the appropriate chart to find the stress concentration factor.

3. Determine the nominal stress by normal methods.

4. Use the relation, $\sigma_{max} = K_t\sigma_{nom} = K_t(P/A)$.

4.1.5 *Static Stress Concentration Factors in Ductile Materials*

So far this chapter has discussed how to use stress concentration factors. However, there are occasions when not to use the factors. Stress concentration factors do not apply in the plastic region, and therefore are useless in calculating the stresses in a statically loaded ductile member deforming plastically. Since most load bearing members are ductile, static stress concentration factors are seldom used. A more important use of stress concentration factors is in fatigue loading, as will be shown later in this chapter.

To understand why stress concentration is not significant in a ductile material, see Figs. 4.12 and 4.13. Fig. 4.12 shows an idealized perfectly brittle material, with the stresses at points A and B shown on the stress-strain diagram. If the load is applied as shown, the stress at point A(σ_{max}) is **K** times the stress at point B(σ_{nom}). When $\sigma_{max} = \sigma_{fracture}$, brittle failure will occur at point A and spread, destroying the member. Fig. 4.13 shows the same loading process with a perfectly ductile material. In the elastic region the maximum stress is K times the nominal stress, as was shown before. However, when $\sigma_{max} = \sigma_{yield}$, the maximum stress cannot increase with the load and the material around point A begins to yield. This yielding leads to a permanent deformation of the material around the point A. This deformation acts as a stress reliever by altering the flow lines of the stress in the area of the stress concentration. The deformation is permanent, so that after the initial present load is removed the stress concentration has no real effect. If the stress concentration at point A is highly localized, the deformation occurs over a very limited area and is not noticeable to normal inspection. 'Highly localized' refers to an area which is relatively small part of the overall member, such as small holes in a plate.

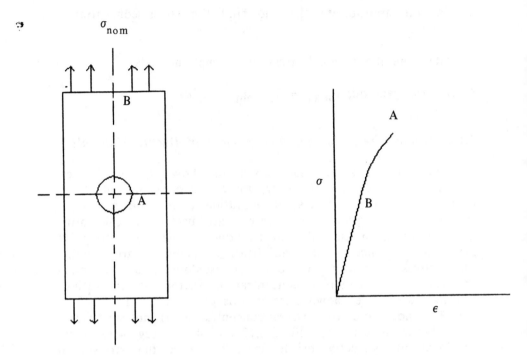

Fig. 4.12 Stress-strain diagram for a brittle material

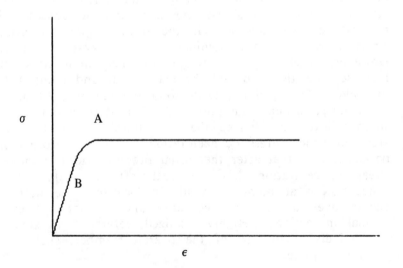

Fig. 4.13 Stress-strain diagram for a ductile material

The models of Figs. 4.12 and 4.13 are examples of ideal situations. No actual material is perfectly brittle or perfectly ductile. Real materials show combinations of brittleness and ductility. In any event, it is safe to disregard the effect of stress concentration on any ductile material such as mild steel as long as the effect is highly localized.

4.1.6 Fatigue Loading

4.1.6.1 Fatigue

Up to now this chapter has dealt only with stress concentrations on statically loaded systems. A static load is one in which the load is applied gradually and once only, so that the strain has time to develop. The tensile strength test is a good example of this, in which the sample is loaded slowly to failure. The failure stress for a static system is called the ultimate strength. Many engineering problems can be modelled as a static system, such as a rivet that supports a constant stress of 15 kpsi (103 MPa).

However, many loads are not static, they are dynamic or cyclic. Cyclic loading can either be the repeated removal and reapplication of a load or many complete reversals of a load. A good example of the first type of cyclic loading is a rivet in a pressure vessel that is constantly being filled and emptied. The fibres of a shaft under a bending moment experience complete reversal of stress, going from tension to compression with every revolution. Fig. 4.14 shows various patterns of time-dependent cyclic stresses.

Materials experiencing cyclic loading fail at much lower stresses than the ultimate strength of material, and they fail suddenly without much warning. This phenomenon was first noticed in the early days of the railroads. Axles that were loaded far below their yield point began failing in great numbers, so steps were taken to find the cause. The early researchers called this dynamic failure *fatigue*. The study of fatigue led to an understanding that dynamically stressed material had a finite life that was dependent on the amount of alternating stress present. As an aid to design engineers, an *S-N* diagram can be used to find the life span of a 1member experiencing totally reversed stresses, but different methods must be used when alternating stresses exist in

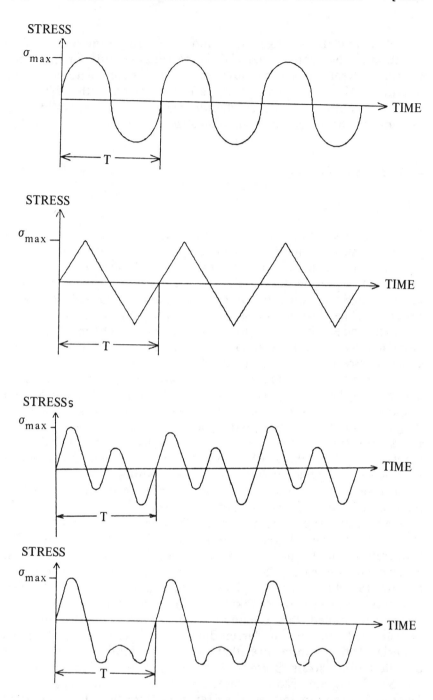

Fig. 4.14 Various cyclic stresses. (a) Completely reversed
sinusoid. (b) Completely reversed ramp. (c) Sinusoid with
high frequency ripple. (d) Secondary peaks.

combination with a static component, such as an axially loaded shaft supporting a bending moment. These methods will be explored in detail later in this chapter.

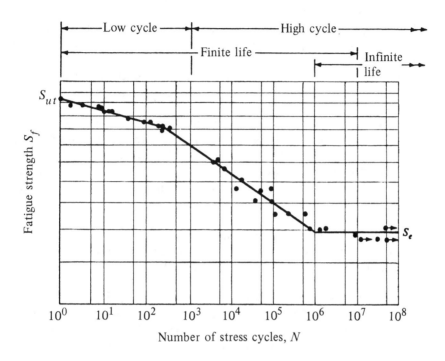

Fig. 4.15 S-N Diagram for UNS G41300 Steel, S_{ut}=125 kpsi (862 MPa)

As can be seen from the figure, there is a point where the S-N curve flattens out at at a certain stress. This stress is called the endurance limit for the material. If no alternating stresses are larger than the endurance limit, the member will theoretically have infinite life. Most non-ferrous materials, such as aluminium, have no flattening of the S-N curve. There is no alternating stress small enough to allow members made from these materials infinite life, but for practical purposes the endurance limit of these materials is taken to be the stress that would allow the material to have a life of from 10^8 to 5×10^8 cycles.

Sometimes members are required to have infinite life, so the member is designed such that no stress exceeds the endurance limit, including stresses in regions with stress concentrations. Other members are designed for finite life.

As Fig. 4.15 indicates, there are two regions of finite life: the high and low cycle regions. The low cycle region is in an area from zero to around 10^3 cycles. Sometimes a statically loaded part is designed to survive low cycle fatigue, especially if there is a good chance the part will experience some overloading. The study of fatigue and of designing parts to survive under dynamic loading is a major part of the engineering discipline.

Fatigue is not as well documented or understood as static failure, but the following basic principles are known:

1. Fatigue is the sudden, unexpected and complete failure of a member. Fatigue starts with the development of small, unnoticeable cracks in a region of stress concentration, followed by the slow propagation of these cracks until they reach a stage when they rapidly enlarge, causing failure.

2. Fatigue is caused by the accumulation of damage with each cycle of loading. Mechanical fatigue does not disappear or lessen with a period of rest or non-usage.

3. Fatigue always occurs first in a region of stress concentration. This concentration could take the form of a stress raiser, a weld or a small machining or forging flaw that would go unnoticed in a normal inspection.

4. Fatigue life cannot be accurately estimated as can the failure or yielding of statically loaded parts. To obtain a really accurate life estimate for a member it is necessary to construct a full scale model, but S-N curves give a good general indication of when fatigue failure will occur.

4.1.7 Fatigue Stress Concentration Factors

As has been mentioned before, ductile materials are not very sensitive to stress concentration factors but fatigue stressed members always fail in regions of stress concentration, regardless of how ductile the material is. Experimental work has shown that even ductile materials are sensitive to stress concentration in the fatigue mode, but this sensitivity is not as great as the theoretical stress concentration factor. Ductile members in fatigue loading experience the fatigue stress concentration factor, K_f, given by:

$$\sigma_{max} = K_f \sigma_{nom} = K_f \, (P/A) \tag{4.4}$$

The factor K_f is related to K_t by the following relationship,

$$K_f = 1 + q(K_t - 1) \tag{4.5}$$

where q is the notch sensitivity, which can have any value between 0 and 1, given by:

$$q = (K_f - 1)/(K_t - 1) \tag{4.6}$$

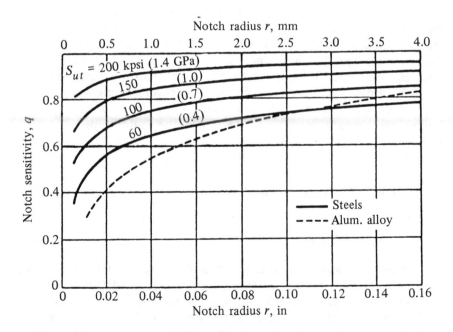

Fig. 4.16 Notch sensitivity factors

The notch sensitivity can be found from Fig. 4.16. The fatigue stress concentration factor is simply a theoretical stress concentration that has become a function of a material. It can be seen that higher strength alloys are more sensitive to stress concentrations, especially when they have large notches. As a general rule the notch sensitivity is unity for high strength, case hardened or low ductile materials, meaning that for these materials:

$$K_f = K_t \qquad (4.7)$$

4.1.8. *Fatigue Failure*

As shown earlier, the mechanics of materials that experience purely alternating stresses are very different from the mechanics of materials under the action of purely static stresses. A material experiencing components of both static and alternating stresses shows characteristics that are distinct from either of the cases. *Fracture mechanics* is the term used to describe fatigue failure, but here it will be used more specifically to denote fatigue under conditions where both alternating and static stresses are present. Much more is understood about static conditions than simple alternating stress. Again, more is known about simple alternating stress than about the fracture mechanics of static and alternating stresses combined, owing to the fact that more experiments in the laboratory have been conducted for reversed stresses.

When considering fracture mechanics, it is important to know the mean stress as well as the amplitude of the alternating stress. See Fig. 4.17.

$$\sigma_{mean} = (\sigma_{max} + \sigma_{min})/2 \qquad (4.8)$$

$$\sigma_{amp} = (\sigma_{max} - \sigma_{min})/2 \qquad (4.9)$$

When the stress amplitude of a fracture is plotted against the mean stress of the fracture, an interesting diagram occurs as shown in Fig. 4.18. The failures seem to follow a pattern. It is known that the endurance limit is the limit for an alternating stress (life span is not infinite). It is also known that the ultimate strength or the yield point could be the limit for a purely static stress. Several empirical relationships have

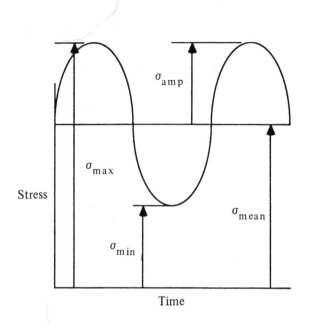

Fig. 4.17 Alternating stress with static component

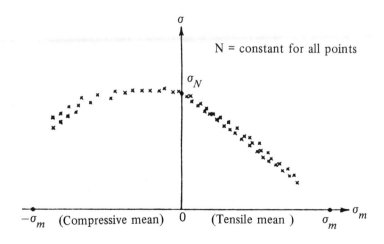

Fig. 4.18 Plot of mean stress vs alternating stress for failure

been presented to connect these points in the region of combined static and alternating stresses, thereby estimating the failure points of materials in this region. Four of the most important relationships are the Goodman line, the Soderberg line, the Elliptic Relationship and Gerber's Parabolic Relationship.

These relationships are plotted in Fig. 4.19 and are presented in equation form in equations (4.10) to (4.13)

Goodman line

$$\frac{\sigma_a}{\sigma_e} + \frac{\sigma_m}{\sigma_u} = 1 \tag{4.10}$$

Soderberg line

$$\frac{\sigma_a}{\sigma_e} + \frac{\sigma_m}{\sigma_y} = 1 \tag{4.11}$$

Elliptic line

$$\left[\frac{\sigma_a}{\sigma_e} \right]^2 + \left[\frac{\sigma_m}{\sigma_u} \right]^2 = 1 \tag{4.12}$$

Gerber's Parabolic

$$\frac{\sigma_a}{\sigma_e} + \left[\frac{\sigma_m}{\sigma_u} \right]^2 = 1 \tag{4.13}$$

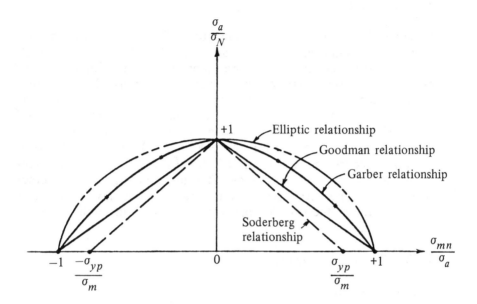

Fig. 4.19 Empirical relationships for fatigue failure

The way these graphs are used is simple, even if the calculations to find the mean and alternating stresses in the member are not. Once these stresses are known, they are plotted on the diagram. If the plotted point is above or outside the bounds of the graph, it is unsafe. If the plotted point is inside the bounds of the relationships then the member is safe by the criteria of which particular test was chosen.

These graphical relationships were made for design engineers who need a simple method to design parts with a complex loading pattern. The Goodman line is the relationship most often used because it is conservative and very simple. The Soderberg line is also very simple, but it is too conservative. This method merely draws a straight line from the endurance limit to the yield point. These relationships are also used for the analysis of existing members. There are other methods of designing or analysing members that have alternating and static stresses, but most of them are very complex and outside the scope of this book.

4.2 FRACTURE MECHANICS

4.2.1 *Introduction*

Fracture is a non-homogeneous process of deformation that causes the material to separate, usually forming a crack. Consequently the load carrying capacity of the material will decrease. Fracture may occur without a warning. It is possible in materials which have rather high ductility. The fracture can occur at various levels of dimensions: atomistic (10^{-8} inches)(25×10^{-8} mm), microscopic (10^{-4} inches)(25×10^{-4} mm) and macroscopic (10^{-1} inches)(2.5 mm). At atomistic level fracture occurs as a result of breakage of bonds between atoms across a fracture plane thereby creating a new crack surface. At the microscopic and macroscopic levels, fracture occurs mainly due to the passage of cracks through a region of material.

4.2.2 *Types of Fracture*

4.2.2.1 *Cleavage*

When a cleavage crack in a solid under a tensile component of externally applied stress spreads, it results in cleavage

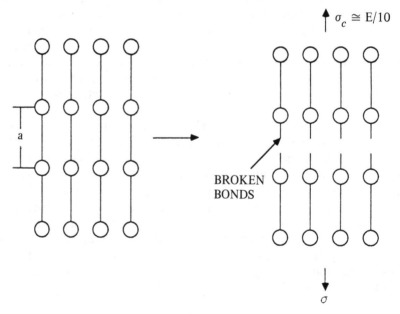

Fig. 4.20 Cleavage fracture viewed at atomistic level

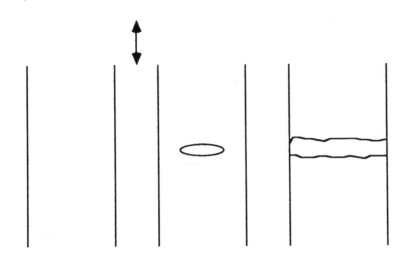

Fig. 4.21 Cleavage fracture viewed at macroscopic level

fracture. The material fractures due to the breakage of atomic bonds by the concentrated tensile stresses at the crack tip. In most materials certain planes of atoms are more easily subject to this type of fracture and these are called cleavage planes. In brittle materials these fractures propagate continuously from one grain to another. On the other hand, in some materials such as mild steel the fracture is mostly discontinuous. But the continuity is maintained since some of the grains fail in shear.

4.2.2.2 Shear

Shear fracture is a process of non-homogeneous plastic deformation that occurs by the shearing of atomic bonds. In solids there are some planes of atoms which have a low resistance to shear. These planes are called slip planes. Shear fracture occurs in pure single crystals when the two halves of the crystal slip apart on the crystallographic glide planes that have the largest amount of shear stress resolved across them.

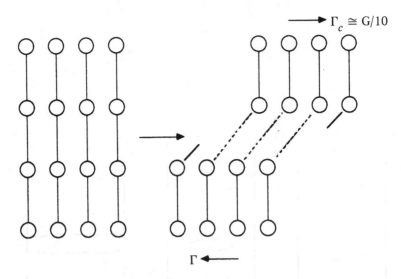

Fig. 4.22 Shear fracture viewed at the atomistic level

There are two types of shear fracture: slant fracture and chisel point fracture. Slant fracture is caused by the shear that occurs on only one set of parallel planes whilst the chisel point fracture is the resultant of the shear that takes place in two directions.

4.2.2.3   Cup-Cone

In polycrystalline materials the advancing shear crack follows the path of minimum resolved shear stress and this path is

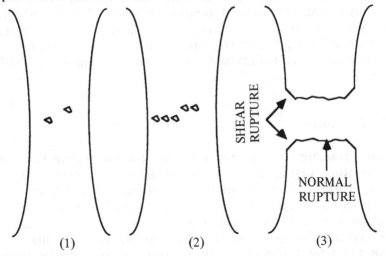

Fig. 4.23 Cup-cone fracture viewed at the macroscopic level

determined by both the applied stress and the presence of stress concentrations such as voids. The crack is the resultant of the formation of voids and their coalescence by localized plastic strains. Shear fracture normally begins at the centre of the structure and gradually grows outwards. The fracture path is usually normal to the tensile axis and the fracture surface has a fibrous appearance. Hence this form of fracture is usually called *normal rupture* or *fibrous rupture*.

4.2.3 *Cyclic Loading*

Fracture, known as fatigue, can occur at stress levels below the tensile strength. This fatigue is a result of plastic deformation in the initiation and propagation of cracks. The overall fatigue process can be subdivided into three main stages: crack initiation, crack propagation and rupture of the remaining section. Since fatigue failure involves the effect of the number of cycles under alternating tensile and compressive stresses, it is difficult to make a prediction of the fatigue life time. However, the low cycle fatigue and fatigue crack propagation can be used to determine quantitatively the fatigue life time.

4.2.3.1 *Initiation and Growth*

Although there are a number of ways by which fatigue cracks can be initiated, they usually start at a free surface.

a. Stage 1 Growth:

After initiation at a surface slip band in a single crystal, the crack will continue to advance into the material along the primary slip planes and later veer onto a plane at right angles to the principal tensile stress. Crack growth before transition is stage 1 growth and the crack growth after the transition is stage 2 growth. Transition is influenced by the magnitude of the tensile stress. Fig. 4.24 shows clearly the two stages of fatigue crack growth.

b. Stage 2 Growth:

In stage 2 the plastic deformation taking place at the tip of a crack can be observed. The crack advances a finite increment in each loading cycle and a mark called striation is formed on

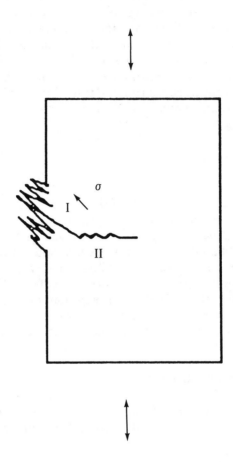

Fig. 4.24 Schematic diagram of stage 1 and stage 2 growth.

the fracture surface in each load cycle. These two aspects are related to each other as shown in Fig. 4.24. At the begining of the cycle the crack tip is sharp but as it advances it becomes much more blunt and the plastic zones at the tip expand. At the crack tip both these effects interact between the applied stress and the amount of plastic deformation. A new fracture surface is created by this plastic shearing process during the loading stage. During the unloading stage the extended material at the tip exerts a back stress leading to deformation marking or striation on the fracture surface as the crack closes.

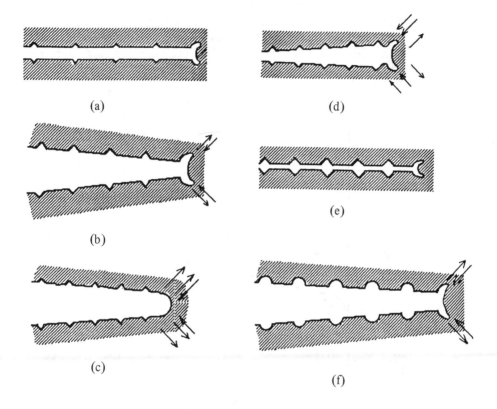

(a)

(b)

(c)

(d)

(e)

(f)

Fig. 4.25 Fatigue crack propagation in the stage 2 mode (a) Zero load. (b) Small tensile load. (c) Maximum tensile load. (d) Small compressive load. (e) Maximum compressive load. (f) Small tensile load

The *S-N* curve shows the dependence of the number of cycles required to cause failure on the applied alternating stress.

4.2.4 *Static Loading*

Fracture under static loading is generally called the *creep fracture*. These creep fractures are intergranular and appear as normal and shear ruptures. In the creep range at high stresses and low temperatures these fractures occur at voids at the grain boundary, known as *triple points*. On the other hand rupture can result from the formation of a multiplicity of voids along the grain boundaries at lower stresses and higher temperatures. This process is normally referred to as *cavitation*. But in certain metals where grain boundary voids are not observed, fracture normally occurs by shear.

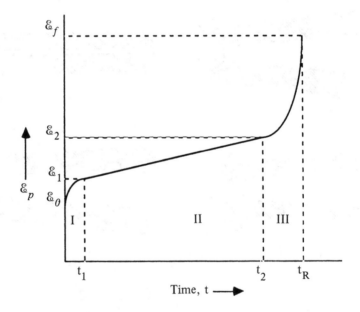

Fig. 4.26 Schematic creep curve

Fig. 4.26 shows a creep curve. The steady state creep rate is governed by the process of dislocation climb. The third stage is characterized by an increased creep rate and results in rupture. This increase in strain rate is either due to an increase in stress because of loss in cross-sectional area as internal voids grow or due to the onset of necking.

4.2.5 *Stress Corrosion Cracking*

Stress corrosion cracking refers to the embrittlement of a metal or alloy as a result of the simultaneous action of the corrodent and stress. In a given environment a material under stress may be:

a) completely passive,
b) subjected to an overall corrosive attack and
c) subjected to a local attack such that cracks
   form and grow ending in failure

Not all the alloys suffer embrittlement. Only the last type of behaviour results in stress corrosion cracking. The stress that is required for this type of cracking is usually tensile, but shear stress also results in cracking.

4.2.6 *Modes of Fracture*

There are three distinct modes of separation that can occur at the crack tip as shown in the Fig. 4.27.

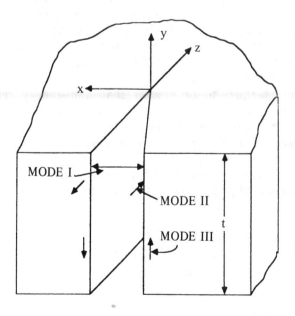

Fig. 4.27 Modes of fracture

Mode 1:

> Mode 1 occurs owing to the tensile component of the stress normal to the faces of the crack.

Mode 2:

> Mode 2 is the result of the shear component of stress normal to the loading edge of the crack.

Mode 3:

> When the shear component of the stress is applied parallel to the loading stage of the crack it results in mode 3.

4.2.7. Griffith's Theory

In 1920, Griffith proposed the first explanation of the discrepancy between the observed fracture strength of crystals and the theoretical cohesive strength. Although Griffith's original theory was only applicable to perfectly brittle materials such as glass, his ideas have subsequently helped in the understanding of the phenomenon of fracture in metals.

Griffith analysed the fracture behaviour of components that contain sharp discontinuities, based on the assumption that incipient fracture in brittle materials occurs when the elastic energy supplied at the tip of the crack during an increase in crack length is equal to or greater than the energy required to create the new crack surface. This criteria can be used to determine the magnitude of the tensile stress which will just cause a crack of a certain size to propagate.

Consider an infinite plate of unit thickness with a crack of length 2a subjected to uniform tensile stress σ. The total potential energy of the system is given by:

$$U = U_0 - U_a + U_\gamma \qquad (4.14)$$

where,

> U_0 — the elastic energy of the uncracked plate

U_a — the decrease in the plastic energy due to
   introduction of the crack in the plate
U_γ - the increase in the elastic surface energy
   caused by the formation of the crack surfaces

Griffith used the stress analysis of Inglis to show that

$$U_a = \frac{\Pi\sigma^2 a^2}{E} \qquad (4.15)$$

Now the elastic surface energy, U_γ is equal to the product of
the elastic energy of the material, γ_e, and the new surface
area of the crack is given by:

$$U_\gamma = 2(2a\gamma_e) \qquad (4.16)$$

Hence Equation (4.14) is written as:

$$U = U_0 - \frac{\Pi\sigma^2 a^2}{E} + 4a\gamma_e \qquad (4.17)$$

On differentiating this equation with respect to the crack
length, a and equating it to zero, the following form of
equation is obtained:

$$\sigma\sqrt{a} = \left[\frac{2\gamma_e E}{\Pi} \right]^{1/2} \qquad (4.18)$$

This indicates that crack extension in brittle material is
governed by the product of the applied nominal stress and the
square root of the crack length.   Equation (4.18) can be
rearranged as:

$$\frac{\Pi\sigma^2 a}{E} = 2\Upsilon_e \qquad (4.19)$$

The left-hand side of the equation is the energy release rate G and the right-hand side of the equation represents the material's resistance to crack extension, R.

The stress required to propagate a crack in a brittle material can be given as a function of the size of the crack as:

$$\sigma = \left[\frac{2\,\gamma_e\,E}{\pi\,a} \right]^{1/2} \qquad (4.20)$$

This equation does not strictly apply for metals, because it does not consider the plastic deformation before fracture, which would blunt the tip of the crack and increase the effective radius at the tip and hence the frature stress. Orowan suggested that the Griffith's equation should be modified for metals by including a term expressing the plastic work required to extend the crack, as follows:

$$\sigma = \left[\frac{2\,E\,(\gamma_e + \gamma_p)}{\pi\,a} \right]^{1/2} \qquad (4.21)$$

CHAPTER 5

Properties of materials

5.1 INTRODUCTION

The final strength of any material used in an engineering component depends on its mechanical and physical properties after it has been subjected to one or more different manufacturing processes. Also, there are several properties that determine the suitability of the material in its initial state for any particular manufacturing process. The initial strength of the virgin material is important, because that strength will affect the ease with which it can be deformed into its required shape and final- ly, its ability to resist loads during service. Factors which increase or decrease the strength of the starting material may be equally important. It may be desirable either to reduce its strength sufficiently to allow it to be formed into shape easily with the available machines, or alternatively to increase the final strength of the manufactured component and render it more serviceable. **Strength** is an imprecise term which may here be understood to indicate the ability of a material either to accept or to resist deformation.

A similar argument applies to a rather more elusive property of any material — namely, its **ductility**, which is usually understood to mean the ability of a material to accept large amounts of deformation (mainly tensile) without fracture. Again considering manufacturing processes, a large value of this parameter will obviously be beneficial. Many metal-working processes are limited only by the available ductility of the material being worked, so that the amount of deformation which can be imposed on the material has to be restricted to avoid **fracture**. There are, however, some manufacturing processes for which the opposite of ductility is beneficial. A suitable generic term for this property might be **brittleness**; for example, it is well-known that certain brittle materials are much easier to machine or shear than are ductile materials.

It is mainly the interplay of properties such as strength and ductility during fabrication which has influenced the technology of production. For example, it is common knowledge that most metals when heated will become softer and easier to deform. If the speed of deformation is too great, however, this benefit will be lost and the material may become either too hard or so brittle that fast deformation will lead to fracture. The occurrence and magnitude of such effects as these depend in some way on the microstructure of the material, so a knowledge of the metal-lurgy of metals or the corresponding microscopic structure of non—metals is necessary for any understanding of the broad subject of this book, namely the **strength of materials**. The aim of the initial discussion in this chapter is, in fact, to indicate those properties of materials which are important both during and after manufacture, to see <u>why</u> they are important and <u>how</u> they influence the manufacturing process. It is clearly necessary to have more precise terms than 'strength' and 'ductility', and in this chapter some of the **standard mechanical tests** will be cons-idered to see whether it is possible to define such concepts with more precision. Of course, to do this it is necessary also to have some knowledge of the **mathematical theory of the plasticity** or **rheology** of **ideal substances**. These topics were discussed in great detail in the companion volume (Vol.1) but will be extend-ed and applied throughout this chapter.

Once the various properties of importance in manufacture have been defined and understood, it is then possible to consider how this knowledge may be used to control the process and the product, and how these properties are affected by different production processes. In this way it should be easier to decide the method of manufacture most able to suit a given component and material so as to give it the final shape, strength and properties required. Thus it can be understood why the subject traditionally entitled **strength of materials** is so important, not only as it relates to the <u>final</u> condition of the materials found in any engineering artefact (an approach which categorizes most previous books with that title) but also as it relates to the materials before they are formed into their final shapes.

For example, it might be relevant to consider changing the shape or material of a manufactured component to suit the available production technique. Such questions are outside the scope of this book, and properly belong to the more specialized realms of **design for manufacture** or **manufacturing engineering**.

In the final analysis any successful manufacturing process must be economically sound and high priority should always be given to economic factors. The costs of manufacture are important from the outset i.e. from the time a component is <u>specified</u> to fulfil a certain function for a certain lifetime until its final inspection, testing, and <u>guarantee</u>. The whole manufacturing process entails both design and production of the component, particularly in the manner in which they affect the final strength of the material.

There are several physical and chemical properties which influence the choice and treatment of materials in manufacture. An example of a physical property is **thermal conductivity** which will affect the flow of heat within the body of the material whilst it is being deformed and therefore its rate of cooling and hardening. Similarly, a well-known example of an important chemical property is that of corrosion resistance. Its importance in the final product is obvious, and it may well be important during the manufacturing process too, because it can sometimes influence the formation of surface films which affect lubrication, or thermal and electrical conduction. A very brief discussion of the more important physical and chemical properties is therefore included in this chapter. Finally, since the subject "properties of materials" is common to both "strength of materials" and "manufacturing technology", the content of this chapter is based upon sections of the book *"Manufacturing Technology"*, by Alexander et al.

5.2 STANDARD MECHANICAL TESTS

To summarize the previous discussion, it is very important to know the strength of a material, both for its eventual use and also to determine the forces required to shape it. Since it is impracticable to test every article after it has been designed and made, several simple general tests are used to measure the mechanical properties of the stock material before, during and after manufacture of the final product.

5.2.1 *Tensile Tests*

The simplest and most widely accepted **tensile test** (as discussed in Vol.1, Ch.2, Sec.2.3) requires a cylindrical (or flat) bar with enlarged ends. This tensile specimen is subjected to a steadily increasing tensile force along its axis, and the extension of a gauge length is accurately measured, according to the appropriate standard. The results usually required are the maximum tensile stress, the **yield stress** (Y) (or **proof stress** in some cases as discussed in Vol.1), the **percentage elongation to fracture** and the **reduction of cross–sectional area at fracture**. In addition, the **Young's Modulus of Elasticity**, or **Young Modulus** (E) may be measured. Some typical **load–extension curves** for various important metals are sketched in Fig. 5.1.

It should be recalled from Vol.1 that the stresses recorded are customarily **nominal**, obtained by dividing the actual load by the <u>original</u> area of the cross–section. For accurate calculations it is preferable to use the **true stress** σ_1, but this requires continuous evaluation of the <u>actual</u> cross–sectional area, which decreases during the test. Similarly the **nominal strain** depends on the original gauge length. A better measure is the **incremental true strain** $d\epsilon_1 = dl/l$, which can be integrated for simple <u>uniaxial</u> stress to give the **logarithmic strain** or **true strain** ϵ_1.

5.2.2 Compression Tests

It is important for metal forming calculations to know the yield stress at much higher strains than can be obtained in tension. **Axial compression** of a short cylinder may be used, with suitable correction for the frictional resistance on the flat ends, but a more accurate result is obtained by the transverse **plane strain compression** of a well–lubricated strip, as illustrated in Fig 5.2 and discussed comprehensively in Vol.1.

5.2.3 Hardness Testing

Hardness testing was also discussed at length in Vol.1 (Ch.2 Sec.2.4). Tensile and compressive tests are destructive of the sample, but it is often important to check the strength properties of stock material or finished components, <u>without</u> destruction. There are several types of hardness test for this purpose, which make only a small indentation in the surface.

The oldest and best known hardness tests in the U.K. are the **Brinell test** in which a standard ball (usually 10 mm dia.) is pressed into a metal under a prescribed load, typically 3000 kgf (= 29.42 kN or 6615 lbf), and the **Vickers test**. The **Brinell Hardness Number** (BHN or HB) is then defined as the load <u>in kgf</u> divided by the actual spherical surface area of the indentation in <u>mm²</u>. Likewise, the **Vickers Hardness Number** (VHN or HV) is the load in kgf divided by the pyramidal surf-ace area (again in mm²) of the indentation. Note that 1 kgf/mm² $\doteq$ 9.807 MPa. It is found empirically, with support

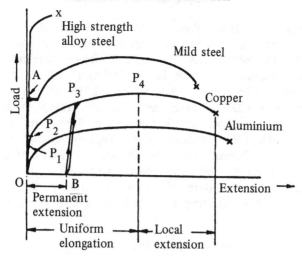

Fig. 5.1 Typical shapes of tensile test curve
obtained at low speed and room temperature

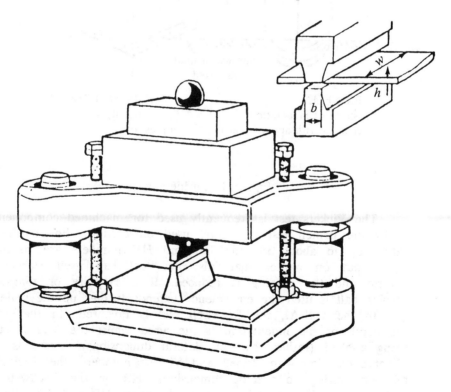

Fig. 5.2 Plane strain compression test

from slip–line field theory (to be discussed later), that the Brinell Hardness Number is directly related to the **uniaxial yield stress** Y. In fact, BHN $\simeq 2.9Y$, it being remembered that, to use this equation, Y must be in kgf/mm². If Y is in S.I. units of MN/m² (N/mm², or MPa) then BHN in its usual units of kgf/mm² $\triangleq 0.102Y$. These tests are illustrated in Fig. 5.3. The Vickers test involves an indenter in the form of a diamond pyramid with 136° included angle (approximating to the contact angle of the Brinell ball). Much smaller loads (from 2 to 120 kg) are applied and the total surface area of the actual contacting faces is determined by measuring the length of the diagonals of the resulting small lateral square indentation.

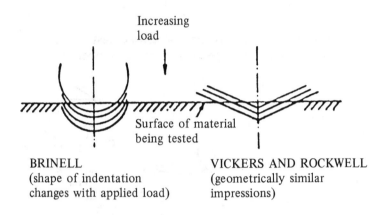

Increasing
load

Surface of material
being tested

BRINELL
(shape of indentation
changes with applied load)

VICKERS AND ROCKWELL
(geometrically similar
impressions)

Fig. 5.3 Hardness tests

The Vickers test is frequently used for machined components whilst the Brinell test is mainly used on castings. For hardness values up to about 350 the HB and HV numbers are roughly equal, but on harder materials the steel ball itself starts to deform and the reading is reduced. If a sapphire or tungsten carbide ball is used the equivalence extends further up the scale.

In the U.S.A., the **Rockwell test** is favoured. In that test the depth of the indentation is measured whilst the load is still being applied (rather than the lateral dimensions). There is no direct correlation with Brinell and Vickers numbers, the Rockwell numbers having no specific dimensions. HR on the Rockwell C scale is approximately 0.1 HB over the range HB = 200 to 400.

5.2.4 *Fatigue Tests*

Another very important subject which was not discussed in Vol.1 is the phenomenon of **fatigue**. It has been recognized for many years that static tensile or compressive testing is not adequate for predicting the strength of components subjected to vibration or repeated loading. These can fail at much lower stress levels, and there is a general relationship (due to Goodman) which shows the allowable oscillating stress level for a given mean stress, as illustrated in Fig. 5.4 for various surfaces produced by different modes of grinding.

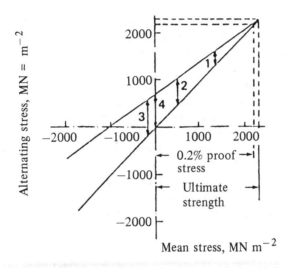

Fig. 5.4 A Goodman diagram for different ground surfaces
1 abusive; 2 conventional; 3 gentle; 4 annealed abusive

Fatigue testing needs considerable time, since each point on the final graph of applied stress S against the number N of cycles to failure requires a new specimen and N is usually between 10^6 and 10^8 (see Fig. 5.5).

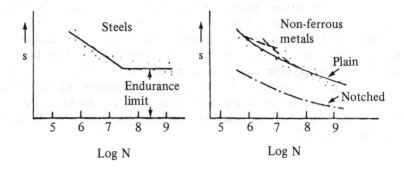

Fig. 5.5 Typical *S–N* curves

For many non-ferrous alloys the *S–N* curve falls steadily, but for steels there is often a levelling-off after some 10^6 to 10^7 cycles. If the stress does not exceed this endurance limit, the specimen will last indefinitely (see the book by Crane and Charles). The cyclic loading is usually applied in **rotating bending**, or **reversed bending**, although for fatigue research a **tensile** or **push–pull** machine is preferred, with very carefully prepared surfaces. Van Vlack in his definitive book shows that any surface cracks or other defects can reduce the **fatigue life**, as can **residual tensile stresses**.

Another very important failure phenomenon is that of **high-stress low-cycle fatigue** which is potentially dangerous in materials as disparate as animal bone and aerospace components.

5.2.5 *Impact Testing*

Another important subject not mentioned in Vol.1 is that of the behaviour of relatively brittle materials such as cast iron, which may fail under even a single impact. Since it may be very important to avoid this type of fracture, **impact tests** have been devised in which a **notched specimen** is hit by a heavy pendulum. (The notch is to induce tensile stress.) The energy absorbed is measured from the height of follow-through of the pendulum. The **Izod test** involves using specimens of square section (see Rollason, for example). The **Charpy test** is easier to conduct, especially at elevated or lowered temperatures (since the specimen is simply laid on top of an anvil), and requires the use of notched cylindrical bars.

These tests are sketched in Fig. 5.6. The significance of temperature is that some alloys, especially common steels, become very susceptible to a phenomenon called **notch–brittleness** at sub–zero temperatures.

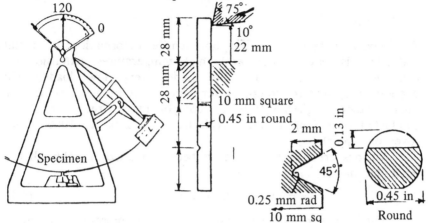

Fig. 5.6 Izod and Charpy specimens

5.2.6 *High–temperature Tests*

At high temperatures the plastic deformation of materials is dominated by **diffusion processes** which, for metals, become evident above about 2/3 of the **absolute melting temperature** T_m; i.e. the **homologous temperature** (a ratio defined as $T_H = T/T_m$) exceeds about 0.67 for hot working. The **diffusion rate** increases as the temperature increases, so the yield stress diminishes, but the latter also depends inversely on the strain rate as illustrated in Fig. 5.7.

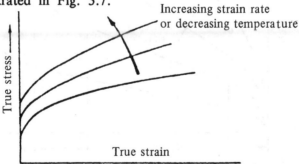

Fig. 5.7 The influence of strain rate and temperature on yield stress

Tension, compression or hardness tests may all be used at elevated temperatures, but it is very important to record all time–dependent parameters since, for example, creep may occur and vary with the applied stress, as will now be described.

5.2.7 *Creep Tests*

An important feature of the hot tensile deformation of metals and alloys is that, at sufficiently high temperatures, extension will continue at a very slow rate under very low loads. This phenomenon, termed **creep,** is very important in gas turbines and many other high–temperature components. Creep tests are conducted over long periods, typically from 1000 to 10000 hours (over one year).

Various creep equations have been suggested, notably that of Andrade, namely

$$\epsilon = A + B \ln t + C \, e^{\eta t}. \tag{5.1}$$

The results of a creep test can best be presented graphically, generally exhibiting three distinct régimes (**primary, secondary** and **tertiary creep**) as shown in Fig. 5.8. The secondary creep is usually recorded as the steady (minimum) creep rate. In the tertiary stage the decrease in cross–sectional area produces increasing stress, thus causing the creep to accelerate until fracture occurs.

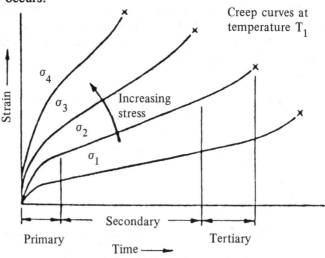

Fig. 5.8 Typical creep curves

Because of the length of time involved in creep testing, a shorter method is often used in which only approximate measurements of strain are made during the test, the main purpose of which is to determine the time to rupture at a given temperature and stress. These **stress–rupture tests** can be further speeded up by testing a string of specimens in series in a long furnace. The specimens are all subjected to the same end load but differing temperatures (which must be accurately measured, of course).

5.2.8 *Fracture Toughness*

In recent years much attention has been given to the **fracture toughness** of certain brittle materials, which is related to the ease with which a crack, once started, will propagate. A simple view of this process is that the opening of a crack releases elastic deformation energy but also requires the supply of surface energy to the two newly created areas of crack surface. If, in a brittle material, the released strain energy U is sufficient for this, the crack will propagate.

For a plane crack of length $2c$ and an assumed cylindrical volume of $\pi c^2.1$ surrounding it, this condition is given by:

$$U = \frac{\sigma^2}{2E}\pi c^2 \geqslant \gamma.2c \qquad (5.2)$$

where σ is the applied stress and γ is the **total surface energy per unit length of crack** (for both surfaces of the crack).

This leads to the definition of a **critical stress** at which the crack may propagate, namely:

$$\sigma_c \triangleq \sqrt{\frac{4E\gamma}{\pi c}} \triangleq \sqrt{\frac{E\gamma}{c}}. \qquad (5.3)$$

Typical values for iron are $E = 210$ kN/mm^2 (GPa), $\gamma = 2\times10^3$ N/mm and $c = c_0 = 2.5\times10^{-7}$ mm (the smallest possible crack, equal to one atomic spacing). Then:

$$(\sigma_c)_{\max} = 40.99\times10^3 \text{ kN/mm}^2 \text{ (GPa)} \triangleq \frac{E}{5}. \qquad (5.4)$$

Strengths approaching this extremely high value can in fact be found for **metallic whiskers** which are free from **mobile dislocations**.

Metals in their common form are usually ductile, so further energy p will be absorbed in plastic deformation. Equation (5.3) should then be written to include this factor and any other energy loss, such as acoustic emission.

$$\sigma_c \overset{\simeq}{=} \sqrt{\frac{E(\gamma+p)}{c}}. \tag{5.5}$$

This makes the material less brittle, or tougher. The real situation is more complex, but a **Strain Energy Release Rate** G can be determined from an experimental **Crack Opening Displacement** (COD) test.* The force necessary to extend the crack from length c_1 to c_2 is determined (Fig. 5.9), from which the value of G (per unit area) can be found for a specimen of large width w, from the following equation

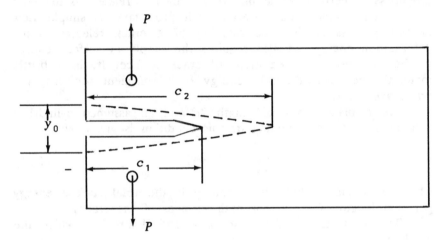

Fig. 5.9 COD determination

$$G = -\frac{dU}{dA} = -\frac{dU}{da} \cdot \frac{1}{w} = -\frac{y_0}{2w} \cdot \frac{\Delta P}{\Delta a} \tag{5.6}$$

where y_0 is a pre-selected mouth opening of the crack. At a critical value G_c, the crack will propagate, as in equation (5.5):

$$\sigma_c = \sqrt{\left[\frac{EG_c}{c}\right]}. \tag{5.7}$$

*   Sometimes referred to in the literature on fracture mechanisms as the "Crack Tip Opening Displacement" (CTOD)

The real situation is often complex and reference is made to the stress field, described by a **Stress Intensity Factor** K. This is not the stress concentration at a notch or other change in section, but describes the stress distribution and can be calculated for special conditions.

The critical value K_c for the growth of a crack in a thin sheet under **plane stress** is determined by the shape and size of the plastic zone around the crack and is known as the **Fracture Toughness**. This decreases as the thickness of a test plate is increased, having a minimum value K_{1c} for **plane strain condit-ions** because of the constraint imposed on the plastic deform-ation zone of a thick plate.

K_{1c} is therefore a material property from which the value of K_c for any other known configuration can in theory be found. As described by Hertzberg (see References) the **toughness** K_{1c} is determined from the experimental measurement of force required to extend the crack by a certain length. For a given COD, the work required will depend upon the length of the original crack and its extended length.

Table 5.1 gives some comparative values:

TABLE 5.1

Comparative fracture toughness values

	Strength MPa	Toughness MPa.m$^{\frac{1}{2}}$
Medium carbon steel	260	55
Pressure-vessel steel	500	200
Aluminium alloy 2024-T4	350	55
7075-T6	460	40
7178-T6	560	23
Titanium alloy Ti-6A1-4V	830	55
Ti-4A1-4Mo-4S	1095	35
Polymethacrylate	30	1
Polycarbonate		2.2
Glass		0.4

Recently, Atkins and Mai have written a comprehensive text book on both elastic and plastic fracture, giving considerable attention to the problems of fracture in materials subjected to large deformation plastic flow, which is therefore of great interest in the context of all manufacturing processes involving plastic deformation.

5.2.9 *Plastic Anisotropy*

In sheet metal forming in particular it is important to recognize that the properties of rolled sheet may differ substantially in the rolling and transverse directions, as well as in the "through–thickness" direction.

   This feature can be measured in terms of the now well-known so–called *r*–value, which is the ratio of the transverse to the longitudinal strain in a tensile test on wide, flat strip using techniques described by Hosford and Caddell. Volume is always approximately conserved in plastic deformation, so the thickness strain is also dependent upon the *r*–value.

5.3 THE BASIC STRESS–STRAIN CURVE

5.3.1 *General Concept*

Consideration of the standard mechanical tests described above reveals that it should be possible to predict the behaviour of any given metal from the results of a single test, such as the tensile test, for example. Unfortunately, it is not possible to predict the precise stress–strain pattern which will exist in any given situation.

   Nevertheless, the general concept that any given material possesses a basic stress–strain curve is useful, especially in pred-icting the behaviour of the material when subjected to plastic flow. In most metal–working processes the stresses are predom-inantly compressive, and fracture does not occur readily. Thus, although a tensile specimen will fracture after only about 40 per cent extension in the tension test, it is possible to roll or forge the same specimen until it has extended many times its original length, simply because the stresses in rolling or forging are predominantly compressive. During the deformation the metal must have been strain hardened, and the question arises as to how to determine the stress–strain curve at large strains. The results of the tensile test are useful only up to relatively small strains, even if measurements are made right up to fracture.

Also, the method of defining strain for such large deformations is important; this is discussed in more detail by Alexander et al. (Vol.2)

There is ample evidence to show that the stress–strain curve for a given metal in uniaxial tension corresponds closely with the curve for a metal in uniaxial compression, provided the true stress (load/current cross–sectional area) and true strain (logarithmic strain) are used as co–ordinates (see Section 5.2.1). To obtain a stress–strain curve up to the large strains required, the best method is to test a specimen of the material in compression, using either the axial compression of a short cylinder or the plane strain compression test illustrated in Fig. 5.2. In the plane strain compression test the strain is substantially zero in the width direction of the specimen, owing to the constraint of the material outside the platens (see Fig. 5.2). Since in simple uniaxial tension or compression there is no such restraint on the sideways movement of the material, more general definitions of stress and strain are required. These definitions are the **effective** or **generalized** stress (or strain).

As shown in Vol.1, the simplest definitions of these quantities in terms of the principal stresses or strain increments are, using the yield criterion of Maxwell (or von Mises), as follows:

Effective stress

$$\bar{\sigma} = \frac{1}{\sqrt{2}} \sqrt{(\sigma_1 - \sigma_2)^2 + (\sigma_2 - \sigma_3)^2 + (\sigma_3 - \sigma_1)^2}. \qquad (5.8)$$

Effective strain increment

$$\delta\bar{\epsilon} = \sqrt{\frac{2}{3}\left[\delta\epsilon_1{}^2 + \delta\epsilon_2{}^2 + \delta\epsilon_3{}^2\right]}. \qquad (5.9)$$

In the plane strain compression test $\delta\epsilon_2 = 0$, say, and therefore $\sigma_2 = \frac{1}{2}(\sigma_1 + \sigma_3)$, as discussed later in this chapter. Also, since the volume remains substantially constant during plastic flow, $\delta\epsilon_1 + \delta\epsilon_2 + \delta\epsilon_3 = 0$ and $\delta\epsilon_3 = -\delta\epsilon_1$.

Substituting these values in equations (5.8) and (5.9) gives the following relationships between effective stress and strain, and <u>actual</u> stress σ_1 and strain ϵ_1 in the plane strain compression test:

$$\bar{\sigma} = \frac{\sqrt{3}}{2}\sigma_1 \qquad (5.10)$$

$$\bar{\epsilon} = \frac{2}{\sqrt{3}} \epsilon_1 . \tag{5.11}$$

In uniaxial tension or compression $\bar{\sigma} = \sigma_1$ and $\bar{\epsilon} = \epsilon_1$, σ_1 and ϵ_1 being the <u>actual</u> stress and strain in the specimen.

Thus, as can be derived from the discussion in Alexander et al. Vol.1, Ch.4, Sec.4.7, to derive the stress–strain curve in uniaxial tension or compression from the stress–strain curve obtained from a plane strain compression test (σ_1 vs. ϵ_1), it is necessary to divide the ordinates of the curve by the factor $2/\sqrt{3}$ ($\triangleq 1.155$) and also to multiply the abscissae by the same factor.

In the case of high–temperature testing, where the <u>rate</u> of straining is of paramount importance, the definition of **effective strain rate** is obtained simply from the expression for the incremental strains [equation (5.9)] as:

$$\dot{\bar{\epsilon}} = \sqrt{\frac{2}{3}\left[\dot{\epsilon}_1^2 + \dot{\epsilon}_2^2 + \dot{\epsilon}_3^2\right]} . \tag{5.12}$$

5.3.2 Constitutive Equations

Several comprehensive discussions about constitutive equations can be found in the book edited by Argon. Approximate ranges of homologous temperature T_H and strain rate were investigated for many metal forming processes and are summarized in Table 5.2 opposite.

It may be concluded from those discussions that one possible constitutive relation which might be useful for representing the dependence of flow stress $\bar{\sigma}$ upon effective strain $\bar{\epsilon}$, effective strain rate $\dot{\bar{\epsilon}}$ and absolute temperature T is:

$$\bar{\sigma} = D\bar{\epsilon}^{n_1}\dot{\bar{\epsilon}}^{n_2}\exp\left(Q/RT\right) \tag{5.13}$$

where Q is an **activation energy**
R the **universal gas constant**
D, n_1 and n_2 constants.

Since equation (5.13) gives a zero value for $\bar{\sigma}$ if either $\bar{\epsilon}$ or $\dot{\bar{\epsilon}}$ is zero, which is certainly <u>not</u> the situation in practice, a more realistic equation might be

$$\bar{\sigma} = \sigma_0(1+B\bar{\epsilon})^{n_1} \times (1+D\dot{\bar{\epsilon}})^{n_2} \exp\left(Q/RT\right). \tag{5.13a}$$

TABLE 5.2
Typical ranges of T_H and strain rate

Process	Range of T_H	Approximate range of strain rate (s^{-1})
Casting	0·9–1·0	0·0–0·4
Hot forging	0·7	100–200
Hot extrusion	0·7	10–20
Cold forging and extrusion	0·16–0·32	10–100
Warm forging	0·55	10–100
Creep forming	0·63+	10^{-2}–10^{-4}
Sheet metal forming	0·16–0·32	10–100
Powder forming	Not known	Not known
Deposition forming	0·9	200–1000
Machining, shearing	0·16–0·32	$\geqslant$1000
Explosive forming	0·16–0·32	100–1000
Hot rolling	0.7	40–100
Cold rolling	0.16–0.32	100–1000

A different approach, more geared to the **dislocation mechanics** of physical metallurgy, is that described by Ashby and Frost in the book edited by Argon. Those researchers developed the very useful concepts of **deformation mechanism maps** and **fracture mechanism maps**, typically as depicted in Fig. 5.10 overleaf.

The strain rate of a plastically deforming crystalline solid is dependent upon a number of basic atomic processes such as: **dislocation motion, diffusion, grain boundary sliding, twinning,** or a **phase transformation.** Such processes can combine to give at least 12 distinctive deformation mechanisms, e.g. **yielding** (or slip), **power law creep, Nabarro–Herring creep, Coble creep, super–plastic flow,** etc.

Deformation mechanism maps for any **polycrystalline** material can be constructed in a **stress–temperature space,** showing the area of dominance of each **flow mechanism** for each of which a **rate equation** exists, linking shear strain rate γ to shear stress τ, temperature T and to structure (including all parameters describing its **atomic structure,** e.g. **bonding, crystal class, defect structure, grain size, dislocation density** and **arrangement, solute** or **precipitate concentration,** etc.). An example of such a map is shown in Fig. 5.10 on which typical

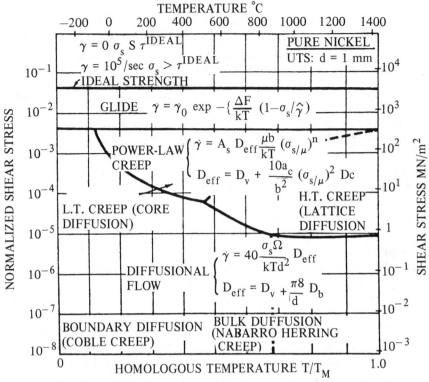

Fig. 5.10 The fields in which a particular mechanism dominates.
Boundaries are revealed by equating flow rates

equations have been shown, relating the **normalized shear stress**
τ/μ (where μ = **shear modulus**), to **homologous temperature**. In
that figure, boundaries are found by equating flow rates. In fact,
Kelly has shown that τ cannot exceed τ_{ideal} ($\approx \mu/20$), and shown
by the line marked **ideal strength** in the figure, which is for
pure nickel, but polycrystalline metals deform at stresses much
less than this by **dislocation glide** and/or **movement of vacancies**.
Examples of other maps and methods of using them are
described by Ashby and Frost. In some forming processes, where
very large strain rate ranges may be encountered, the description
of steady state behaviour given in equation (5.13) is inadequate.
At very high stress, an exponential stress dependence on strain
rate is more appropriate and the power law and exponential
variation can be combined as:

$$\dot{\epsilon} = A \, \{\sinh \, (a\bar{\sigma})\}^{m} \, \exp \, (-Q/RT). \tag{5.14}$$

This relationship has been shown to give a good fit with experimental data for aluminium alloys over many orders of magnitude in strain rate.

However, the description is still of **steady state** behaviour and there appears to be no good evidence that it can be used as an **equation of state**† relationship when stress or temperature change sharply. This problem has been approached in two ways.

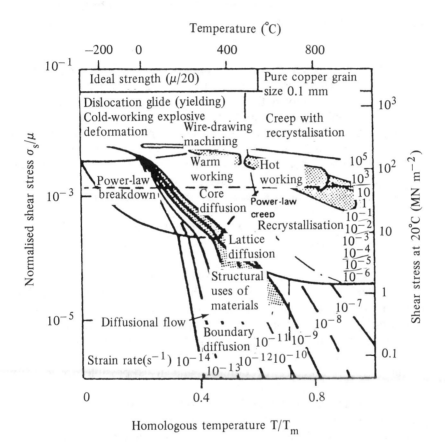

Fig. 5.11 Forming operations on a
deformation map for pure copper

† An **equation of state** implies a fixed one-to-one relationship between dependent and independent variables, e.g. as for the equation relating the pressure, volume, and absolute temperature of ideal gases in thermodynamics, viz. $PV = RT$.

One procedure is to define an **equation of state** which contains an internal variable describing the structure of the material. If the variation of this internal variable with change in stress, temperature and time can be established, an adequate **constitutive equation** may result. Numerous attempts have been made to provide such a description but they all suffer from the disadvantage that large numbers of sophisticated tests have to be conducted to establish the internal variables. Thus, not only constant stress and constant strain rate tests are required, but also complex relaxation and other stress and temperature cycling.

Ashby's **deformation and fracture mechanism maps** have made an important contribution to the way of describing and illustrating both deformation and fracture phenomena. It is possible to delineate various regions for different deformation processes on a **deformation mechanism map** as was illustrated in Fig. 5.11 (for pure copper). It is worth noting here that Ashby plots curves of **normalized shear stress** vs. **homologous temper-ature,** for various strain rates, derived from **dislocation mechanisms.** The same approach can be used for fracture.

5.3.3 *Modelling of Dynamic Material Behaviour in Hot Deformation*

Another (similar) approach has been adopted by H. Gegel and his co-researchers at the Wright Patterson Air Force Base in America. Their approach, however, is based on thermodynamic considerations and does not rely on the microstructural behavioural description of the material. It is 'similar' only in so far as the final presentation is of areas in which deformation is possible without fracture or defects, and it is possible to identify optimum regions in which deformation is most efficient. This approach seems to be superior to that of Ashby in that it is more general (not necessarily tied to metals) and is easier to understand, although it does present a completely new approach, the validity of which many other researchers find difficult to accept. Attention is drawn towards the **instantaneous power** being dissipated, viz.:

$$\overline{\sigma}\overline{\epsilon} = \int_0^{\dot{\overline{\epsilon}}} \overline{\sigma}\mathrm{d}\dot{\overline{\epsilon}} + \int_0^{\overline{\sigma}} \overline{\epsilon}\mathrm{d}\overline{\sigma} \qquad (5.15)$$

or

$$P = G + J. \tag{5.16}$$

where the power dissipated by plastic work is denoted by the area G under the curve in Fig. 5.12 and given the apt title **dissipator content** whilst the area J above the curve is termed the **dissipator co-content** and is related to the structural (metallurgical or non-metallic) mechanisms which occur dynamically to dissipate power.

From equation (5.15) it follows that, at any given temperature and effective strain, the partitioning of power between J and G is given by

$$\left[\frac{\partial J}{\partial G}\right]_{T,\bar{\epsilon}} = \left[\frac{\partial \ln \bar{\sigma}}{\partial \ln \dot{\bar{\epsilon}}}\right]_{T,\bar{\epsilon}} \tag{5.17}$$

which is the **strain rate sensitivity** of a material conventionally evaluated as:

$$m = \frac{\partial(\log \bar{\sigma})}{\partial(\log \dot{\bar{\epsilon}})}_{T,\bar{\epsilon}} \; \stackrel{\Omega}{\simeq} \; \frac{\Delta(\log \bar{\sigma})}{\Delta(\log \dot{\bar{\epsilon}})}_{T,\bar{\epsilon}}. \tag{5.18}$$

[Note: $m = n_2$ of the discussion relating to equation (5.13).]

Determining the values of m as a function of the effective strain rate $\dot{\bar{\epsilon}}$ for any given material and assuming that the exponential equation:

$$\bar{\sigma} = A\dot{\bar{\epsilon}}^m \tag{5.19}$$

is valid for each value of m, the values of J are then calculated as a function of $\dot{\bar{\epsilon}}$ at a given T and $\bar{\epsilon}$. A **piece-wise quadratic fit** is used to determine the various values of m over the range of $\dot{\bar{\epsilon}}$ used in the tests.

Thus, the power dissipated in changing the material structure is:

$$J = \int_0^{\dot{\bar{\epsilon}}} \dot{\bar{\epsilon}} \, d\bar{\sigma} = \int_0^{\bar{\sigma}} (\bar{\sigma}/A)^{1/m} d\bar{\sigma} = \frac{1}{A^{1/m}} \cdot \frac{\bar{\sigma}^{(1/m+1)}}{(1/m+1)} = \frac{\dot{\bar{\epsilon}} \cdot \bar{\sigma} \cdot m}{m+1}. \tag{5.20}$$

Gegel and his colleagues have shown that m must lie between 0 and 1 if the material deformation is not to become unstable. For values of $m < 1$ the curve of $\bar{\sigma}$ versus $\dot{\bar{\epsilon}}$ appears similar to that shown in Fig. 5.12(a). When $m = 1$, the curve is a straight line giving a **maximum dissipator co-content** (of area $J_{max} = \dot{\bar{\epsilon}}.\bar{\sigma}/2$), as shown in Fig. 5.12(b).

Hence, it seems sensible to define an **efficiency of**

deformation in terms of a **non–dimensional parameter,** in line with other **stability criteria,** namely

$$\eta = \frac{J}{J_{max}} = \frac{2m}{m+1}.$$ (5.21)

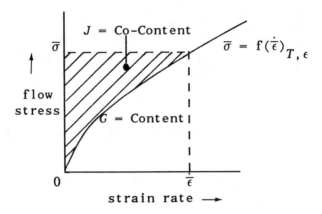

Fig. 5.12(a) Schematic representation of G content and J co–content for workpiece having a constitutive equation represented by curve $\bar{\sigma} = f(\dot{\bar{\epsilon}})$. The total power of dissipation is given by the enveloping rectangle.

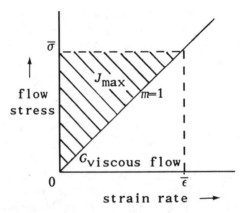

Fig. 5.12(b) Schematic representation illustrating the J_{max} which occurs when the strain rate sensitivity (m) of the material is equal to one

These researchers then proceed to determine curves of $\overline{\sigma}$ versus $\dot{\overline{\epsilon}}$ for gas turbine disc materials such as Ti–6242 at various temperatures and strains, by cross–plotting and interpolation of data obtained from conventional stress–strain compression testing up to high values of strain. A three–dimensional plot of efficiency η vs temperature T and logarithmic strain rate $\ln\dot{\overline{\epsilon}}$ can then be constructed and **two–dimensional maps** showing η **contours** derived for various values of **effective strain** $\overline{\epsilon}$ as shown typically in Figs. 5.13 and 5.14.

It is also possible to delineate regions on these maps where fracture or defects are likely to occur as shown in Fig. 5.15. For example, the parameter $\sigma_m/\overline{\sigma}$ (where σ_m = the mean or hydrostatic stress) can be determined everywhere throughout the volume of material being deformed (e.g. by finite element methods).

Ti-6242, BETA, 0.6 STRAIN

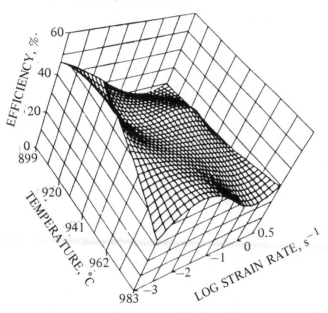

Fig. 5.13 Three–dimensional plot showing variation of efficiency of dissipation with temperature and strain rate for Ti–6242 preform at 0.6 strain

It must be ensured that $\sigma_m/\overline{\sigma}$ does not exceed algebraically a certain limiting value, typically –2/3 (i.e. compressive) for extrusion at the exit, to avoid **internal cracking (centre burst)**. This parameter, taken together with **dynamic recrystallization,** phase transformations, spheroidization of acicular structures, precipitation mechanisms, edge cracking and local kinking of β–platelets in Ti–6242 can all contribute to prediction of zones on these maps which should be avoided.

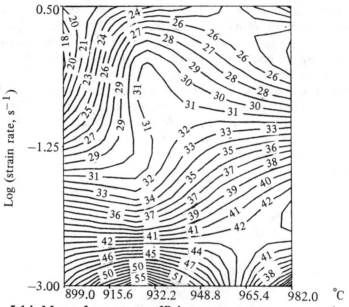

Fig. 5.14 Map of constant efficiency contours on a strain–rate vs. temperature frame for Ti–6242 β–preform at 0.6 strain

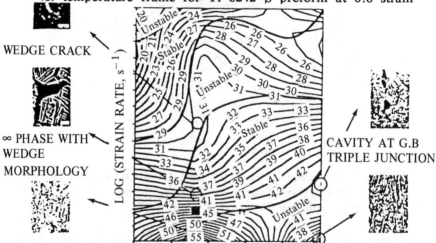

WEDGE CRACK

∞ PHASE WITH
WEDGE
MORPHOLOGY

CAVITY AT G.B
TRIPLE JUNCTION

Fig. 5.15 Processing map for Ti–6242 β microstructure
with stable regions identified (courtesy of
Wright Patterson Air Force Base)

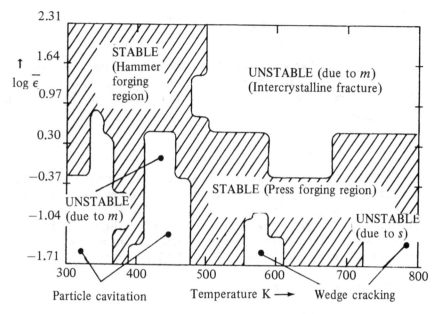

Fig. 5.16 Stability map showing safe processing regions for
Al–5Si alloy, at $\bar{\epsilon}$ = 0.6 (courtesy of Gopinath)

It is very interesting to observe that the predicted **optimum
forging conditions** of both $\alpha + \beta$ and β **preform micro–structures**
of Ti–6242 are 927°C and $10^{-3}s^{-1}$, corresponding almost exactly
with the **creep–forming conditions** which have been arrived at in
recent manufacturing practice by laborious and expensive trial
and error methods. More recently, by applying **Lyapunov function
stability criteria** (well–known in control theory, e.g. as described
by Schulz and Melsa) to material processing variables such as m
and introducing an analogous **entropy coefficient** s related to
stress σ and absolute temperature T by the equation:

$$s = \frac{1}{T} \frac{\partial(\log \bar{\sigma})}{\partial(1/T)}\bigg|_{\bar{\epsilon},\dot{\epsilon}} \qquad (5.22)$$

Gegel and his colleagues have pointed out that:

$$s = \dot{s}_{sys}/\dot{s}_{app} \qquad (5.23)$$

where $\dot{s}_{sys}$ = the rate of entropy production in the system and
$\dot{s}_{app}$ = the rate of entropy input into the system.

The analogy between m and s can best be seen by
comparing equation (5.18), viz.:

$$m = \frac{\partial(\log\overline{\sigma})}{\partial(\log\dot{\overline{\epsilon}})}_{T,\overline{\epsilon}} \stackrel{\simeq}{=} \frac{\Delta(\log\overline{\sigma})}{\Delta(\log\dot{\overline{\epsilon}})}_{T,\overline{\epsilon}} \qquad (5.18)$$

with equation (5.22).

Using the **Lyapunov function stability criteria** then leads to the condition that, in stable regions, the two second order partial differential coefficients of m and s with respect to log $\dot{\overline{\epsilon}}$ should be <u>negative</u>, viz.:

$$\frac{\partial m}{\partial(\log\dot{\overline{\epsilon}})} < 0 \; ; \; \frac{\partial s}{\partial(\log\dot{\overline{\epsilon}})} < 0. \qquad (5.24)$$

If the rate of change of the slope of a function is negative, this implies that the function is at its maximum. For the parameters m and s, therefore, the condition for stability is that they should be as large as possible, with the proviso that they must both be less than unity. These conditions have been used in developing the maps shown in Figs. 5.15 and 5.16.

5.3.4   Summary and Conclusions

It seems that the mapping techniques being developed by Gegel and his group offer the most useful approach for developing better mathematical modelling of material behaviour. Data–bases have been established for several materials of interest, using these techniques developed by Gegel and his team.

As has already been mentioned, many researchers find difficulty in accepting the concepts of dissipator content and co–content. Also, it has been observed that, in this approach, variables such as T and $\overline{\epsilon}$ are being treated as state variables (much as pressure, volume and temperature in gas thermo–dynamics), whereas they must be dependent on the history of their development. For example, if a material is taken to a certain value of temperature and total strain without passing through any phase transformation it will then have a different stress–strain–rate relationship than if it <u>had</u> passed through a phase change.

However, the method manifestly works. It is not actually necessary to use the concept of **efficiency**, viz. $\eta = J/J_{max}$. However, it is easier to correlate the parameter η with the microstructures evaluated over various regions of the processing maps and thereby arrive at more meaningful conclusions for understanding the intrinsic **workability** of materials. Similar results could be obtained simply by maximizing the m–value of equation

(5.18) (up to its maximum stable value of unity), i.e. by developing three–dimensional plots and mappings with m as the ordinate rather than $\eta = 2m/(m+1)$, since:

$$\eta m + \eta = 2m, \therefore m(2-\eta) = \eta, \text{ and } m = \eta/(2-\eta). \quad (5.25)$$

There is, in fact, almost a linear relationship between η and m for values of η and m between 0 and 1 as shown in Fig. 5.17 below.

Since the limiting value of $m = 1$ corresponds with **Newtonian viscosity**, i.e. the viscosity possessed by most simple liquids, this is obviously the most efficient way to deform a material (by making it approximate to the liquid state). The higher the value of m, the easier (more efficient) is the process of deformation going to be (e.g. metals becoming **super–plastic**). However, it is not possible to correlate the value of m so easily with observed or predicted microstructures as is the efficiency η.

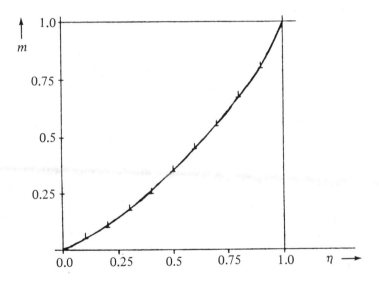

Fig. 5.17 Variation of m with η for equation (5.25)

It may be concluded that the best approach to **constitutive equations** may be that of developing **deformation and fracture**

mechanism maps. The **deformation efficiency** concept is attractive and has demonstrably led to a better understanding of the important parameters to be developed to give a meaningful data−base for any given material of interest. **Lyapunov function stability criteria** seem to provide powerful tools for delineating stable and unstable regions on these mappings of deformation behaviour. Apparently, these regions so discovered do actually correspond to material instabilities well−known to materials scientists dealing with both metals and non−metals. One remaining problem, however, is that of adequately accounting for the history of the deformation.

Modelling of dynamic material behaviour in hot deformation can most easily be included by using this mapping approach. The data−base so developed could then be used to provide data as a function of temperature, strain, strain rate and time (to include the history of deformation) which can then be fed into existing comprehensive computer solutions for rolling, for example.

However, many computer solutions are of the **slab** type (see later discussion of plastic flow) and do not include either temperature or strain rate effects. Much work has been done in industry on this problem, but it has not been published widely.

5.4 PHYSICAL PROPERTIES OF IMPORTANCE IN MANUFACTURE AND USE

Analysis of processes such as hot forming, casting or others used at high temperatures requires a knowledge of **thermal conductivity** κ, **specific heat** c, and **thermal expansion** α.

The last is especially important in resistance to thermal shock. A material such as silicon−aluminium−oxy−nitride has a much lower thermal expansion coefficient than, say, aluminium oxide and it is much more resistant to cracking when suddenly heated or cooled. A high thermal conductivity is also valuable in reducing any large **temperature gradients** which might otherwise occur in a material or composite.

When composites are used it is necessary to consider the compatibility of expansion coefficients, otherwise high stresses and even plastic deformation may occur.

5.5 CHEMICAL PROPERTIES

Oxidation is clearly important in hot working operations although in an enclosed process like extrusion, for example, there is very little access of air, so oxidation is substantially avoided.

The book edited by Braithwaite explains how **chemical reaction with lubricants** plays a major role in the effectiveness of lubricants. Titanium, for example, is very difficult to lubricate because of its thin inert oxide skin. Schey points out that stainless steels present the same problem.

Corrosion is also a critical feature. A high proportion of all metallic scrap arises from **rusting and related electrochemical degradation processes.**

Much attention has been given to **plating** and other **protective coatings to avoid corrosion** and also to improve other surface properties such as **hardness and wear resistance** (see the handbooks by Swann, Ford and Westwood and also by Peterson and Winer.

5.6 WORKABILITY

Although industrial producers will readily recognize some alloys as being easier to work than others, there is no simple test or clearly defined set of properties.

Ductility is a related property, and it is generally considered that a greater reduction of area to failure in a tensile test indicates better workability. While this is true in general terms, the actual fracture in a tensile test depends upon the complex stress state in the neck region for all but the most brittle alloys.

In the comprehensive book by Atkins and Mai, various criteria are mentioned which have often been proposed on the supposition that fracture in tension depends upon the total plastic work or the tensile plastic work expended in the deformation. It is very difficult to obtain accurate experimental data because the final stress state before fracture is frequently ill–defined. Recent analytical work using **finite–element plasticity** has given a better insight into these problems.

Hot workability is often described in terms of the number of revolutions to failure in a torsion test, but again the analytical significance is obscure. Torsion testing is also a valuable indicator of **cold workability,** but the results depend strongly upon the axial constraint, possibly because fracture and rewelding occur during the test, so the conditions must be carefully controlled.

5.7 ELEMENTS OF THE THEORY OF PLASTICITY

5.7.1 *General Theory*

To understand in more detail the mechanics of the forming
processes which are being discussed, it is necessary to understand
the fundamentals underlying the theory of the plastic flow of
materials. This is the one unifying theme that underlies all
forming processes. A complete treatment of the **mathematical
theory of plasticity** is beyond the scope or intent of this book;
comprehensive early texts have been written by Nadai, Hill, and
Prager and Hodge. Later treatments of the theory of plasticity,
discussed specifically with the needs of engineers in mind, are
given by Ford with Alexander, Johnson and Mellor, and
Alexander. A simpler, more practical account is given by Rowe.
What it is intended to do here is to consider the basic fundamentals on which the theory rests and thereby to show how a
better understanding of the behaviour of materials subjected to
these processes can be achieved.

In the first place the question may be asked: 'What is the
main difference between the behaviour of metal when subjected
to large deformation in metal-working processes and its
behaviour when stressed within the elastic range?' The simple
answer is that in metal-working processes the amount of straining or deformation is many times larger. It is this essential
difference between metal-working and elastic straining which is
all-important in formulating a workable theory. The solution of a
problem in elasticity is often achieved by making the assumption
that the external shape is unaltered during elastic straining. That
this is justified for most metals can be realized when it is
remembered that their elastic range never exceeds about 0.4 per
cent strain, corresponding with a change in external dimensions
of 0.4 per cent, and is often only 0.1 per cent.

By their very nature the processes of metal-working demand
that strains of more than one hundred times this value be
imposed. Indeed, in the case of forging or extrusion, reductions
in cross-sectional area of the order of 95 per cent are found.
For such large deformations a more precise definition of strain is
necessary than simply that of "(change in length)/(original
length)". For example, if a tensile specimen is stretched to

double its length the **engineer's strain** would be unity, or 100 per cent. To achieve 100 per cent compressive strain (engineer's strain) it would obviously be necessary to compress the specimen until it had zero thickness.   Clearly, then, this measure of strain is unsuitable for such large deformations, since 100 per cent compression certainly represents much more straining than does 100 per cent extension. Thus can be realized the first result of having such large deformations, namely the necessity for specifying some better measure of strain. Another consequence of having large strains of this order is that, as far as determining the changes of shape and the forces required is concerned, it will be permissible to neglect elastic strains.   This considerably simplifies the theoretical approach, (but it should be remembered that elastic deformation is very important when considering residual stresses).

Bearing in mind the large deformations which are imposed in metal–working, another question can be asked: 'Why is it that, in a tensile test, the metal will rarely withstand more than about 30 per cent elongation without fracture, whereas in metal-working processes much larger strains can be imposed?' Consideration of the major processes reveals the answer to this question – they are all predominantly compressive in nature. Any stress system can be regarded as the sum of an all–round **hydrostatic stress** (usually taken as being equal to the **mean stress**) and stresses equal to the differences between the actual stresses and this mean stress. It was shown experimentally by Bridgman, and later again by Crossland, that there is no plastic flow occasioned by this hydrostatic stress. The plastic flow is caused by the **deviatoric** stress system, so–called because the stresses take the values by which the actual stresses 'deviate' from this mean stress. Although the mean stress is not responsible for any plastic flow it has a profound influence on fracture, and the more compressive its value the more deformation can be imposed before ′failure occurs owing to fracture. The exact dependence is not properly understood, but the tests of Bridgman and Crossland have given us much information. A corollary of the observation that the mean stress occasions no plastic flow is that there can be no permanent change of volume. Thus, if elastic strains are neglected, it may reasonably be assumed that there is no change of volume <u>at all</u> during plastic flow.

Now, almost all metal-working processes take place under conditions of complex stressing. In other words, if uniaxial tension or compression is regarded as a simple system of stress–

ing, in which there is only one principal stress acting along the axial direction of the prismatic specimen, then most metal-working processes involve more complicated systems of stress. Thus, in formulating a theory, a general three-dimensional system of stress must be introduced and yet another question must be asked: 'How can the behaviour of metal subjected to a complex system of stress be correlated with its behaviour in simple tension or compression?' To answer this question it is convenient to invoke the concepts of **equivalent stress (or strain)** and **effective stress (or strain).** These are parameters which are functions of the imposed complex stresses or strains which quantify their effectiveness in causing the plastic flow of materials.

Having discussed the principles on which the theory rests, it is now possible to develop the equations of plastic flow. Before so doing, it is well to recall the equations of three-dimensional elasticity, to compare them with those for plastic flow. Initially, what may be called the **laws of elasticity** and the **laws of plasticity** can be set up, as follows:

Laws of Elasticity
    Hooke's Law
    Equations of Equilibrium (of stresses and forces)
    Equations of Compatibility (of strains and displacements)

Laws of Plasticity
    Stress-Strain Relations
    Equations of Equilibrium
    Equations of Compatibility
    Yield Criterion (function of stresses necessary to initiate and maintain plastic flow)

To these must be added the **boundary conditions** for both stresses and displacements.

Because of the magnitude of the strains and also the non-linearity of the relationships involved in metal-working processes it is necessary to formulate the equations in terms of incremental strains or strain rates. This difficulty does not arise for elastic straining, since the strains are always sufficiently small for the total strains to be used. Thus, Hooke's Law for a three-dimensional system of stress, referred to x,y,z cartesian coordinates, is as follows:

$$\epsilon_{xx} = \frac{1}{E}[\sigma_{xx} - \nu(\sigma_{yy} + \sigma_{zz})]$$

$$\epsilon_{yy} = \frac{1}{E}[\sigma_{yy} - \nu(\sigma_{zz} + \sigma_{xx})]$$

$$\epsilon_{zz} = \frac{1}{E}[\sigma_{zz} - \nu(\sigma_{xx} + \sigma_{yy})]$$

$$\gamma_{xy} = \frac{1}{G}\tau_{xy}$$

$$\gamma_{yz} = \frac{1}{G}\tau_{yz}$$

$$\gamma_{zx} = \frac{1}{G}\tau_{zx}.$$

(5.26)

In these equations, τ_{xy} is the stress in the y direction acting on a plane normal to the x direction, γ_{xy} is the corresponding engineering shear strain, E is **Young's Modulus**, G is the **shear modulus**, ν is **Poisson's ratio**, and $E = 2G(1+\nu)$. Equations (5.26) are the well-known equations of elasticity representing Hooke's law. Considering the first one of these equations, a positive (tensile) stress σ_{xx} in the x direction produces a strain of σ_{xx}/E in that direction, whilst the positive tensile stress σ_{yy} will produce a negative strain (contraction) of $\nu(\sigma_{yy}/E)$ in the x direction, and σ_{zz} similarly produces a contraction. By adding together the first three equations of equations (5.26) the elastic volume change is determined as:

$$(\epsilon_{xx} + \epsilon_{yy} + \epsilon_{zz}) = \frac{1-2\nu}{E}(\sigma_{xx} + \sigma_{yy} + \sigma_{zz}).$$

(5.27)

(The factor $E/(1-2\nu)$ is three times the **bulk modulus** K of the material, since $(\sigma_{xx} + \sigma_{yy} + \sigma_{zz})/3$ is the mean stress, or hydrostatic component of the stress system.)

Thus, if the volume change is to be zero in plastic flow, Poisson's ratio ν must be replaced by the factor $1/2$ in the stress–strain relations, since the right–hand side of equation (5.27) then becomes zero. Young's Modulus E has no meaning for plastic flow in which the elastic strains are neglected, and must be replaced by an analogous parameter which will be considered later.

In any element of material subjected to a complex system of stress there are three mutually perpendicular directions in which the local direct stresses attain either maximum or minimum values. These direct stresses act on planes on which the shear stresses are zero, and these planes are known as 'principal' planes, the direct stresses as 'principal' stresses. If these principal directions are denoted by 1, 2, and 3, equations (5.26), for example, reduce to three equations:

$$\epsilon_1 = \frac{1}{E}[\sigma_1 - \nu(\sigma_2 + \sigma_3)]$$

$$\epsilon_2 = \frac{1}{E}[\sigma_2 - \nu(\sigma_3 + \sigma_1)] \qquad (5.28)$$

$$\epsilon_3 = \frac{1}{E}[\sigma_3 - \nu(\sigma_1 + \sigma_2)].$$

5.7.1.1 *Simple Matrix Notation*

Equations (5.28) may be written in matrix notation as follows:

$$\begin{Bmatrix} \epsilon_1 \\ \epsilon_2 \\ \epsilon_3 \end{Bmatrix} = \frac{1}{E} \begin{bmatrix} 1 & -\nu & -\nu \\ -\nu & 1 & -\nu \\ -\nu & -\nu & 1 \end{bmatrix} \begin{Bmatrix} \sigma_1 \\ \sigma_2 \\ \sigma_3 \end{Bmatrix} \qquad (5.28a)$$

or $$\epsilon = \mathbf{B}\sigma. \qquad (5.28b)$$

To express the stresses in terms of the strains, we should have

$$\sigma = \mathbf{B}^{-1}\epsilon \qquad (5.28c)$$

where:

$$\mathbf{B}^{-1} = \frac{\text{adj } \mathbf{B}}{\det \mathbf{B}} = \frac{\begin{bmatrix} 1-\nu^2 & \nu(1+\nu) & \nu(1+\nu) \\ \nu(1+\nu) & 1-\nu^2 & \nu(1+\nu) \\ \nu(1+\nu) & \nu(1+\nu) & 1-\nu^2 \end{bmatrix}E}{1(1-\nu^2)-(-\nu)(-\nu-\nu^2)+(-\nu)(\nu^2+\nu)}$$

$$= \frac{\begin{bmatrix} 1-\nu^2 & \nu(1+\nu) & \nu(1+\nu) \\ & 1-\nu^2 & \nu(1+\nu) \\ \text{SYM} & & 1-\nu^2 \end{bmatrix}E}{(1-2\nu)(1+\nu)^2} = \frac{\begin{bmatrix} 1-\nu & \nu & \nu \\ & 1-\nu & \nu \\ \text{SYM} & & 1-\nu \end{bmatrix}E}{(1-2\nu)(1+\nu)}.$$

Written out in full, therefore,

$$\begin{Bmatrix} \sigma_1 \\ \sigma_2 \\ \sigma_2 \end{Bmatrix} = \begin{bmatrix} 1-\nu & \nu & \nu \\ \nu & 1-\nu & \nu \\ \nu & \nu & 1-\nu \end{bmatrix} \frac{E}{(1-2\nu)(1+\nu)} \begin{Bmatrix} \epsilon_1 \\ \epsilon_2 \\ \epsilon_3 \end{Bmatrix}. \qquad (5.28d)$$

For example:

$$\sigma_2 = \frac{E}{(1-2\nu)(1+\nu)} \{(1-\dot{\nu})\epsilon_2 + \nu(\epsilon_1 + \epsilon_3)\}.$$

It will be seen later that the stress–strain equations of plastic flow will be similar to equations (5.28), but with ν replaced by 1/2 and with strains such as ϵ_1 replaced either by the incremental strain $\delta\epsilon_1$, or by strain rate $\dot{\epsilon}_1 = d\epsilon_1/dt$, where t is time or some quantity which increases in proportion to time. The parameter $1/E$ will disappear, however, being replaced by a term which will have the same value in each of the equations (5.28). The actual value of this term, determining the relationship existing between the plastic strain and the stresses causing plastic flow will depend on both the work–hardening characteristics of the material and/or the constraints of neighbouring elements or boundaries of the element of material under consideration.

5.7.1.2 *The principal stress cubic*

To give an actual value to this term, consider in more detail the concepts of equivalent stress and equivalent strain already referred to. Firstly, considering the equivalent stress, if the material is isotropic and homogeneous, then this equivalent stress must be a function of the 'invariants' of the stress system. Without considering the mathematical theory in detail, these invariants may be described as those functions of the stresses at a point in the material which do not vary when referred to different sets of cartesian coordinate axes through the point. One of the easiest ways of deriving them is to obtain the cubic equation whose roots give the principal stresses at the point. In full, this cubic equation is as follows:

$$p^3 - p^2(\sigma_{xx} + \sigma_{yy} + \sigma_{zz}) - p(\tau_{xy}{}^2 + \tau_{yz}{}^2 + \tau_{zx}{}^2 - \sigma_{xx}\sigma_{yy} - \sigma_{yy}\sigma_{zz} - \sigma_{zz}\sigma_{xx}) -$$
$$(\sigma_{xx}\sigma_{yy}\sigma_{zz} + 2\tau_{xy}\tau_{yz}\tau_{zx} - \sigma_{xx}\tau_{yz}{}^2 - \sigma_{yy}\tau_{zx}{}^2 - \sigma_{zz}\tau_{xy}{}^2) = 0. \qquad (5.29)$$

Equation (5.29) can be found from the eigenvector equation for the stress tensor. It may be conveniently derived by matrix algebra using a 3×3 matrix definition of the stress tensor σ_{ij} (where i,j can take any of the values 1,2,3), namely:

$$\sigma = \begin{bmatrix} \sigma_{11} & \sigma_{12} & \sigma_{13} \\ \sigma_{21} & \sigma_{22} & \sigma_{23} \\ \sigma_{31} & \sigma_{32} & \sigma_{33} \end{bmatrix}.$$

If the direction cosines of one of the principal planes are denoted by a (3×1) column matrix **a**, i.e. the eigendirection:

$$\mathbf{a} = \begin{bmatrix} a_{11} \\ a_{12} \\ a_{13} \end{bmatrix},$$

for example, then the eigenvector equation is:

$$\sigma \mathbf{a} = p\mathbf{a} \tag{5.29a}$$

p being the multiplicative scalar associated with the eigen-equation, from which the eigenvalues are determined. (p_1, p_2, and p_3 associated with a_{11} a_{12} a_{13}, a_{21} a_{22} a_{23}, and a_{31} a_{32} a_{33}).

$$\text{Thus } (\sigma - \mathbf{I}p)\mathbf{a} = 0 \tag{5.29b}$$

$$\text{or} \qquad \mathbf{Ka} = 0 \tag{5.29c}$$

where **I** is the *unit matrix* and $\mathbf{K} = \sigma - \mathbf{I}p$ is the **characteristic matrix** of σ.

Equations (5.29b) are a set of **homogeneous equations** which can only have solutions for the unknown **eigendirections a** if the **determinant of the coefficients** is zero, i.e. if:

$$\det \mathbf{K} = \begin{vmatrix} \sigma_{11}-p & \sigma_{12} & \sigma_{13} \\ \sigma_{21} & \sigma_{22}-p & \sigma_{23} \\ \sigma_{31} & \sigma_{32} & \sigma_{33}-p \end{vmatrix} = 0. \tag{5.29d}$$

When expanded, this gives the cubic equation, equation (5.29), from which the principal stresses are determined and which is more conveniently expressed for the present purpose with the cartesian axes x,y,z, replaced by 1,2,3, i.e.

$$p^3 - p^2(\sigma_{11}+\sigma_{22}+\sigma_{33}) - p(\sigma_{12}{}^2+\sigma_{23}{}^2+\sigma_{31}{}^2-\sigma_{11}\sigma_{22}-\sigma_{22}\sigma_{33}-\sigma_{33}\sigma_{11})$$
$$-(\sigma_{11}\sigma_{22}\sigma_{33}+2\sigma_{12}\sigma_{23}\sigma_{31}-\sigma_{11}\sigma_{23}{}^2-\sigma_{22}\sigma_{31}{}^2-\sigma_{33}\sigma_{12}{}^2)=0 \tag{5.29e}$$

In general, for both **singular** and **non-singular** matrices (a singular matrix is one whose determinant is zero so that it cannot be inverted, such as **K** the **characteristic matrix** of σ), the following equation applies:

$$K \text{ adj } K = I \det K = 0, \text{ since } K \text{ is singular.} \qquad (5.29f)$$

Comparing equations (5.29e) and (5.29f), it may be deduced that the three elements of **a** must be proportional to the three elements of any column of adj **K**, i.e.

$$\text{adj } K = \begin{bmatrix} a_{1j} \\ a_{2j} \\ a_{3j} \end{bmatrix} =$$

$$\begin{array}{ccc} p_1\text{-direction} & p_2\text{-direction} & p_3\text{-direction} \end{array}$$

$$\begin{bmatrix} (\sigma_{22}-p_j)(\sigma_{33}-p_j)-\sigma_{23}{}^2 & -\sigma_{12}(\sigma_{33}-p_j)+\sigma_{13}\sigma_{23} & \sigma_{12}\sigma_{23}-\sigma_{13}(\sigma_{22}-p_j) \\ -\sigma_{12}(\sigma_{33}-p_j)+\sigma_{23}\sigma_{31} & (\sigma_{11}-p_j)(\sigma_{33}-p_j)-\sigma_{13}{}^2 & -\sigma_{23}(\sigma_{11}-p_j)+\sigma_{13}\sigma_{21} \\ \sigma_{12}\sigma_{23}-\sigma_{13}(\sigma_{22}-p_j) & -\sigma_{32}(\sigma_{11}-p_j)+\sigma_{12}\sigma_{31} & (\sigma_{11}-p_j)(\sigma_{22}-p_j)-\sigma_{21}{}^2 \end{bmatrix}$$

5.7.1.3 Determination of the principal stresses and their directions

The problem of solving cubic equations was apparently first studied by the Italian mathematician Cardan in the 16th century, as pointed out by Johnson, who has extensively researched the history of science and engineering. Cardan's method consists essentially of replacing the variables of the cubic equation with trigonometrical functions, thus allowing an analytical solution to be found. A detailed discussion of his method, including some computer and programmable calculator programs was given in Vol.1. The computer program set out in that reference is rather lengthy for the present purpose, because it was written for application to both stress and strain matrices. A much simpler computer program based on Cardan's method follows, namely Program 5.1, which gives the principal stresses and their direction cosines for any given stress tensor σ_{ij}.

Shown in Solution 5.1(a) and Solution 5.1(b) are two printouts from this program, the first of which relates to a set of values for a stress tensor given as an example in Vol. 1. The second is the solution for virtually the same stress state but with the sign of the shear stress σ_{23} changed from $+5.4$ to -5.4, to illustrate how greatly the principal stresses and their directions are changed, by what is apparently a quite small change in the stress state.

PROGRAM 5.1 Cardan's method

```
1:*         PROGRAM CARDAN
2:          WRITE (*,1)
3: 1        FORMAT(1X,' CARDAN DETERMINES PRINCIPAL STRESSES S1,S2,S3,',/1X,
4:       1  ' FROM THE CUBIC SJ**3 -J1*SJ**2 -J2*SJ-J3=0, ',/1X,
5:       2  ' THE GIVEN STRESS TENSOR IS S11,S22,S33,S12,S23,S31, ',/1X,
6:       3  ' DEVIATORIC STRESSES ARE D11=S11-SM ETC, D12=S12 ETC, ',/1X,
7:       4  ' SM=(S11+S22+S33)/3. NON ZERO INVARIANTS OF DEVIATORIC ',/1X,
8:.      5  ' STRESS TENSOR ARE XI2,XI3, DIRECTION COSINES OF S1 ARE ',/1X,
9:       6  ' P11,P21,P31, OF S2 P12,P22,P32, OF S3 P13,P23,P33',/1X,
10:      7  ' TYPE IN S11,S22,S33,S12,S23,S31 ')
11:         READ(*,*) S11,S22,S33,S12,S23,S31
12:         SM = (S11+S22+S33)/3.
13:         D11 = S11-SM
14:         D22 = S22-SM
15:         D33 = S33-SM
16:         XI2=S12**2 +S23**2 +S31**2 -D11*D22-D22*D33-D33*D11
17:         XI3=D11*D22*D33+2.*S12*S23*S31-D11*S23**2 -D22*S31**2 -D33*S12
18:      1  **2
19:         R = 2.*SQRT(XI2/3.)
20:         THETA = (ACOS(4.*XI3/R**3 ))/3.
21:         PI = 3.141592654
22:         X = 2.*PI/3.
23:         S1 = R*COS(THETA)+SM
24:         S2 = R*COS(THETA-X)+SM
25:         S3 = R*COS(THETA+X)+SM
26:         WRITE(*,2) S11,S22,S33,S12,S23,S31,S1,S2,S3
27: 2       FORMAT(/1X' STRESS TENSOR WAS S11=',E12.6,' S22=',E12.6,/1X,
28:      1  ' S33=',E12.6,' S12=',E12.6,' S23=',E12.6,' S31=',E12.6,/1X,/1X,
29:      2  ' PRINCIPAL STRESSES ARE S1=',E12.6,' S2=',E12.6,' S3=',E12.6,/)
30:         A11=S12*S23+S31*(S1-S22)
31:         A12=S12*S23+S31*(S2-S22)
32:         A13=S12*S23+S31*(S3-S22)
33:         A21=S12*S31+S23*(S1-S11)
34:         A22=S12*S31+S23*(S2-S11)
35:         A23=S12*S31+S23*(S3-S11)
36:         A31=(S1-S11)*(S1-S22)-S12**2
37:         A32=(S2-S11)*(S2-S22)-S12**2
38:         A33=(S3-S11)*(S3-S22)-S12**2
39:         RMS1=SQRT(A11**2 +A21**2 +A31**2)
40:         RMS2=SQRT(A12**2 +A22**2 +A32**2)
41:         RMS3=SQRT(A13**2 +A23**2 +A33**2)
42:         P11=A11/RMS1
43:         P21=A21/RMS1
44:         P31=A31/RMS1
45:         P12=A12/RMS2
46:         P22=A22/RMS2
47:         P32=A32/RMS2
48:         P13=A13/RMS3
49:         P23=A23/RMS3
50:         P33=A33/RMS3
51:         WRITE(*,3) P11,P21,P31,P12,P22,P32,P13,P23,P33
52: 3       FORMAT(1X,' PRINCIPAL STRESS DIRECTION COSINES ARE ',/1X,
53:      1  ' P11=',E12.6,' P21=',E12.6,' P31=',E12.6,/1X,
54:      2  ' P12=',E12.6,' P22=',E12.6,' P32=',E12.6,/1X,
55:      3  ' P13=',E12.6,' P23=',E12.6,' P33=',E12.6,/)
56:         END
```

SOLUTION 5.1(a)

```
CARDAN
CARDAN DETERMINES PRINCIPAL STRESSES S1,S2,S3,
FROM THE CUBIC SJ**3 -J1*SJ**2 -J2*SJ-J3=0,
THE GIVEN STRESS TENSOR IS S11,S22,S33,S12,S23,S31,
DEVIATORIC STRESSES ARE D11=S11-SM ETC, D12=S12 ETC,
SM=(S11+S22+S33)/3. NON ZERO INVARIANTS OF DEVIATORIC
STRESS TENSOR ARE XI2,XI3, DIRECTION COSINES OF S1 ARE
P11,P21,P31, OF S2 P12,P22,P32, OF S3 P13,P23,P33
TYPE IN S11,S22,S33,S12,S23,S31
    -17.3 -8.7 -6.1 12.5 5.4 3.8

STRESS TENSOR WAS S11=-.173000E+02 S22=-.870000E+01
S33=-.610000E+01 S12= .125000E+02 S23= .540000E+01 S31= .380000E+01

PRINCIPAL STRESSES ARE S1= .437934E+01 S2=-.102603E+02 S3=-.262190E+02

PRINCIPAL STRESS DIRECTION COSINES ARE
P11= .490797E+00 P21= .689152E+00 P31= .533092E+00
P12= .311493E+00 P22= .432625E+00 P32=-.846054E+00
P13= .813694E+00 P23=-.581289E+00 P33= .230501E-02
```

SOLUTION 5.1(b)

```
CARDAN DETERMINES PRINCIPAL STRESSES S1,S2,S3,
FROM THE CUBIC SJ**3 -J1*SJ**2 -J2*SJ-J3=0,
THE GIVEN STRESS TENSOR IS S11,S22,S33,S12,S23,S31,
DEVIATORIC STRESSES ARE D11=S11-SM ETC, D12=S12 ETC,
SM=(S11+S22+S33)/3. NON ZERO INVARIANTS OF DEVIATORIC
STRESS TENSOR ARE XI2,XI3, DIRECTION COSINES OF S1 ARE
P11,P21,P31, OF S2 P12,P22,P32, OF S3 P13,P23,P33
TYPE IN S11,S22,S33,S12,S23,S31
    -17.3 -8.7 -6.1 12.5 -5.4 3.8

STRESS TENSOR WAS S11=-.173000E+02 S22=-.870000E+01
S33=-.610000E+01 S12= .125000E+02 S23=-.540000E+01 S31= .380000E+01

PRINCIPAL STRESSES ARE S1= .105466E+01 S2=-.515013E+01 S3=-.280045E+02

PRINCIPAL STRESS DIRECTION COSINES ARE
P11=-.474706E+00 P21=-.805133E+00 P31= .355549E+00
P12=-.426443E+00 P22=-.142983E+00 P32=-.893142E+00
P13=-.769935E+00 P23= .575601E+00 P33= .275468E+00
```

5.7.2 Simple Tensor Notation

5.7.2.1 Some definitions

The equations of transformation of the coordinates of any spatial point from one set of cartesian coordinate axes (x_1, x_2, x_3) to another set (x'_1, x'_2, x'_3) are easily derived by projecting their individual lengths onto each of the various axes in turn, the generic direction cosine between the ith and the jth axes being defined by the tensor quantity a_{ij}, therefore:

$$x'_1 = a_{11}x_1 + a_{21}x_2 + a_{31}x_3 = \Sigma a_{i1}x_i$$

$$i = 1,2,3$$

$$x'_2 = a_{12}x_1 + a_{22}x_2 + a_{32}x_3 = \Sigma a_{i2}x_i$$

$$x'_3 = a_{13}x_1 + a_{23}x_2 + a_{33}x_3 = \Sigma a_{i3}x_i .$$

Coordinates referred to original x_1 x_2 x_3 and new x'_1 x'_2 x'_3 axes

In general, $x'_j = a_{ij}x_i$, where the repeated (or dummy) suffix (i) means summation over all three values 1,2,3; similarly, $x'_i = a_{ij}x_j$.

Dummy Suffix   (Repeated suffix).

Free Suffix   (Occurs once in a term).

The **Kronecker Delta** or **Substitution Tensor** δ_{ij}:
($\delta_{ij} = 1$ when $i = j$; $\delta_{ij} = 0$ when $i \neq j$, therefore $\delta_{ik}x_k = \delta_{i1}x_1 + \delta_{i2}x_2 + \delta_{i3}x_3$. If $i = 1$, $\delta_{1k}x_k = x_1$. Similarly $\delta_{2k}x_k = x_2$, hence $\delta_{ik}x_k = x_i$).

The <u>definition</u> of a **Vector** is $u'_j = a_{ij}u_i$

The <u>definition</u> of a **Tensor** is $\sigma'_{kl} = a_{ik}a_{jl}\sigma_{ij}$

5.7.2.2 Proof of invariance, using tensors

The proofs follow below, and are self-evident, relying mainly on the use of the Kronecker Delta.

$$\sigma'_{kk} = a_{ik}a_{jk}\sigma_{ij} = \delta_{ij}\sigma_{ij} = \sigma_{ii}$$
$$\sigma'_{ij}\sigma'_{ij} = (a_{ik}a_{jl}\sigma_{kl})(a_{im}a_{jn}\sigma_{mn}) = a_{ik}a_{im}a_{jl}a_{jn}\sigma_{kl}\sigma_{mn} =$$
$$\delta_{km}\delta_{ln}\sigma_{kl}\sigma_{mn} = \sigma_{mn}\sigma_{mn}$$

5.7.3 The Yield Criterion

The three roots of equation (5.29) (p being the unknown quantity) give the principal stresses $p_1 = \sigma_1$, $p_2 = \sigma_2$ and $p_3 = \sigma_3$ at the point considered. Now it should not matter how the original set of x,y,z axes is orientated with regard to evaluating σ_1, σ_2, and σ_3 from this equation. These principal stresses should always have the same value for the particular state of stress concerned. For this to be true the coefficients of p^3, p^2 and p^1 should always be the same, in equation (5.29). In other words, the functions of the stresses in the terms within the brackets in equation (5.29) are **invariant** with respect to rotation of the x,y,z axes in the element considered. This is a most important concept to understand, since it can then be said that any property of the **idealized isotropic homogeneous material**, such as the **criterion of yielding, criterion of fracture, criterion of creep,** or **fatigue** even, should be a function of the invariants of the stress system. The invariants are usually denoted as follows:

5.7.3.1 Invariants of the stress tensor

$$J_1 = \sigma_1 + \sigma_2 + \sigma_3 = \sigma_{xx} + \sigma_{yy} + \sigma_{zz}$$
$$J_2 = -\sigma_1\sigma_2 - \sigma_2\sigma_3 - \sigma_3\sigma_1 = \tag{5.29}$$
$$\tau_{xy}^2 + \tau_{yz}^2 + \tau_{zx}^2 - \sigma_{xx}\sigma_{yy} - \sigma_{yy}\sigma_{zz} - \sigma_{zz}\sigma_{xx}$$
$$J_3 = \sigma_1\sigma_2\sigma_3 = \sigma_{xx}\sigma_{yy}\sigma_{zz} + 2\tau_{xy}\tau_{yz}\tau_{zx} - \sigma_{xx}\tau_{yz}^2 - \sigma_{yy}\tau_{zx}^2 - \sigma_{zz}\tau_{xy}^2.$$

Thus, for this particular problem, the yield criterion and consequently the effective stress, must be a function of the invariants J_1, J_2, and J_3. The invariant J_1 is seen to be three times the arithmetic mean stress $\sigma = (\sigma_1 + \sigma_2 + \sigma_3)/3$, and, since plastic flow has been shown experimentally to be independent of the mean stress, the effective stress should not depend on J_1, at least as far as yielding is concerned. As stated before, J_1 would be important in establishing a criterion of fracture. J_2 and J_3, in contrast, involve the **deviatoric stresses,** now to be discussed, and are therefore directly related to the criterion of yielding.

5.7.3.2 *Hydrostatic and deviatoric stresses*

The easiest way of making the yield criterion independent of the mean stress is by making it a function of the shear stress components of the stress system, since these are unaffected by the magnitude of the mean stress in the system. This is most easily seen from the Mohr's circles of stress as shown in Fig. 5.18 below. No matter how the mean stress σ may be altered, the origin of the deviatoric stresses remains in the same position relative to the Mohr's circles.

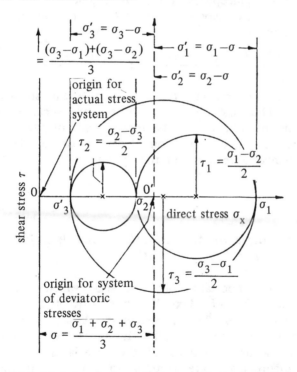

Fig. 5.18 Mohr's circles of stress, illustrating the origins of both total and deviatoric stress systems

As mentioned before, the given system of stress represented by the principal stresses σ_1, σ_2 and σ_3 may be regarded as the sum of a system of hydrostatic stresses equal to the arithmetic mean stress:

$$\sigma_m = \frac{\sigma_1 + \sigma_2 + \sigma_3}{3}$$

plus the deviatoric stresses $\sigma_1', \sigma_2', \sigma_3'$. The deviatoric stresses are so called because they represent the amount by which the actual stresses <u>deviate</u> from the mean stress. Thus:

$$\begin{aligned}
\sigma_1' &= \sigma_1 - \sigma_m \\
\sigma_2' &= \sigma_2 - \sigma_m \\
\sigma_3' &= \sigma_3 - \sigma_m.
\end{aligned} \qquad (5.31)$$

The stresses which are effective in causing plastic flow may be regarded either as functions of the deviatoric stresses (a concept preferred by mathematicians) or as functions of the shear stresses (a concept preferred by engineers). That the two concepts are identical may be realized from the fact that σ_1', for example, is equal to:

$$\frac{(\sigma_1 - \sigma_2) + (\sigma_1 - \sigma_3)}{3} \qquad \text{or} \qquad \frac{2(\tau_1 + \tau_3)}{3}$$

where τ_1, τ_2, and τ_3 are the maximum shear stresses in the stress system. Thus the deviatoric stresses are shear stresses and any yield criterion which is to be independent of the mean stress may be expected to be a function of the maximum shear stresses or principal stress differences in the system.

5.7.3.3 *Yield criteria of Maxwell (Huber–von Mises) and Tresca (Guest)*

There are two important yield criteria which are used in practice, namely that of Maxwell, often attributed to von Mises or Huber, and that of Tresca or Guest. A list of references is to be found in the book by Hill. It is easy to show that both of these criteria are functions of the invariants of the stresses and many theoreticians are content to leave it at that. It is possible to attach some physical significance to the functions involved and this is often preferred by practising engineers.

The Maxwell criterion, in its simplest form, may be expressed in terms of the equivalent stress, given by the invariant function:

$$\bar{\sigma} = \frac{1}{\sqrt{2}} \sqrt{(\sigma_1-\sigma_2)^2+(\sigma_2-\sigma_3)^2+(\sigma_3-\sigma_1)^2}. \qquad (5.32)$$

This quantity is readily shown to be proportional to either the root mean square shear stress, or the shear stress which exists on the octahedral planes (the planes whose normals make equal angles with the three principal stress directions in the element: there are, in fact, eight 'families' of such planes, each family being made up of parallel planes); or the square root of the energy of distortion. The factor $1/\sqrt{2}$ has been chosen so that for uniaxial stressing ($\sigma_2 = \sigma_3 = 0$), $\bar{\sigma}$ is equal to the applied stress σ_1. When $\bar{\sigma}$ attains a certain value (the **yield** or **flow** stress, often denoted by Y, σ_y or σ_f and usually derived experimentally), it is supposed that yielding or plastic flow will occur.

The Tresca criterion may be expressed in an analogous way, in terms of the invariant function:

$$\bar{\sigma} = \sigma_{max}-\sigma_{min}. \qquad (5.33)$$

This quantity is seen to be proportional to the maximum shear stress operating and can be seen to be inferior to the Maxwell criterion in that no 'weight' is given to the intermediate stress. In spite of this the difference between the two criteria never exceeds about $15\frac{1}{2}$ per cent in terms of the stresses and can be reduced still further by multiplying the right−hand side of equation (5.33) by a judiciously chosen factor, its value lying between 1 and 0.866 ($\sqrt{3}/2$).

The equivalent or effective strain increment may again be expressed in terms of invariant functions of the strain increments, either as:

$$\delta\bar{\epsilon} = \sqrt{\frac{2}{9}[(\delta\epsilon_1-\delta\epsilon_2)^2+(\delta\epsilon_2-\delta\epsilon_3)^2+(\delta\epsilon_3-\delta\epsilon_1)^2]} \qquad (5.34)$$

or as:

$$\delta\bar{\epsilon} = \frac{2}{3}(\delta\epsilon_{max}-\delta\epsilon_{min}). \qquad (5.35)$$

Again, the factors $\sqrt{(2/9)}$ and $2/3$ in these equations have been chosen so that for uniaxial stressing (i.e. $\delta\epsilon_2 = \delta\epsilon_3 = -\frac{1}{2}\delta\epsilon_1$, since $\delta\epsilon_1+\delta\epsilon_2+\delta\epsilon_3 = 0$) $\delta\bar{\epsilon}$ is approximately equal to the applied strain increment $\delta\epsilon_1$.

It is of interest to apply these equations to another system of complex stresses and investigate their predictions. Consider the case of pure shear stress, which is of great importance in practice, being found in manufacturing processes involving shearing such as machining, cropping and blanking. It is also of great significance in the theory of plane plastic deformation, which is essentially a process of pure shear. Then the relationship between the principal stresses will be $\sigma_2 = -\sigma_1$, $\sigma_3 = 0$, and between the principal strains $\delta\epsilon_2 = -\delta\epsilon_1$, $\delta\epsilon_3 = 0$.

Substituting these values in equations (5.32) and (5.34) for the Maxwell criterion the effective stress and strain are found to be $\bar{\sigma}_M = \sqrt{3}\sigma_1$ and $\delta\bar{\epsilon}_M = (2/\sqrt{3})\delta\epsilon_1$. Similarly, by substituting in equations (5.33) and (5.35) for the Tresca criterion gives $\bar{\sigma}_T = 2\sigma_1$ and $\delta\bar{\epsilon}_T = (4/3)\delta\epsilon_1$.

Thus it is seen that, although the two criteria predict the same effective stress and strain for the case of simple tension, they predict quite different values for the case of pure shear. Furthermore the ratios of the respective predictions for effective stress and effective strain are:

$$\frac{\bar{\sigma}_T}{\bar{\sigma}_M} = \frac{\delta\bar{\epsilon}_T}{\delta\bar{\epsilon}_M} = \frac{2}{\sqrt{3}} \simeq 1.155.$$

In other words, to make the Tresca criterion correspond with the Maxwell criterion in pure shear it is necessary to multiply both the effective stress [equation (5.33)] and effective strain [equation (5.35)] by the factor $\sqrt{3}/2$. Now why should it be desirable to make the Tresca criterion correspond to Maxwell's and not vice versa? In other words, which is the more realistic criterion of plastic flow?

A partial answer to this question has already been given, in that the Tresca criterion clearly gives no weight to the intermediate stress. An even more striking argument may be found if the work done during plastic flow is considered, since this quantity can be used to test the concepts of effective stress and strain which have been chosen. For it is obvious that if the values specified are in fact the effective stress and effective strain, their product should represent the work done per unit volume in the complex stress system.

Now, this work done will be given by the equation:

$$\delta W = \sigma_1\delta\epsilon_1 + \sigma_2\delta\epsilon_2 + \sigma_3\delta\epsilon_3. \tag{5.36}$$

For uniaxial stressing the last two terms are zero since σ_2 = σ_3 = 0 and the work done per unit volume is simply $\sigma_1 \delta \epsilon_1$. Since the factors in equations (5.32) to (5.35) for both the Maxwell and Tresca criteria have been chosen to make the effective stress and strain correspond with the values of uniaxial stressing, either criterion will give the same result, namely $\delta W = \sigma \delta \epsilon = \sigma_1 \delta \epsilon_1$.

Considering the case of pure shear:

$$\delta W = \sigma_1 \delta \epsilon_1 + \sigma_2 \delta \epsilon_2 + \sigma_3 \delta \epsilon_3 = \sigma_1 \delta \epsilon_1 + (-\sigma_1)(-\delta \epsilon_1) + 0 = 2\sigma_1 \delta \epsilon_1$$

For the Maxwell criterion the product of effective stress and effective strain gives the result: $\bar{\sigma}_M \delta \bar{\epsilon}_M = \sqrt{3}\sigma_1 \times (2/\sqrt{3})\delta \epsilon_1 = 2\sigma_1 \delta \epsilon_1$, which _does_ in fact represent the work done per unit volume. For the Tresca criterion: $\bar{\sigma}_T \delta \bar{\epsilon}_T = 2\sigma_1 \times (4/3)\delta \epsilon_1 = (2+2/3)\sigma_1 \delta \epsilon_1$, which is _2/3 more_ than the correct result.

In extensive experiments carried out long ago by Taylor and Quinney, Hohenemser, Morrison and Shepherd, and many others, it has been found that the Maxwell criterion does, in general, represent the yielding behaviour of most ductile materials. An extensive discussion of all these tests can be found in the book by Hill.

Summarizing then, the Maxwell criterion is the more acceptable criterion for the isotropic homogeneous material of the theory but the Tresca criterion is often easier to apply in practice because of its simpler mathematical form. Consideration of the difference between the predictions of the two criteria for pure shear (which gives the biggest difference to be found) shows that for states of complex stress approximating to pure shear, the Tresca criterion can be made to approximate to Maxwell's by the use of an appropriate factor, applied to both the effective stress and the effective strain.

5.7.3.4 Stress–strain relations for plastic flow

To return to the question of determining the stress–strain relations in plastic flow, the following relations may be written, by analogy with the elastic equations (5.28):

$$\delta \epsilon_1 = \delta \lambda [\sigma_1 - \tfrac{1}{2}(\sigma_2 + \sigma_3)]$$
$$\delta \epsilon_2 = \delta \lambda [\sigma_2 - \tfrac{1}{2}(\sigma_3 + \sigma_1)] \qquad (5.37)$$
$$\delta \epsilon_3 = \delta \lambda [\sigma_3 - \tfrac{1}{2}(\sigma_1 + \sigma_2)].$$

The assumption of a linear multiplier $\delta\lambda$ was originally proposed by Reuss.

Equations (5.37) may be written in matrix notation as follows:

$$\begin{Bmatrix} \delta\epsilon_1 \\ \delta\epsilon_2 \\ \delta\epsilon_3 \end{Bmatrix} = \delta\lambda \begin{bmatrix} 1 & -\frac{1}{2} & -\frac{1}{2} \\ -\frac{1}{2} & 1 & -\frac{1}{2} \\ -\frac{1}{2} & -\frac{1}{2} & 1 \end{bmatrix} \begin{Bmatrix} \sigma_1 \\ \sigma_2 \\ \sigma_3 \end{Bmatrix} \qquad (5.37a)$$

by analogy with equations (5.28a), or:

$$\delta\epsilon = \mathbf{D}\sigma. \qquad (5.37b)$$

It might be thought possible to express the stresses in terms of the strain increments by inverting this equation, thus:

where:
$$\sigma = \mathbf{D}^{-1}\delta\epsilon \qquad (5.37c)$$

$$\mathbf{D}^{-1} = \frac{\text{adj } \mathbf{D}}{\det \mathbf{D}}.$$

Referring back to equation (5.28b), it may be noted that $\mathbf{D}$ is simply $\mathbf{B}$ with $\nu = \frac{1}{2}$, and $E = 1/\delta\lambda$. However,

$$\mathbf{D}^{-1} = \frac{\begin{bmatrix} 1-\frac{1}{2} & \frac{1}{2} & \frac{1}{2} \\ & 1-\frac{1}{2} & \frac{1}{2} \\ \text{SYM} & & 1-\frac{1}{2} \end{bmatrix}}{\delta\lambda(1-1)(1+\frac{1}{2})} \rightarrow \frac{0}{0} \text{ (Indeterminate)}.$$

For example:

$$\sigma_2 = \frac{1}{\delta\lambda(1-1)(1+\frac{1}{2})}\{\tfrac{1}{2}\delta\epsilon_1 + (1-\tfrac{1}{2})\delta\epsilon_2 + \tfrac{1}{2}(\delta\epsilon_3)\} \rightarrow \frac{0}{0},$$

$$(\text{since } \delta\epsilon_3 = -\delta\epsilon_1 - \delta\epsilon_2).$$

It may be concluded that $\mathbf{D}$ cannot be inverted because it is a singular matrix whose determinant is zero. In fact,

$$\det \mathbf{D} = 1(1-\tfrac{1}{4})-(-\tfrac{1}{2})(-\tfrac{1}{2}-\tfrac{1}{4})-\tfrac{1}{2}(\tfrac{1}{4}+\tfrac{1}{2}) = \tfrac{3}{4}-\tfrac{3}{8}-\tfrac{3}{8} = 0.$$

As stated before, $\delta\lambda$ is assumed to be a constant for the particular element and particular moment in time considered, and is analogous to $1/E$ in the elastic equations, the factor $\frac{1}{2}$ being analogous to ν. By substituting the values for the strain increments given by equations (5.37) in equation (5.34), which may be written in a simpler form (using the fact that $\delta\epsilon_1+\delta\epsilon_2+\delta\epsilon_3 = 0$) as:

$$\delta\overline{\epsilon} = \sqrt{\frac{2}{3}[\delta\epsilon_1{}^2+\delta\epsilon_2{}^2+\delta\epsilon_3{}^2]} \qquad (5.34a)$$

it is easily shown that:

$$\delta\overline{\epsilon} = \delta\lambda\overline{\sigma}. \qquad (5.37d)$$

Thus $\delta\lambda = \delta\overline{\epsilon}/\overline{\sigma}$, and equations (5.37) may be written as follows:

$$\left.\begin{array}{l}\delta\epsilon_1 = \dfrac{\delta\overline{\epsilon}}{\overline{\sigma}} \, [\sigma_1-\tfrac{1}{2}(\sigma_2+\sigma_3)] \\[2mm] \delta\epsilon_2 = \dfrac{\delta\overline{\epsilon}}{\overline{\sigma}} \, [\sigma_2-\tfrac{1}{2}(\sigma_3+\sigma_1)] \\[2mm] \delta\epsilon_3 = \dfrac{\delta\overline{\epsilon}}{\overline{\sigma}} \, [\sigma_3-\tfrac{1}{2}(\sigma_1+\sigma_2)]\end{array}\right\}. \qquad (5.38)$$

It can be seen from this that the parameter in the stress–strain relations of plastic flow which is analogous to Young's Modulus in elasticity is the quantity $\overline{\sigma}/\delta\overline{\epsilon}$. The analogy is now complete in that Poisson's ratio ν is replaced by the factor $\tfrac{1}{2}$ required for zero volume change and Young's Modulus E is replaced by the ratio of the equivalent stress divided by the equivalent strain increment, namely $\overline{\sigma}/\delta\overline{\epsilon}$.

If the material work–hardens the effective stress and effective strain increment will be interdependent, through the basic stress–strain curve of the material. This may be written as $\overline{\sigma} = H(\int d\overline{\epsilon})$ where $\overline{\sigma}$ and $\int d\overline{\epsilon}$ are simply the true stress and integrated plastic strain increment measured in a tension or compression test. The slope of this plastic stress–strain curve will be $H' = d\overline{\sigma}/d\overline{\epsilon}$, so that $\delta\overline{\epsilon}$ in equations (5.38) may be replaced by the quantity $\delta\overline{\sigma}/H'$. Clearly in many cases it will not be possible to solve these equations analytically, and a step–by–step method of solution must be adopted. If the material does not work–harden the effective strain increment is not so important, and the factor $\delta\overline{\epsilon}/\overline{\sigma} = \delta\lambda$ is generally eliminated between equations (5.38) by division. Such problems can often be solved analytically for work–hardening materials by using a constant mean value of the yield stress, Y or Y_m.

5.7.3.5 *Incremental and large strains*

It is now possible to consider what is meant by strain when considering the case of large deformations. To fix ideas, consider the uniaxial stressing of a solid round rod in tension. If the current length of the specimen is l, and there is imposed on it an increment of plastic strain $\delta \epsilon_1$ in the axial direction (still neglecting elastic strains), then $\delta \epsilon_1 = \delta l/l$, where δl is the small increase of length of the specimen. The total strain in a large extension of the specimen can be defined as:

$$\epsilon_1 = \int_0^{\epsilon_1} d\epsilon_1$$

and if l_0 is the original length of the specimen, l_1 the final length, then:

$$\epsilon_1 = \int_0^{\epsilon_1} d\epsilon_1 = \int_{l_0}^{l_1} \frac{dl}{l} = \ln \frac{l_1}{l_0}. \qquad (5.39)$$

This is the well-known **logarithmic or true strain**. If, as in traditional engineering handbooks, the symbol $e_1 = (l_1 - l_0)/l_0$ is used ($100e_1$ = per cent extension), then $e_1 = (l_1/l_0)-1$ and:

$$\epsilon_1 = \ln(1+e_1). \qquad (5.40)$$

Similarly:

$$\epsilon_2 = \int_0^{\epsilon_2} d\epsilon_2 = \int_{D_0}^{D_1} \frac{dD}{D}$$

where D is the current diameter of the specimen, etc. Thus:

$$\epsilon_2 = \epsilon_3 = \ln (D_1/D_0). \qquad (5.41)$$

Now since there is no volume change, $d\epsilon_2 = d\epsilon_3 = -\frac{1}{2}d\epsilon_1$, and by integration therefore, $\epsilon_2 = \epsilon_3 = -\frac{1}{2}\epsilon_1$. From equations (5.39) and (5.41) it is seen that:

$$\ln \frac{D_1}{D_0} = -\frac{1}{2} \ln \frac{l_1}{l_0} \quad \text{or} \quad \frac{D_1}{D_0} = \sqrt{\frac{l_0}{l_1}}$$

or:

$$l_1 D_1^2 = l_0 D_0^2 \qquad (5.42)$$

as is required for constancy of volume. Thus, if true strains are used, the equation of constancy of volume for incremental strains, $\delta \epsilon_1 + \delta \epsilon_2 + \delta \epsilon_3 = 0$, may be extended to large strains, as:
$$\epsilon_1 + \epsilon_2 + \epsilon_3 = 0.$$

Considering again the meaning to be ascribed to large strains of the order encountered in metal–working, it is clear that extending a prismatic specimen in uniaxial tension to double its length would impose an equal and opposite strain to that obtained by compressing a similar specimen to half its length. For the tension specimen:

$$\epsilon_1 = \int_{l_0}^{2l_0} \frac{dl}{l} = \ln 2$$

and for the compression specimen, where h is the current length:

$$\epsilon_1 = \int_{h_0}^{\frac{1}{2}h_0} \frac{dh}{h} = \ln \frac{1}{2} = -\ln 2.$$

Thus the logarithmic strain is a true measure of the strain. Alternatively, these logarithmic strains could have been obtained from equation (5.42), since $e_1 = 1$ for the tension specimen subjected to 100 per cent extension, and $-\frac{1}{2}$ for the compression specimen subjected to 50 per cent compression.

Clearly in any system of complex stress, to obtain a measure of the total amount of plastic straining suffered by any element of the material it is necessary to add up all the increments of effective strain which the element has received, and the total effective strain may be written as:

$$\bar{\epsilon} = \int_0^{\bar{\epsilon}} d\bar{\epsilon}. \tag{5.43}$$

{If strain <u>rates</u> are used, equation (5.34), for example, would be written as:

$$\dot{\bar{\epsilon}} = \sqrt{\frac{2}{9}[(\dot{\epsilon}_1 - \dot{\epsilon}_2)^2 + (\dot{\epsilon}_2 - \dot{\epsilon}_3)^2 + (\dot{\epsilon}_3 - \dot{\epsilon}_1)^2]} \tag{5.34a}$$

and equations (5.38) as:

$$\dot{\epsilon}_1 = \frac{\dot{\bar{\epsilon}}}{\bar{\sigma}}[\sigma_1 - (\sigma_2 + \sigma_3)] \quad \text{etc.} \tag{5.38a}$$

It may or may not be possible to express the integral of equation (5.43) in terms of logarithmic functions of the dis–placements. For simple geometrical arrangements in which the principal axes remain fixed in direction in the elements it usually is possible. Uniaxial stressing has already been discussed. For axial symmetry the incremental circumferential strain $\delta \epsilon_\theta$ is related to the incremental radial displacement δr by the equation $\delta \epsilon_\theta = \delta r/r$. This may be integrated to give the total circumferential strain as $\epsilon_\theta = \ln(r/r_0)$ where r is the current radius, r_0 the original radius to the element concerned.

5.7.4 The Prandtl–Reuss Equations

These then, are the basic concepts underlying the theory of plasticity. It should be re-emphasized here that the theory considered above applies only to a material in which elastic strains are neglected and there is zero volume change. This is permissible if it is desired only to estimate the deforming forces and displacements during plastic flow. If a more detailed picture of the deformation is required, as for example, an estimate of the residual stresses developed after some particular metal-working process, it would be necessary to include elastic strains. The problem then becomes more complicated, although the basic equations are easy enough to establish. In effect, the total strain increment is now simply the elastic strain increment plus the plastic strain increment and the basic stress–strain relations (due to Prandtl and Reuss) may be stated as follows:

$$\left.\begin{aligned}
\delta \epsilon_1 &= \frac{\delta \bar{\epsilon}}{\bar{\sigma}}[\sigma_1 - \tfrac{1}{2}(\sigma_2 + \sigma_3)] + \frac{1}{E}[\delta \sigma_1 - \nu(\delta \sigma_2 + \delta \sigma_3)] \\
\delta \epsilon_2 &= \frac{\delta \bar{\epsilon}}{\bar{\sigma}}[\sigma_2 - \tfrac{1}{2}(\sigma_3 + \sigma_1)] + \frac{1}{E}[\delta \sigma_2 - \nu(\delta \sigma_3 + \delta \sigma_1)] \\
\delta \epsilon_3 &= \frac{\delta \bar{\epsilon}}{\bar{\sigma}}[\sigma_3 - \tfrac{1}{2}(\sigma_1 + \sigma_2)] + \frac{1}{E}[\delta \sigma_3 - \nu(\delta \sigma_1 + \delta \sigma_2)]
\end{aligned}\right\} . (5.44)$$

It should be noted that the plastic strain increments such as:

$$\delta \epsilon_1{}^P = \frac{\delta \bar{\epsilon}}{\bar{\sigma}}[\sigma_1 - \tfrac{1}{2}(\sigma_2 + \sigma_3)]$$

are associated with the **total stresses,** whilst the elastic strain increments such as:

$$\delta \epsilon_1{}^e = \frac{1}{E}[\delta \sigma_1 - \nu(\delta \sigma_2 + \delta \sigma_3)]$$

are associated with **incremental changes of the stresses.** These concepts are in accord with experimental observation, but contain the inherent assumption that elastic and plastic strains (or stresses) are separable.

5.8 PLANE STRAIN − THE SLIP−LINE FIELD

In all the preceding discussion the plastic deformation of an element of material has been assumed to occur on a microscopic scale, as though it were a single fibre disconnected from the rest of the material. In any working process the geometrical configuration of the tools, frictional condition at the boundaries, and the constraint of neighbouring material will result in the plastically deforming material being subjected to a 'field' of stress and strain which differs from point to point in the region of plastic deformation. The existence of this 'field' of differing stress−strain states enormously complicates the task of obtaining a solution, and it is only by making drastic assumptions that any information can be obtained in this way. Modern finite element analysis can provide detailed predictions, as discussed later, but it requires large computing power which can be costly.

For the purpose of estimating the deforming loads it is often adequate in some processes such as strip rolling to neglect the variation in distortion over the deforming cross−section, and assume that **plane sections remain plane.** For example, in the theory of wire drawing, it can be assumed that the distortion is not as observed in practice and shown diagrammatically in Fig. 5.19(a), but as in Fig. 5.19(b). In other words the original flat plane transverse cross−sections such as X in Fig. 5.19(b) remain flat after the distortion although spaced further apart as shown by the lines marked Y. By making such assumptions it is possible to allow for the effects of friction at the material−tool interfaces and of work−hardening of the material passing through the process, and thereby to obtain approximate estimates of the deforming loads required. These can be made more accurate by allowing for the **redundant deformation.** For example, an allowance for the redundant internal deformation can often be made by multiplying the result by a factor φ based on slip−line field theory or experimental data as discussed by Rowe.

To obtain information about the severity of the distortion, or of the effect of the process on the material, it is necessary

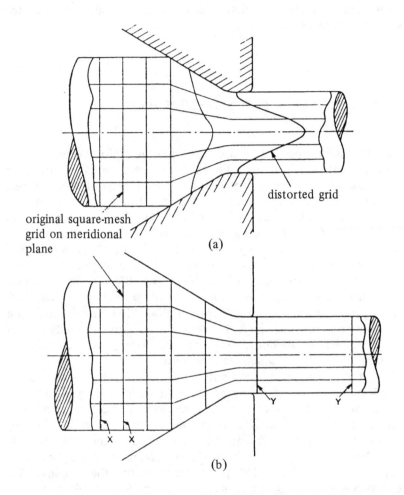

original square-mesh
grid on meridional
plane

distorted grid

(a)

(b)

Fig. 5.19 Wire drawing:
(a) Diagrammatic illustration of exaggerated distortion of square
mesh grid on meridional plane
(b) Diagrammatic illustration of distortion assumed in 'plane
sections remain plane' types of theory

to develop a **field type of theory,** which will give a picture of
the deformation from point to point in the material. Up to the
present the only rigorous means of developing such a field
theory has been by making rather sweeping assumptions about
the material and the way in which it is deformed.

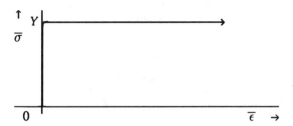

Fig. 5.20 Stress–strain curve of 'plastic–rigid' material

In addition to neglecting elastic strains (equivalent to assuming that the material has an infinite Young's Modulus), work–hardening is neglected, so that the material is assumed to flow under a constant yield stress, its stress–strain curve being as shown in Fig. 5.20. This hypothetical material is often referred to as a 'plastic–rigid' material, since it is either being plastically deformed, or it is rigid in regions where the stresses are below those necessary to cause plastic flow. The constant yield strength is generally denoted by Y. In addition to these assumptions about the material, it is also assumed that deformation takes place under conditions such that the strain is zero in one direction (i.e. under 'plane strain' conditions), because the theory can deal accurately only with two–dimensional flow.

Although at first sight these assumptions appear rather restrictive, in fact they are not, since many materials and processes approximate to these conditions. Since the strains are generally large it is permissible to neglect the small elastic strain, and the constant flow stress is closely representative of the behaviour of all metals during hot working and fairly well of work–hardened metals in cold working. Plane strain conditions are realized approximately in hot and cold strip rolling and machining and closely approximated in the extrusion of rectangular bar, ironing of cups, and other similar processes such as tube drawing, for example. The reason that the assumptions have to be made is to produce a workable 'field' theory, so it is fortunate that these assumptions are not too unrealistic.

Considering equations (5.37) (the stress–strain relations for the plastic–rigid material), if the principal strain increment $\delta\epsilon_2$ is zero, the principal stress σ_2 in that direction is seen to be the mean of the other two principal stresses, and therefore equal to the mean stress $\sigma_m = (\sigma_1 + \sigma_2 + \sigma_3)/3$. The Mohr's circles for the states of stress and strain rate are therefore as shown in Fig.

time), then $\dot{\epsilon}_3 = -\dot{\epsilon}_1$ in order that the condition of no volume change ($\epsilon_1 + \epsilon_2 + \epsilon_3 = 0$) be satisfied.

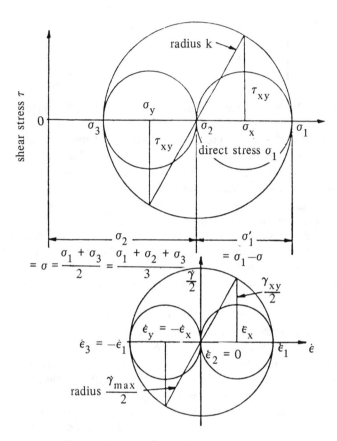

Fig. 5.21 Mohr's circles of stress and strain-rate under plane strain conditions for plastic-rigid material

The maximum shear stress in the plane of flow is $k = (\sigma_1 - \sigma_3)/2$ and if σ_x, σ_y and τ_{xy} are the direct and shear stresses acting at a general point in the plane of flow, it can be seen from the Mohr's circle that:

$$\left[\frac{(\sigma_x - \sigma_y)^2}{4} + \tau_{xy}^2\right]^{\frac{1}{2}} = k. \tag{5.45}$$

Applying the Maxwell criterion of flow, we find from equation (5.32) that the effective stress is:

$$\overline{\sigma} = \frac{\sqrt{3}}{2}(\sigma_1 - \sigma_3) = \sqrt{3}k. \qquad (5.46)$$

Thus, if the stresses increase until the radius k of Mohr's circle attains $Y/\sqrt{3}$, yielding will then occur.

Applying the Tresca criterion of flow, equation (5.33) shows that the effective stress is:

$$\overline{\sigma} = \sigma_1 - \sigma_3 = 2k \qquad (2.22)$$

and yielding will occur when the radius of the Mohr's circle attains the value $Y/2$.

Put in another way, there is no difference between the functional relationship between the stresses representing either the Maxwell or the Tresca criterion under plane strain conditions because the deformation occurs in pure shear. This can be summarized by using equation (5.45) for the yield criterion, noting again that yielding will occur when:

$$\left[\frac{(\sigma_x - \sigma_y)^2}{4} + \tau_{xy}^2\right]^{\frac{1}{2}} = k$$

where $k = Y/\sqrt{3}$ for Maxwell (von Mises), $Y/2$ for Tresca.

Referring again to the Mohr's circles of stress and strain rate (Fig. 5.21), it is clear that the direct stress acting on the plane which has the maximum shear stress k acting on it is the mean stress σ_m. Moreover, this plane will have associated with it the maximum shearing strain and zero direct strain. In fact, in the deforming material the planes of maximum shear form an orthogonal curvilinear network made up of elements such as that shown in Fig. 5.22. Each element is subjected to the maximum shear stress k and the mean normal stress σ on its faces. The directions of the principal stresses σ_1 and σ_3 at the point A, say, are at $45°$ to the lines of maximum shear stress (shown as α and β lines in the figure) and are as indicated. The third principal stress σ_2 acts normal to the plane, in the direction of zero strain.

Now why are the lines of maximum shear being considered and not the principal planes? Since the stress system at any point in the deforming material is made up of the sum of the two systems $(\sigma,\sigma,\sigma) + (k,0,-k)$ and the hydrostatic stress system causes no plastic flow, the deformation may be regarded as that of a simple pure shear as stated before. Thus the planes of maximum shear are the important unknowns in any problem; it can be shown that they are the characteristics of the equations governing the flow of the material. Furthermore, since there can be no volumetric change in the hypothetical plastic–rigid material, the hydrostatic component of the stress system can cause no dimensional change at all, and the distortion consists of a network of simple shear. The element shown in Fig. 2.5, for example, might be sheared into the shape shown diagrammatically by the broken lines.

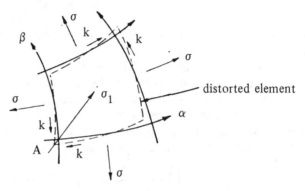

Fig. 5.22 Curvilinear element in the slip–line field

The orthogonal network of characteristics or lines of maximum shear has come to be called the **slip–line field**. This is considered by some to be rather an undesirable terminology, since lines of slip have a special meaning for the metallurgist, and **shear–line field** is a better term, as often used by Prager. Nevertheless, metallurgical slip tends to occur in directions of maximum resolved shear stress. In a polycrystalline material the **slip lines** are reliably indicative of the deformation. It can be demonstrated by considering the equilibrium of the forces on the curvilinear element shown in Fig. 5.23, (it will be noticed that the shear stress has been allowed to vary in the diagram) that:

$$\left.\begin{array}{l}\dfrac{\partial \sigma}{\partial \alpha} + \dfrac{\partial k}{\partial \beta} - \dfrac{2k}{r_\alpha} = 0 \ \ (\text{along an } \alpha\text{-line}) \\[4mm] \dfrac{\partial \sigma}{\partial \beta} + \dfrac{\partial k}{\partial \alpha} + \dfrac{2k}{r_\beta} = 0 \ \ (\text{along a } \beta\text{-line})\end{array}\right\} . \quad (2.23)$$

If φ is the angular displacement of a characteristic from some datum and is positive *anti–clockwise*, as shown, then the α and β lines in the diagram have positive curvature, since in

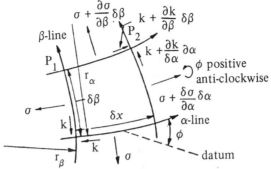

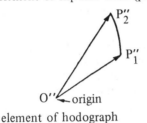

element of slip line field (physical plane)

element of hodograph

Fig. 5.23 Element of physical plane (slip–line field)
and hodograph (velocity diagram)

moving along the lines in the positive direction, the angle φ
increases. Thus the respective curvatures are $1/r_\alpha = \partial\varphi/\partial\alpha$ and
$1/r_\beta = \partial\varphi/\partial\beta$ and equations (5.48) become, after integration,

$$\left.\begin{array}{l} \sigma - 2k\varphi + \int\dfrac{\partial k}{\partial\beta}d\alpha + \int 2\varphi\dfrac{\partial k}{\partial\alpha}d\alpha = \text{Constant}_1, \quad \text{along an } \alpha\text{-line} \\[2ex] \sigma + 2k\varphi + \int\dfrac{\partial k}{\partial\alpha}d\beta + \int 2\varphi\dfrac{\partial k}{\partial\beta}d\beta = \text{Constant}_2, \quad \text{along a } \beta\text{-line} \end{array}\right\}.(5.49)$$

These equations, without the integrals and with σ replaced
by $-p$, (p being the pressure on a slip–line), are known as
Hencky's Equations. The additional integral terms were first
suggested by Christopherson, Oxley and Palmer, who found it
necessary to allow the material to work–harden, in order to
obtain a solution to the machining problem.

Hencky's equations are as follows:

$$\left.\begin{array}{l} \delta p + 2k\delta\varphi = 0, \quad \text{along an } \alpha\text{-line} \\ \delta p - 2k\delta\varphi = 0, \quad \text{along a } \beta\text{-line} \end{array}\right\}.(2.24a)$$

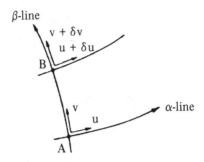

Fig. 5.24 Velocities of neighbouring points in slip–line field

Analogous equations for the velocities were derived by Geiringer (see Hill) from the condition of zero extension along the shear lines. If the velocity of the point A in Fig. 5.24 is u along the α–line and v along the β–line, and $u+\delta u$, $v+\delta v$ at point B, in order that there should be no extension of AB, the projections of the velocities of each point in the v direction must be equal. Thus, if $\delta\varphi$ is the small angular change in the β–line between points A and B, then $(v+\delta v)\cos\delta\varphi + (u+\delta u)\sin\delta\varphi = v$, or $\delta v+u\delta\varphi = 0$, along the β–line, and $\delta u-v\delta\varphi = 0$, along the α–line. Summarizing, the Geiringer equations are:

$$\left.\begin{array}{l} \dfrac{\partial u}{\partial\varphi} - v = 0 \text{ along an } \alpha\text{-line} \\[2mm] \dfrac{\partial v}{\partial\varphi} + u = 0 \text{ along a } \beta\text{-line} \end{array}\right\} \quad (5.50)$$

and are simply a statement of the fact that there is zero extension along the slip–line.

The theory of the slip–line field has existed for many years, and major advances have been made, notably the realization by Prager and Green, independently, that the distortion could be 'mapped' as a velocity diagram; followed by a geometrical method of construction developed by Prager, who more realistically called the velocity diagram the **hodograph**; and later the inclusion of strain–hardening by Christopherson et al.

Prager's geometrical method allows the solution of problems (involving mixed boundary conditions) hitherto intractable by numerical methods, and also gives a convenient representation of the solution to any problem.

5.9 VELOCITY DISCONTINUITIES

One important consequence of the assumption of a plastic–rigid material is that the displacements need not be continuous through the material. In other words, it is possible for there to be **block slipping** between neighbouring sectors in the deforming material. Thus if the curve C illustrated in Fig. 5.25 is a line along which such a discontinuity occurs let it be assumed that at any point P on the curve, the material on one side of C in the region A has velocity with components v and u normal and tangential to the curve respectively, and on the other side of the region B it has velocity with corresponding components v' and u'.

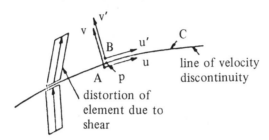

Fig. 5.25 Velocity discontinuity

It is easily seen that if the material is neither to 'pile up' nor form a cavity at the point P, $v' = v$. There is no such restriction in the tangential components, and it is possible for there to be a discontinuity of amount $u'-u$. Furthermore, since such a discontinuity must correspond with an infinitely large rate of shearing strain, the line of velocity discontinuity must also be a slip–line, and Geiringer's appropriate equation for zero extension must be satisfied along both sides of it. That the discontinuity corresponds to an infinite rate of shearing strain can be realized by considering the distortion of an element approaching the discontinuity, as shown in the same figure. It will suffer a finite shear strain in an infinitesimal interval of time, i.e. $\delta u = v \, \delta\varphi$ in region A and $\delta u' = v' \, \delta\varphi$ in region B, in moving along an elemental length of the slip–line. Since $v'=v$, and $\delta\varphi$ is common to both sides of the curve, $\delta u'=\delta u$, and the discontinuity in tangential velocity $u'-u$ must be constant along the slip-line. In a real metal, of course, such discontinuities are not possible and the velocity discontinuity may be imagined as the limiting state of a band or layer of very intense shear strain.

5.10 STRESS DISCONTINUITIES

Discontinuities in the stress components are also possible. Assume that the line C in Fig. 5.26 is such a discontinuity.

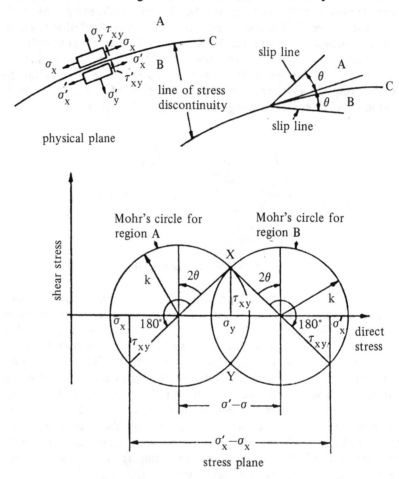

Fig. 5.26 Stress discontinuity

Consider the equilibrium of the two similar adjacent elements shown, one on each side of the line. Clearly $\sigma'_y = \sigma_y$, $\tau'_{xy} = \tau_{xy}$, but it is not necessary for σ'_x to be equal to σ_x. In the figure are also illustrated the Mohr's circles for the two states of stress in regions A and B. It is not necessary for the material to be deforming plastically for a stress discontinuity to exist, although if it is plastic, each of the radii of the two circles then equals the maximum shear stress k.

It is apparent from this that a line of stress discontinuity can <u>never</u> coincide with a slip–line, since the only way in which the two equal circles can intersect at their top (or bottom) points is by being coincident. One of the most obvious examples of a stress discontinuity occurs at a free surface such as that of a bent beam.

It is easy to show that the slip–lines are 'reflected' at a stress discontinuity as illustrated in the figure, since the radius of each Mohr's circle must be rotated through the same angle 2θ to concide with the maximum shear stress, where θ is the angle between the slip–line and the stress discontinuity. Moreover, the change in mean stress between corresponding slip–lines in the A and B regions, $\sigma'-\sigma$, is seen to be:

$$\sigma'-\sigma = 2k \sin 2\theta. \qquad (5.51)$$

(If the discontinuity is very small, $\sigma'-\sigma \rightarrow \delta\sigma$ and $2\theta \rightarrow \delta\varphi$, and Hencky's equation $\delta\sigma-2k\,\delta\varphi = 0$ is recovered.)

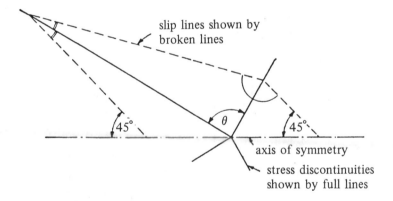

slip lines shown by broken lines

θ

45°   45°

axis of symmetry

stress discontinuities shown by full lines

Fig. 5.27 Intersection of stress discontinuities
at axis of symmetry

Since slip–lines are reflected at a stress discontinuity, it is easy to prove that, if a stress discontinuity meets an axis of symmetry (which must be a principal plane), it is effectively 'reflected' at 90°, as shown in Fig. 5.27. Thus the point of incidence becomes the meeting–point of four lines of stress dis-continuity, since it lies on an axis of symmetry. It is left as an

exercise for the reader to prove that, in order that the shear
lines should cross the axis of symmetry at 45° and also be
reflected at the stress discontinuities, it is not possible for the
angle θ in Fig. 5.27 to have any value other than 90°.

In the real material, the plastic stress discontinuity may be
envisaged as the last vestige of an elastic layer coinciding with
it. For, in moving from region A to region B of Fig. 5.26 (it
being imagined that the line C is replaced by a band), equilibrium conditions necessitate that the Mohr's circles for states of
stress in the layer should pass through both points X and Y in
that figure. Since these circles must lie between those shown in
the figure, they must all have radii less than k, implying that
the material within the layer is elastic.

5.11 PRAGER'S GEOMETRICAL METHOD

Prager's method of geometrically constructing slip–line fields   will
now be described briefly. Before considering this, however, it
may be helpful to remind the reader about the concept of the
pole of the Mohr's circle as used by Prager and explained at
length in Vol.1.

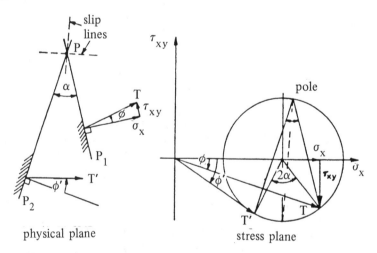

physical plane   stress plane

Fig. 5.28 Illustrating the 'pole' of Mohr's circle

In Fig. 5.28 is shown a generic point P in a field of plastic
stress, and the components of stress σ_x and τ_{xy} giving a total
traction of T in the plane PP_1, are assumed to be known. Thus
it is possible to draw the Mohr's circle, as shown in the figure,
since the radius of the circle is known to be equal to the shear

yield stress k. It is necessary to establish a convention for the sign of shear stresses, and Prager states a suitable convention as follows: 'The traction T acts in the physical plane at an angle φ to the normal to the plane considered, where φ is equal and opposite to the angle made by the traction T and the positive σ_x axis in the stress plane'. An alternative, simpler convention, is that a positive shear stress tries to cause clockwise rotation of the element on which it acts. (This is actually opposite to the convention usually adopted in three-dimensional stress analysis.)

If it is now desired to know the stresses on another plane PP_2 through P, at an angle α to the first plane, then it is well known that the radius of the Mohr's circle is simply rotated through a corresponding angle 2α, to give the required stress T' as shown in the figure. Since the arc of a circle subtends twice the angle at the centre that it does at the circumference, the **pole** of the Mohr's circle may be defined as: **that point from which a line parallel to a chosen *plane* in the material would cut the Mohr's circle at the point representing the state of stress on that plane.** Thus, if T on the plane PP_1 is known, the Mohr's circle can be drawn and a line from T to the point **pole** parallel to PP_1 in the physical plane constructed, hence *locating* the pole. To find the stress on a plane PP_2 at an angle α to PP_1, we simply draw a line from the pole of the Mohr's circle parallel to PP_2. Where it again touches the circle gives the required stresses $(T = T')$. Alternatively, the radius may be rotated by 2α, which is often simpler, especially in computer solutions.

Since the slip-lines are coincident with those planes associated with the maximum shear stresses, by joining the point **pole** to the top and bottom points of the Mohr's circle it is possible to find the slip-line directions as shown by the broken lines in Fig. 5.28. In moving along a slip-line, Hencky's relations [equations (5.49a)] state that $\delta p = \pm 2k\delta\varphi$. Since the abscissa of the centre of the Mohr's circle is given by $\sigma = -p$, it is seen that the Mohr's circle moves by an amount equal to $\pm 2k\delta\varphi$, where $\delta\varphi$ is the angular change in moving along the slip-line. Now for a change of $\delta\varphi$ in the physical plane there is a rotation of the radius of the Mohr's circle by an angle $2\delta\varphi$, and since the radius of the circle is k, $\pm 2k\delta\varphi$ also represents the peripheral movement of the pole of the circle. Thus the pole of the Mohr's circle effectively traces out a cycloid in moving along a slip-line, by rolling without slipping along either top or bottom tangent.

Corresponding elements of the cycloid and slip-line are seen to be at right angles, since the instantaneous centre of rotation is at either the top or the bottom of the circle. For each slip-line in the slip-line field there is a cycloid in the stress plane, representing the trace of the pole of the Mohr's circle as it rolls along either top or bottom tangent.

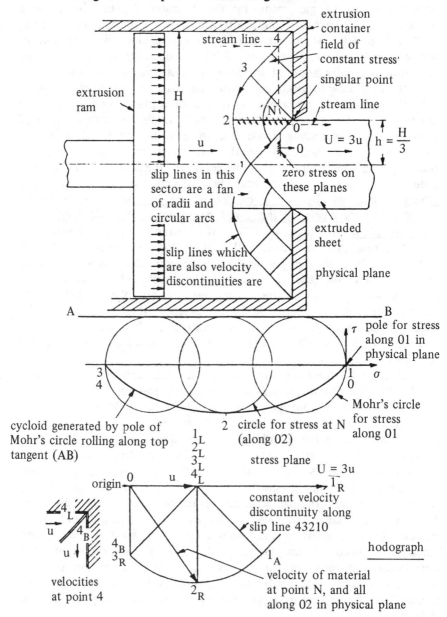

Fig. 5.29 Frictionless extrusion of sheet (3:1 extrusion ratio)

A typical example of a slip–line field, for the special problem of frictionless extrusion in plane strain (3:1 ratio) is shown in Fig. 5.29 opposite.

A third diagram introduced by Prager completes the representation of any slip–line field solution, namely the **hodograph.** If two neighbouring points P_1 and P_2 on a slip–line are considered (see Fig. 5.23), in order for there to be no extension of the slip–line, the velocity component of P_1 along the slip–line must be equal to the velocity component of P_2 along the slip–line. This can only be so if the ends of the vectors $O''P''_1$ and $O''P''_2$ in the hodograph, representing the velocities of the points P_1 and P_2, are such that the line $P''_1P''_2$ is normal to P_1P_2 in the physical plane. Thus, if the ends of all such vectors representing the velocities of points along the slip–lines indicated in the physical plane are joined together, there will be created an orthogonal network of lines having elements normal to their corresponding elements in the physical plane. Velocity discontinuities are represented by vectors *parallel* to the slip–line along which they exist.

The slip–lines cross the axis of symmetry at $45°$ and meet the walls of the container at $45°$, since the container is assumed to be perfectly lubricated (zero shear stress). To determine the magnitude of the velocity discontinuity it is convenient to imagine the material to be split as shown in the inset diagram for the point 4, in which:

4_L means material just to the left of point 4

4_B means material just below point 4

4_A means material just above point 4

4_R means material just to the right of point 4.

Another method of notation, particularly suited to computer drawing, is to number the spaces rather than the nodes of the slip–line field. Each node in the hodograph then locates the velocity vector, drawn from the pole of the hodograph, corresponding to the slip–line field space having the same number.

The example shown is particularly simple in that the slip–line field can be constructed immediately. Many slip–line fields are started from a **singular point** such as 0 in Figs. 5.29 and 5.30, from which a **centred fan** of slip–lines comprising radii and circular arcs is constructed as indicated. The real power of Prager's geometrical method is in the step–by–step solution of more complicated problems. This can be effected by the following technique. Suppose it is desired to extend the solution shown in Fig. 5.30 to the left, i.e. to determine the points 2ˊ and 3ˊ.

The angular orientation of these points (2˝ and 3˝) is known from the stress plane since all the cycloids are of identical size. By drawing a cross as shown at the point 3˝ having this correct angular orientation, the point 3˝ can be located by trial and error, in such a way that the distance 3˝C = 3C, and 3˝B = 2˝B. This is tantamount to replacing elements of the slip–lines with circular arcs, and gives very good results even for quite a large assumed mesh, the solution being carried on until some appropriate boundary condition is satisfied. The method of drawing chords, introduced and described by Johnson and Mellor is a simpler one and is readily adapted to the computer.

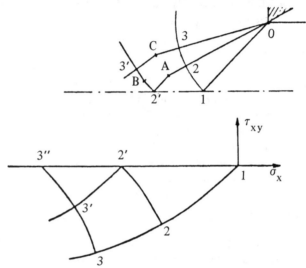

Fig. 5.30 Step–by–step method of extending slip–line field

The directions of flow of the material can be easily derived from the hodograph and indicated on the slip–line field, by drawing in short vectors representing the velocities at a number of points in the physical plane. The streamlines can be shown by drawing sequential contiguous vectors. The distortion of an originally square mesh grid can be determined by integrating velocities along the streamlines, and the deforming forces can be found by integration of the stresses acting on an appropriate boundary.

Having determined a slip–line field solution of the type shown, it is necessary to check that the velocity diagram is consistent with the slip–line field, and does not imply a negative rate of working anywhere. Thus, near the point 2 in the extrusion problem just discussed (Fig. 2.12) the material is subjected to the shear stresses shown in Fig. 5.31, as can be

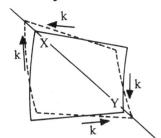

Fig. 5.31 Test for positive
plastic work rate

seen from the stress plane in Fig. 5.29. Clearly, the material
should be deforming in the manner shown by the broken lines in
Fig. 5.31, i.e. points X and Y should be moving apart. By
making a check of the hodograph, it can be seen whether this is
so or not. Velocity discontinuities need separate consideration in
this respect, it being necessary to check that the sense of the
velocity discontinuity is consistent with the direction of shear
stress in the material.

     A further check should also be made to see that those
regions (such as to the left of 4321 and to the right of 01 in
Fig. 5.29) will not in fact deform under the applied stresses on
their boundaries. The problem is difficult, but suitable techniques
have been developed for examining these regions.

     A more detailed account of the geometrical construction of
slip–line fields can be found in the book by Ford and
Alexander. Numerical methods are dealt with by Hill and by
Johnson, Sowerby and Haddow, and have recently returned to
prominence because of their suitability for computer solutions.

5.12 LIMIT ANALYSIS

Limit analysis is a powerful technique which has been developed
to give a more realistic idea of the collapse loads of structures,
as described by Baker, Horne and Hayman. It consists in finding
upper and lower bounds to the collapse load and these can be
brought closer and closer together by assuming more accurate
pictures of the deformations and loads, until the desired degree
of precision is attained. In considering metal–working processes,
it may be of assistance in replacing the 'plane sections remain
plane' type of theory for estimating deforming loads. Particularly
is this true for processes like extrusion in which plane sections
certainly do not remain plane, and Johnson has pioneered the
development of upper bound solutions to many forming problems.

     The science of limit analysis applies generally to the plastic–
rigid non–work–hardening material only, and rests on two theorems, one relating to the lower bound, the other to the upper
bound. These theorems are capable of mathematical proof (see

books by Hill, Prager, and Prager and Hodge); two concepts facilitate the development of the theorems, as follows:

(1) *The Statically Admissible Stress Field:* This is a field of generalized stresses (which may be forces, bending moments, or torques) which is in statical equilibrium within itself and with the externally imposed stresses at the boundaries.

(2) *The Kinematically Admissible Velocity Field:* This is a field of generalized velocities (which may be strains, linear displacements, or angular rotation) which is kinematically compatible within itself and with the externally imposed displacements at the boundaries.

The theorems may be stated as follows:

(1) *The Lower Bound Theorem:* If a statically admissible stress field exists, such that the stresses are everywhere just below those necessary to cause yielding, then the loads associated with that field constitute a lower bound solution.

(2) *The Upper Bound Theorem:* If a kinematically admissible velocity field exists, the loads required to be applied to cause the velocity field to operate constitute an upper bound solution.

The latter can be understood quite simply. If another kinematically admissible velocity field exists with a lower energy requirement, it will always come into play first as the load is increased. The postulated field can therefore never require less force than the real one.

Since the material is rigid, plastic deformation will only be possible if the yield criterion is exceeded at the joints, (or along the slip–lines) of the structure (or the material). Thus Theorem 2 really states that the loads required to cause the structure to deform as a mechanism will constitute an upper bound. It will be seen that many statically admissible stress fields and kinematically admissible velocity fields exist for any given problem. If one of each can be found to give the same answer, then the true collapse load has been found.

It will be realized, by reference to the slip–line field solution of the extrusion problem just discussed, that the slip–line field proposed was a kinematically admissible velocity field, but not necessarily a statically admissible stress field. This is because the velocity of the rigid material outside the deforming region had been considered, but not the stresses in the rigid material. Thus the solution may be only a partial solution, giving an *upper* bound to the extrusion force. In practice, recognizing the approximations involved, upper bound solutions are usually obtained from much more simplified fields, as discussed below. This is generally true of all slip−line field solutions, although Bishop established methods for deriving the complete solution from the partial solution, by extending the slip−line field into the rigid material.

The concept of velocity discontinuities introduced into the theory of the slip−line field gives a ready artifice for developing kinematically admissible velocity fields in plane strain problems. Similarly, stress discontinuities enable the development of stat− ically admissible stress fields. Consequently, upper and lower bound solutions should be readily determined for plane strain deformation. As a matter of fact, most research workers in this field have concentrated mainly on upper bound solutions to metal−working problems, which are easier to visualize, and lower bound solutions have not been extensively determined. A further practical advantage is that the press or other machine allocated on the basis of an upper bound solution will be <u>at least</u> powerful enough to complete the process.

The easiest method of visualizing these concepts is by considering specific examples. To begin with, consider the structure shown in Fig. 5.32(a). To find an upper bound it is necessary to postulate a kinematically admissible velocity field. This has been done in Fig. 5.32(b), hinges effectively forming at points A, B and C, when the bending moment at those points reaches the yield moment of the beam. If the small deflection under the load is δ, then the angular rotation of A is δ/a, at B δ/b, and, at C $\delta/a+\delta/b$. Thus the total work necessary to cause collapse is $2M_0(\delta/a+\delta/b)$, and, equating this to $P_u\delta$, the work done by the upper bound load P_u, gives the result $P_u = 2M_0(1/a+1/b)$. To find a lower bound it is necessary to postulate a statically admissible stress field. Such a field is shown in Fig. 5.32(c), which is a bending moment diagram in which the limit− ing collapse moment has been attained at each of the important points A, B, and C. It can be seen from this that $P_l ab/(a+b) = 2M_0$, and the lower bound solution is $P_l = 2M_0(1/a+1/b)$. In this particular case both lower and upper bounds coincide, the true solution has been found and the actual collapse load of the structure determined, because the static and kinematic fields which would actually exist in practice were assumed.

These techniques are very powerful for finding the collapse load of structures, and have revolutionized design methods in civil engineering practice. They can be applied also to processes of metal-working, as will now be discussed.

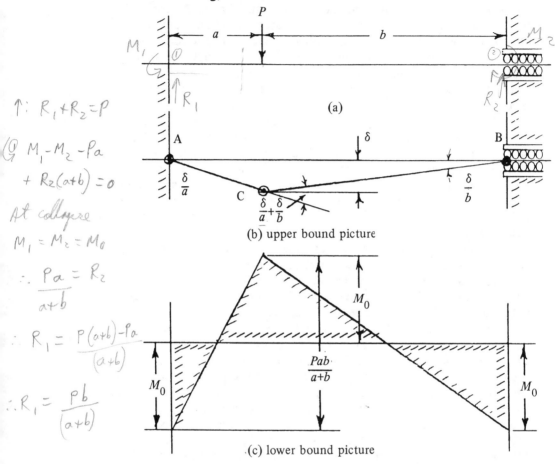

$\uparrow: R_1 + R_2 = P$

$\circled{2}\ M_1 - M_2 - Pa$
$\quad + R_2(a+b) = 0$

At collapse

$M_1 = M_2 = M_0$

$\therefore\ Pa = R_2$
$\quad\overline{a+b}$

$\therefore\ R_1 = \dfrac{P(a+b) - Pa}{(a+b)}$

$\therefore R_1 = \dfrac{Pb}{(a+b)}$

(a)

(b) upper bound picture

(c) lower bound picture

Fig. 5.32 Limit analysis of encastré beam

In Fig. 5.33(a) is shown an upper bound velocity field for the problem of lubricated extrusion, under plane strain conditions. The lines CA and BC are velocity discontinuities and the path of a typical element of material as it flows through the die is also shown. The shear strain in crossing the discontinuity BC is v_T/v_N as shown in the hodograph, Fig. 5.33(b), and this is found to be equal to $1/(\sin\varphi\cos\varphi)$. Similarly, the shear strain in crossing the velocity discontinuity CA is $1/(\sin\theta\cos\theta)$, so that the total work done per unit volume is simply:

$$k\left[\frac{1}{\sin\varphi\cos\varphi} + \frac{1}{\sin\theta\cos\theta}\right].$$

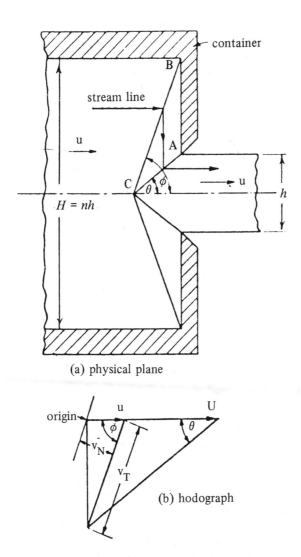

(a) physical plane

(b) hodograph

Fig. 5.33 Upper bound solution for
sheet extrusion (small reduction)

It is easy to show that the work done per unit volume is, in fact, equal to the mean pressure of extrusion $\bar{p}$ (see Alexander et al. Vol. 2 p.168) and, since $H \cot \varphi = h \cot \theta$, or $n \cot \varphi = \cot \theta$, where $n = H/h$, the following relationship applies:

$$\bar{p}_u = \frac{k(1+n)(1+n \tan^2 \theta)}{n \tan \theta}. \tag{5.52}$$

Since this is an upper bound solution, the minimum value of $\bar{p}_u$ in equation (5.52) is required, and is found by differentiation to occur for a value of $\theta = \tan^{-1}(1/\sqrt{n})$, in which case:

$$\frac{\bar{p}_u}{2k} = \frac{1+n}{\sqrt{n}}. \tag{5.53}$$

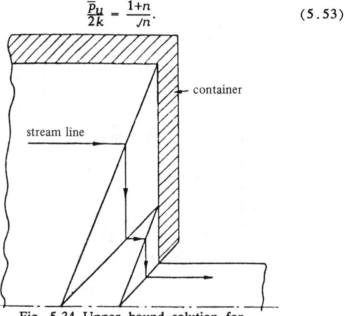

stream line

container

Fig. 5.34 Upper bound solution for
sheet extrusion (large reduction)

As discussed by Johnson, a smaller upper bound can be found for higher extrusion ratios by assuming the pattern of velocity discontinuities shown in Fig. 5.34 above:

A lower bound solution for this problem can be found easily by using a discontinuous stress field suggested for the problem of indentation by Shield and Drucker. It is valid for extrusion ratios (n) greater than 3:1 and is shown in Fig. 5.35. The broken lines are stress discontinuities, separating regions of constant stress, the Mohr's circles for these states of stress being as indicated in the stress plane. The circle B, for example, is the Mohr's circle representing the state of stress in the region B, O_B being the pole of that circle (Prager's definition). It can be seen that the slip-lines in each of the regions of constant stress meeting the container wall will do so at 45°, as required by the condition of zero friction, and the field is in fact a statically admissible stress field provided the radii of the larger Mohr's circles are just less than k.

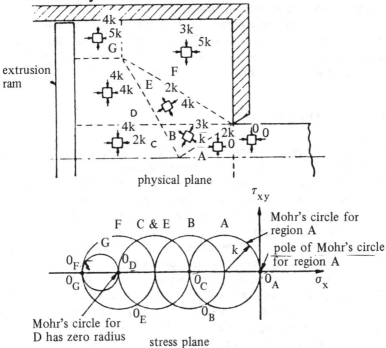

Fig. 5.35 Lower bound solution for sheet extrusion

Thus a lower bound for this problem is found to be (for $n > 3$):

$$\frac{\overline{p}_l}{2k} = 2.5\left[1 - \frac{1}{n}\right]. \tag{5.54}$$

Values of the two bounds are compared in the following table:

n	3	5	10	15	20
$\overline{p}_u/2k$	2.31	2.68	3.48	4.13	4.69
$\overline{p}_l/2k$	1.667	2.0	2.25	2.33	2.38

Clearly, as n increases, equation (5.53) tends to the value $\sqrt{n}$, whilst equation (5.54) tends to the constant value 2.5, so that the divergence increases with increasing extrusion ratio.

Another interesting method of using limit analysis to give upper bounds to metal—working problems, <u>and</u> allow for Coulomb friction on the boundaries is to assume a pattern of velocity discontinuities in the same way as Johnson, and to analyse the forces on each of the deforming blocks of metal by drawing polygons of forces. In this way, Coulomb friction can be taken into account, merely by drawing the reactions at the boundaries at the appropriate friction angle.

A different technique, useful for high friction conditions, assumes that the shear stress is a fraction m of the shear yield stress k, i.e. $\tau = mk$. This factor is then included in the evaluation of the work done (see the book by Rowe).

5.13 INSTABILITY

Instability during plastic deformation can occur in tension, compression, or shear. Probably tensile plastic instability, as manifested in the simple tensile test by the 'necking' phenomenon, is the most important, especially in relation to the forming of sheet metal under the action of tensile stresses in the plane of the sheet. Many engineering components such as automobile panels, refrigerators, washing machines etc. have to be made by deep drawing or pressing in biaxial tension, and tensile plastic instability can occur at relatively small strains. Therefore, this problem will be considered first. The manifestation of compressive plastic instability is found in large-deformation buckling, which will be considered later, as will be the problem of plastic instability in adiabatic shear.

5.13.1 *Tensile Plastic Instability*

To fix ideas, consider the simple tension test again. As the specimen extends under the action of the applied load, the material 'work—hardens' owing to the deformation it is suffering. Thus it is necessary to impose an increasing stress on the cross-sectional area in order to maintain plastic flow. This can be accomplished by increasing the applied load, but there is another effect which must not be forgotten. The requirement for conservation of the plastic volume in the tensile test is such that as the specimen extends it also contracts, thus reducing the cross—sectional area on which the applied load is acting. This also increases the stress as deformation proceeds, so that it is not necessary to increase the load as much as it would have been if the effect were not present. There comes a time, in fact, when it is not necessary to increase the load at all and the specimen simply extends under a constant load, the steadily decreasing area of cross-section ensuring the rise of stress necessary to offset work-hardening of the material. As deformation proceeds, the rise of stress due to the contraction of the specimen under the constant load 'outweighs' that due to work-hardening, and the specimen-load system becomes unstable. Fracture follows in a short while, the detailed behaviour during this phase depending to some extent on the characteristics of the testing machine used in the experiment. 'Necking' occurs in ductile metals, at some location along the specimen which must depend on the local imperfections of either manufacture or material.

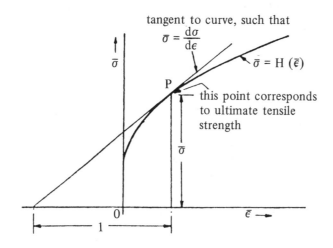

Fig. 5.36 Illustrating location of point of tensile plastic instability on true stress-true strain curve

It is easy to describe what happens in the simple tensile test to cause instability, but in sheet-metal stretch-forming processes in which the metal is subjected to a complex biaxial system of stress, the problem is more complicated. Professors Swift and Mellor have studied this problem in great detail. An introductory approach to the problem may be made, however, as follows. Considering the tensile test as a first step, assume that the relationship between true stress and true strain is of the form $\bar{\sigma} = H(\bar{\epsilon})$ as shown in Fig. 5.36, and that elastic strains may be neglected. If the current area of the specimen is A, and the current load is P, then the true stress is $\sigma_1 = P/A$, due to the applied load. It has already been established that this will be equal to the equivalent stress $\bar{\sigma}$, if the metal is deforming uniformly and plastically. Since elastic strains are being neglected, the equation Al = constant applies, where l is the current gauge length of the specimen, since there is no permanent volume change. Differentiating this equation it is found that:

$$A\mathrm{d}l + l\mathrm{d}A = 0 \tag{5.55}$$

and hence the longitudinal strain increment:

$$\mathrm{d}\bar{\epsilon} = \mathrm{d}\epsilon_1 = \mathrm{d}l/l = -\mathrm{d}A/A. \tag{5.56}$$

An increase of load on the tension specimen of amount $\mathrm{d}P$ is now imposed. The increase of stress $\mathrm{d}\sigma_1$ may be found by differentiating the equation $\sigma_1 = P/A$ to be:

$$\mathrm{d}\bar{\sigma} = \mathrm{d}\sigma_1 = \frac{\mathrm{d}P}{A} - \frac{P\mathrm{d}A}{A^2} = \frac{\mathrm{d}P}{A} - \bar{\sigma}\frac{\mathrm{d}A}{A} \tag{5.57}$$

$$\text{or } \mathrm{d}P/\mathrm{d}A = \bar{\sigma} - \mathrm{d}\bar{\sigma}/\mathrm{d}\bar{\epsilon}. \tag{5.58}$$

Now, instability occurs when the load attains a stationary maximum value, i.e. $\mathrm{d}P/\mathrm{d}\bar{\epsilon} = 0$. The cross–sectional area will still be decreasing, however, so that $\mathrm{d}A$ has a finite negative value, and $\mathrm{d}P/\mathrm{d}A$ will be zero. Thus, from equation (5.58), instability in the tension test occurs when:

$$\bar{\sigma} = \mathrm{d}\bar{\sigma}/\mathrm{d}\bar{\epsilon}. \tag{5.59}$$

This is a very important result, since it allows the determination of another quantity of interest from the true stress–true strain curve, namely the point associated with the maximum load which a tensile specimen can withstand. This is the point giving the **ultimate tensile strength (U.T.S.)** of the material, and it can now be seen that this may be determined from the true stress–true strain curve, by finding the point at which the tangent to the curve has the same magnitude as the true stress at that point. This is shown diagrammatically in Fig. 5.36, which also illustrates a simple geometrical construction for finding the point P corresponding with the U.T.S. of the material.

When a complex state of stress exists in the material, a similar procedure may be adopted to find the conditions at which instability will occur. This was discussed by both Swift and Mellor, the method being essentially to find the equivalent stress and strain associated with the given stress system, and to find the states of stress and strain which will give rise to instability, or unrestrained deformation without increase of load. Since <u>biaxial</u> stress states are important in forming sheet, the problem of instability under biaxial tension was considered by Swift and Mellor, who both introduced the stress ratio $x = \sigma_2/\sigma_1$, and assumed that the stress σ_3 normal to the sheet was zero.

To solve any given problem of this type, it is possibly simpler to consider what tensile plastic instability means in terms of the applied loads associated with that particular problem, rather than introducing the parameter x. Thus, for example, instability in a closed-end cylinder with thin walls subjected to an internal pressure p would occur when $dp/p = 0$. If the radius of the cylinder is r, and the wall thickness t, then the two principal stresses will be the circumferential stress σ_1, and the longitudinal or axial stress σ_2 given by the equations:

$$\left.\begin{array}{c} \sigma_1 = \dfrac{pr}{t} \\[2mm] \sigma_2 = \dfrac{pr}{2t} \end{array}\right\}. \tag{5.60}$$

Differentiating these equations:

$$\left.\begin{array}{c} \dfrac{\delta\sigma_1}{\sigma_1} = \dfrac{\delta p}{p} + \dfrac{\delta r}{r} - \dfrac{\delta t}{t} \\[3mm] \dfrac{\delta\sigma_2}{\sigma_2} = \dfrac{\delta p}{p} + \dfrac{\delta r}{r} - \dfrac{\delta t}{t} \end{array}\right\}. \tag{5.61}$$

Thus, instability of the vessel occurs when $\delta p/p = 0$, or:

$$\frac{\delta\sigma_1}{\sigma_1} = \frac{\delta\sigma_2}{\sigma_2} = \frac{\delta r}{r} - \frac{\delta t}{t}. \tag{5.62}$$

Substituting equations (5.60) into equation (5.32), recognizing that the radial stress σ_3 in a thin-walled cylinder is often assumed to be zero, it is found that $\bar{\sigma} = (\sqrt{3}/2)\sigma_1$, and from equation (5.24a), $\delta\bar{\epsilon} = (2/\sqrt{3})\delta\epsilon_1$. Now $\delta r/r$ is the hoop strain $\delta\epsilon_1$ and $\delta t/t$ is the 'through thickness' strain $\delta\epsilon_3$. From equations (5.38) it may be seen that $\delta\epsilon_2 = 0$ and $\delta\epsilon_3 = -\delta\epsilon_1$. Thus, considering equation (5.62), $\delta r/r - \delta t/t$ is equal to $2\delta\epsilon_1$, or $\sqrt{3}\delta\bar{\epsilon}$, and this equation may be re-written in the form:

$$\frac{\delta\bar{\sigma}}{\bar{\sigma}} = \sqrt{3}\delta\bar{\epsilon} \tag{5.62a}$$

since $\delta\bar{\sigma}/\bar{\sigma} = \delta\sigma_1/\sigma_1$ for a constant stress ratio σ_2/σ_1. In the limit, this condition of instability may be expressed as:

$$\frac{d\overline{\sigma}}{d\overline{\epsilon}} = \sqrt{3}\overline{\sigma}. \qquad (5.63)$$

It is sometimes convenient to refer this equation to the **subtangent** z of the true stress–true strain curve, the condition of instability then being given by the equation:

$$\frac{d\overline{\sigma}}{d\overline{\epsilon}} = \frac{\overline{\sigma}}{z}. \qquad (5.64)$$

In the particular case of the cylindrical closed–end vessel the value of z is $1/\sqrt{3}$, as shown diagrammatically in Fig. 5.37 where the case under discussion is compared with the case of instability in uniaxial tension in which $z = 1$. It can be seen from this diagram that the equivalent strain at instability is considerably less for the cylinder with closed ends than for the tension specimen. The maximum strain in the cylinder is the hoop strain, i.e. $\epsilon_1 = (\sqrt{3}/2)\overline{\epsilon}$, so that this too is considerably less than the longitudinal strain in the tension test. Since rupture of the wall of the thin–walled vessel will follow shortly after instability, the apparent lack of ductility observed in practice is explained by the foregoing analysis.

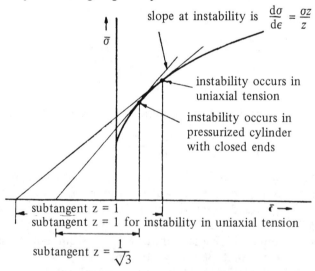

Fig. 5.37 Illustrating location of point of tensile instability on true stress–true strain curve (closed–end cylinder under pressure)

By way of comparison, and to illustrate the necessity for considering each problem *per se*, consider the case of plane–strain extension. It might at first be thought that instability in this case would occur at the same point on the stress–strain curve as for the closed-ended cylindrical shell, in which plane-strain conditions exist. The mode of application of the load is different, however, since it is assumed in this example that the sheet is flat and being stretched in its plane by forces in such a way that the strain ϵ_2 is zero at all times. Thus $\sigma_2 = \frac{1}{2}\sigma_1$, and $\sigma_3 = 0$, as before. If P_1 is the load per unit length in the σ_1 direction, P_2 is the load per unit length in the σ_2 direction, and t is the thickness of the sheet, the stresses and loads are related by the equations:

$$\sigma_1 = \frac{P_1}{t} \quad \text{and} \quad \sigma_2 = \frac{P_2}{t}. \tag{5.65}$$

Now $P_2 = \frac{1}{2}P_1$ at all times, and instability will occur when $dP_1 = 0$. Differentiating the first of equations (5.65):

$$\frac{\delta\sigma_1}{\sigma_1} = \frac{\delta P_1}{P_1} - \frac{\delta t}{t} \tag{5.66}$$

and instability will occur when:

$$\frac{\delta\sigma_1}{\sigma_1} = -\frac{\delta t}{t}. \tag{5.67}$$

As for the previous case $\delta\sigma_1/\sigma_1 = \delta\bar{\sigma}/\bar{\sigma}$, and $-\delta t/t = \delta\epsilon_1 = (\sqrt{3}/2)\,\delta\bar{\epsilon}$. Thus instability occurs when $\delta\bar{\sigma}/\bar{\sigma} = (\sqrt{3}/2)\delta\bar{\epsilon}$, or in the limit, when:

$$\frac{d\bar{\sigma}}{d\bar{\epsilon}} = \frac{\sqrt{3}}{2}\,\bar{\sigma} \tag{5.68}$$

i.e. when the subtangent has the value $2/\sqrt{3}$ ($\stackrel{\triangle}{=} 1\cdot155$).

In this case, although the system of stress is exactly the same as for the cylindrical shell with closed ends, the equivalent strain at instability is higher than for uniaxial tension, although the actual maximum strain in the sheet will be smaller (since $\epsilon_1 = (\sqrt{3}/2)\bar{\epsilon}$). This value is important for the problem of deep drawing cylindrical flat-bottomed cups, in which instability and subsequent failure occur at the junction of the cup wall and punch profile radius, under a system of stresses closely corresponding with plane-strain tension.

Swift and Mellor have also considered several other cases of biaxial stressing, but the principles are the same as for the two problems discussed here. Readers who are sufficiently interested might care to prove for themselves that for equal biaxial tension the subtangent has the value 2, whilst for a thin spherical vessel under internal pressure it has the value 2/3, which can also be shown to be the value for a thin-walled cylinder with open ends under internal pressure.

5.13.2 *Compressive Plastic Instability*

This phenomenon manifests itself in the compressive testing of cylindrical rods, which will buckle if they are too long. Similarly, in the production of axially symmetrical components by the processes of forging or extrusion, buckling of the initial billet will also occur if it is too long in relation to its diameter.

Engineers are generally familiar with the Euler <u>elastic</u> buckling load for a long prismatic column which, as discussed in Chapter 1, is of the form:

$$P_{cr} = C\pi^2 EI/l^2$$

where: P_{cr} = the critical load for elastic buckling,
          E = Young's Modulus,
          I = $\pi r^4/4$ for a cylindrical rod of radius r,
          l = column length, and
          C = 4 for a fixed−ended column (for example).

A comprehensive discussion of plastic buckling is given in the second edition of the book by Johnson and Mellor. They point out that a realistic estimate of the plastic buckling of short columns of the type under discussion here can be obtained by simply replacing the Young's Modulus E with the plastic tangent modulus $E_p = d\bar{\sigma}/d\bar{\epsilon}_p$. The value of E_p chosen from the stress−strain curve must correspond with the value of $\bar{\sigma} = \sigma_{cr}$, of course, which may involve some iteration, unless a simple constitutive relation can be used for the stress−strain curve of the material concerned.

For the plastic buckling of short fixed−ended cylindrical columns, radius r, of the type used as billets in forging or extrusion, $I = \pi r^4/4$, E is replaced by E_p, P_{cr} by $\sigma_{cr} \times \pi r^2$, $C = 4$, and hence Euler's equation takes the form:

$$l/r = \pi\sqrt{(E_p/\sigma_{cr})} \qquad (5.69)$$

where E_p is the value of $d\bar{\sigma}/d\bar{\epsilon}_p$ at the stress σ_{cr}. Since the current values of l, r, σ_{cr} and E_p are inter−related through the stress−strain curve and the constancy of the volume $V = \pi r^2 l$, it is possible to determine σ_{cr}.

5.13.3 *Shear Instability*

Buckling is specifically avoided in large plastic deformation by suitable choice of diameter/height ratio based on analyses of the type described in the previous discussion. It is more common to find failure in shear, due to intense local heating by the localized shear, which causes softening and thus facilitates further shear in the same band. This unstable situation is enhanced by high-speed working and with low-conductivity materials such as titanium. The extreme condition is known as **adiabatic shear instability.** In hot forging the local heating is sometimes visible as 45° cross patterns of brighter colour red on the sides of the billets.

5.14 MATRIX PLASTICITY

In recent years the techniques of finite element analysis have been developed, notably by Zienkiewicz and Argyris, to such an extent that it is necessary to discuss the matrix form of the basic equations of plasticity required for the finite element method. Specific application to metal forming is discussed by Rowe, Sturgess, Hartley and Pillinger.

In the finite element method the structure or continuum is split up into a number of small elements (often triangular or tetrahedral) and the forces and displacements at the nodes of these elements analysed in terms of the boundary conditions and the stiffness or flexibility of each element. For a continuum the forces and displacements can only be connected by consideration of the stresses and strains and their inter-relation, which involves the material properties of the continuum and the dimensions of the element. These considerations lead to extensive sets of simultaneous equations which have to be solved for each element and for the whole body under discussion, generally on a large (main-frame) digital computer. This process is greatly facilitated by the use of matrix algebra and it is, therefore, valuable to consider how the stress-strain relations are affected, particularly for an elastic-plastic continuum.

Using matrix algebra, equations (5.26) can be written for a general three-dimensional system as:

$$\epsilon = \varphi_e \, \sigma \qquad\qquad (5.70)$$

where:

$$\epsilon = \begin{bmatrix} \epsilon_{xx} \\ \epsilon_{yy} \\ \epsilon_{zz} \\ \gamma_{xy} \\ \gamma_{yz} \\ \gamma_{zx} \end{bmatrix} \quad \text{and} \quad \sigma = \begin{bmatrix} \sigma_{xx} \\ \sigma_{yy} \\ \sigma_{zz} \\ \tau_{xy} \\ \tau_{yz} \\ \tau_{zx} \end{bmatrix}$$

and:

$$\varphi_e = \frac{1}{E}\begin{bmatrix} 1 & -\nu & -\nu & 0 & 0 & 0 \\ -\nu & 1 & -\nu & 0 & 0 & 0 \\ -\nu & -\nu & 1 & 0 & 0 & 0 \\ 0 & 0 & 0 & 2(1+\nu) & 0 & 0 \\ 0 & 0 & 0 & 0 & 2(1+\nu) & 0 \\ 0 & 0 & 0 & 0 & 0 & 2(1+\nu) \end{bmatrix}.$$

It is often more convenient to write a column matrix such as ϵ or σ above as a horizontal row, to save space. Remembering that the transpose of a matrix is obtained by converting its columns into rows, a convenient way of writing σ, for example, would therefore be thus:

$$\sigma^t = [\sigma_{xx}\ \sigma_{yy}\ \sigma_{zz}\ \sigma_{xy}\ \sigma_{yz}\ \sigma_{zx}]. \tag{5.71}$$

The matrix φ_e could be described as the **stiffness matrix** for the basic cubic element of an isotropic elastic continuum. The multiplication rule for two matrices involves multiplying each term in a row of the first matrix by the corresponding term of the appropriate column of the second matrix and adding the resulting products. Thus, for example, considering the second term of the column vector ϵ the following equation is derived:

$$\epsilon_{yy} = \frac{1}{E}[-\nu\sigma_{xx}+\sigma_{yy}-\nu\sigma_{zz}+0.\tau_{xy}+0.\tau_{yz}+0.\tau_{zx}]$$

$$= \frac{1}{E}[\sigma_{yy}-\nu(\sigma_{xx}+\sigma_{zz})] \tag{5.72}$$

corresponding with the second of equations (2.1). Two useful references for matrix algebra and its application to the digital computer are the books by Bickley and Thompson, and Pipes and Hovanessian.

The problem which must now be considered is that of developing the **flexibility matrix** (the inverse of the **stiffness matrix**), for the basic cubic element of an isotropic elastic-plastic continuum. Stated in another way, it is now necessary to write the Prandtl–Reuss equations (5.44) in matrix form in such a way that the total strain increment can be related to the stress increment. (It was noted previously that in their original form the Prandtl–Reuss equations contain plastic strain increments which were related to the total stresses.)

To do this it is necessary to consider the quadratic form of the scalar quantities $\bar{\sigma}^2$ and $\delta\bar{\epsilon}^2$ previously defined in equations (5.8) and (5.9). It can be shown (see for example Ford with Alexander) that:

$$\bar{\sigma}^2 = (3/2)[\sigma'_{xx}{}^2+\sigma'_{yy}{}^2+\sigma'_{zz}{}^2+2\tau_{xy}{}^2+2\tau_{yz}{}^2+2\tau_{zx}{}^2] \qquad (5.73)$$

where σ'_{xx}, σ'_{yy} and σ'_{zz} are the deviatoric normal stresses in the general cartesian coordinate system (e.g. $\sigma'_{xx} = \sigma_{xx}-\sigma$). The corresponding form of $\delta\bar{\epsilon}_p{}^2$ is:

$$\delta\bar{\epsilon}_p{}^2 = (2/3)[\,_p\delta\epsilon_{xx}{}^2+_p\delta\epsilon_{yy}{}^2+_p\delta\epsilon_{zz}{}^2+$$
$$2_p\delta\epsilon_{xy}{}^2+2_p\delta\epsilon_{yz}{}^2+2_p\delta\epsilon_{zx}{}^2] \qquad (5.74)$$

where $_p\delta\epsilon_{xx}$, $_p\delta\epsilon_{yy}$, and $_p\delta\epsilon_{zz}$ are the corresponding normal plastic strain increments and $_p\delta\epsilon_{xy}$, $_p\delta\epsilon_{yz}$ and $_p\delta\epsilon_{zx}$ are the plastic <u>tensor</u> shear strain increments (i.e. <u>half</u> the engineering shear strain increments such as $_p\delta\gamma_{xy}$).

The column matrices for stress and strain previously defined in equation (5.70) will not allow these quadratic forms to be developed. It would be possible to define a 9×1 column matrix which, when pre-multiplied by its transpose, would give the '2' factors required in the product defined in equation (5.73). However, such a large matrix would lead to extremely large matrices in finite element analyses and Argyris has suggested using the column vectors:

$$\sigma^{\mathbf{t}} = [\sigma'_{xx} \quad \sigma'_{yy} \quad \sigma'_{zz} \quad \sqrt{2}\tau_{xy} \quad \sqrt{2}\tau_{yz} \quad \sqrt{2}\tau_{zx}] \qquad (5.75)$$

$$_p\delta\epsilon^{\mathbf{t}}=[\,_p\delta\epsilon_{xx} \quad _p\delta\epsilon_{yy} \quad _p\delta\epsilon_{zz} \quad \sqrt{2}_p\delta\epsilon_{xy} \quad \sqrt{2}_p\delta\epsilon_{yz} \quad \sqrt{2}_p\delta\epsilon_{zx}]. \qquad (5.76)$$

With these definitions it is seen that:

$$\bar{\sigma}^2 = (3/2)\sigma^{\mathbf{t}}\sigma' \qquad (5.77)$$

$$\delta\bar{\epsilon}_p{}^2 = (2/3)_p\delta\epsilon^{\mathbf{t}}{}_p\delta\epsilon. \qquad (5.78)$$

The Prandtl-Reuss equations are often written in tensor notation in the form (see Hill):

$$\delta\epsilon'_{ij} = \frac{1}{2G}\,\delta\sigma'_{ij} + \frac{3}{2}\frac{\delta\bar{\epsilon}}{\bar{\sigma}}\,\sigma'_{ij}\ .$$

where i and j take all values from 1 to 3 representing the cartesian axes and the **shear modulus** $G = E/\{2(1+\nu)\}$. Using the definitions given by equations (5.75) to (5.78) above, the corresponding matrix equation is:

$$\delta\epsilon' = \frac{1}{2G}\,\delta\sigma' + \frac{3}{2}\frac{\delta\bar{\epsilon}}{\bar{\sigma}}\,\sigma'\ . \qquad (5.79)$$

In these equations, $\delta\epsilon'$ and $\delta\epsilon'_{ij}$ represent <u>deviatoric</u> strain increments such as $\delta\epsilon'_{xx} = \delta\epsilon_{xx} - \delta\epsilon$, where $\delta\epsilon$ is the increment of mean strain given by $1/3 \times (\delta\epsilon_{xx} + \delta\epsilon_{yy} + \delta\epsilon_{zz})$, i.e.

$$\frac{1}{3}\frac{\delta V}{V}, \quad \text{or} \quad \frac{1-2\nu}{E}(\delta\sigma_{xx} + \delta\sigma_{yy} + \delta\sigma_{zz})$$

from equation (5.27), or $(1/3)(\delta\sigma_m/K)$, where K is the **bulk modulus**, given by $K = E/\{3(1-2\nu)\}$.

Thus, the **total strain increment** may be written as:

$$\delta\epsilon = \delta\epsilon' + \begin{bmatrix}1\\1\\1\\0\\0\\0\end{bmatrix}\delta\epsilon = \frac{1}{2G}\{\delta\sigma - \begin{bmatrix}1\\1\\1\\0\\0\\0\end{bmatrix}\delta\sigma\} + \begin{bmatrix}1\\1\\1\\0\\0\\0\end{bmatrix}\frac{\delta\sigma}{3K} + \frac{3\bar{\sigma}\delta\bar{\sigma}}{2E_p\sigma^2}\sigma'$$

where K is the **bulk modulus** and E_p is the **tangent modulus**, given by $E_p = d\bar{\sigma}/d\bar{\epsilon}$.

Differentiating equation (5.77):

$$2\bar{\sigma}\delta\bar{\sigma} = \frac{3}{2}(\sigma'^t\delta\sigma' + \delta\sigma'^t\sigma') = 3\sigma'^t\delta\sigma' = 3\sigma'^t(\delta\sigma - \begin{bmatrix}1\\1\\1\\0\\0\\0\end{bmatrix}\delta\sigma)$$

$$= 3\sigma'^t\,\delta\sigma.$$

since: $\quad 3\sigma'^t\begin{bmatrix}1\\1\\1\\0\\0\\0\end{bmatrix}\delta\sigma = 3(\sigma'_{xx} + \sigma'_{yy} + \sigma'_{zz})\delta\sigma = 0,$

as can be seen by summing equations (5.31).

Also, $\delta\sigma = \begin{bmatrix}1 & 1 & 1 & 0 & 0 & 0\end{bmatrix}\delta\sigma$ and:

$$\text{putting } I_6 = \begin{bmatrix}1 & 0 & 0 & 0 & 0 & 0\\0 & 1 & 0 & 0 & 0 & 0\\0 & 0 & 1 & 0 & 0 & 0\\0 & 0 & 0 & 1 & 0 & 0\\0 & 0 & 0 & 0 & 1 & 0\\0 & 0 & 0 & 0 & 0 & 1\end{bmatrix}$$

$$\frac{1}{2G} = \frac{1+\nu}{E} \quad \text{and} \quad \frac{1}{2G} - \frac{1}{3K} = \frac{3\nu}{E},$$

therefore:

$$\delta\epsilon = \left\{\frac{1+\nu}{E} \mathbf{I}_6 - \frac{\nu}{E} \begin{bmatrix} 1 \\ 1 \\ 1 \\ 0 \\ 0 \\ 0 \end{bmatrix} \begin{bmatrix} 1 & 1 & 1 & 0 & 0 & 0 \end{bmatrix}\right\}\delta\sigma + \frac{9}{4E_p\bar{\sigma}^2}\,\sigma'\sigma'^t\delta\sigma$$

$$\text{or} \quad \delta\epsilon = (\varphi_e + \varphi_p)\,\delta\sigma = \varphi_t\,\delta\sigma. \tag{5.80}$$

The **elastic flexibility matrix** associated with these particular definitions of the stress and strain column vectors is thus:

$$\varphi_e = \frac{1}{E}\begin{bmatrix} 1 & -\nu & -\nu & 0 & 0 & 0 \\ -\nu & 1 & -\nu & 0 & 0 & 0 \\ -\nu & -\nu & 1 & 0 & 0 & 0 \\ 0 & 0 & 0 & 1+\nu & 0 & 0 \\ 0 & 0 & 0 & 0 & 1+\nu & 0 \\ 0 & 0 & 0 & 0 & 0 & 1+\nu \end{bmatrix} \tag{5.81}$$

and the **plastic flexibility matrix,** although rarely used in this form, is as follows:

$$\varphi_p = \frac{9}{4E_p\bar{\sigma}^2} \times$$

$$\begin{bmatrix}
\sigma'_{xx}{}^2 & \sigma'_{xx}\sigma'_{yy} & \sigma'_{xx}\sigma'_{zz} & \sqrt{2}\sigma'_{xx}\tau_{xy} & \sqrt{2}\sigma'_{xx}\tau_{yz} & \sqrt{2}\sigma'_{xx}\tau_{zx} \\
\sigma'_{yy}\sigma'_{xx} & \sigma'_{yy}{}^2 & \sigma'_{yy}\sigma'_{zz} & \sqrt{2}\sigma'_{yy}\tau_{xy} & \sqrt{2}\sigma'_{yy}\tau_{yz} & \sqrt{2}\sigma'_{yy}\tau_{zx} \\
\sigma'_{zz}\sigma'_{xx} & \sigma'_{zz}\sigma'_{yy} & \sigma'_{zz}{}^2 & \sqrt{2}\sigma'_{zz}\tau_{xy} & \sqrt{2}\sigma'_{zz}\tau_{yz} & \sqrt{2}\sigma'_{zz}\tau_{zx} \\
\sqrt{2}\tau_{xy}\sigma'_{xx} & \sqrt{2}\tau_{xy}\sigma'_{yy} & \sqrt{2}\tau_{xy}\sigma'_{zz} & 2\tau_{xy}{}^2 & 2\tau_{xy}\tau_{yz} & 2\tau_{xy}\tau_{zx} \\
\sqrt{2}\tau_{yz}\sigma'_{xx} & \sqrt{2}\tau_{yz}\sigma'_{yy} & \sqrt{2}\tau_{yz}\sigma'_{zz} & 2\tau_{yz}\tau_{xy} & 2\tau_{yz}{}^2 & 2\tau_{yz}\tau_{zx} \\
\sqrt{2}\tau_{zx}\sigma'_{xx} & \sqrt{2}\tau_{zx}\sigma'_{yy} & \sqrt{2}\tau_{zx}\sigma'_{zz} & 2\tau_{zx}\tau_{xy} & 2\tau_{zx}\tau_{yz} & 2\tau_{zx}{}^2
\end{bmatrix} \tag{5.82}$$

Thus φ_p is singular and cannot be inverted, but the matrix φ_t is non-singular so solutions can be found.

A more detailed discussion of matrix representations of the stress and strain tensors can be found in Vol. 1. The application to large-scale deformation finite element analysis is given by Rowe et al.

5.15 STRESS–STRAIN CURVES – EMPIRICAL
RELATIONSHIPS

A major difficulty in using a computer to obtain solutions to
problems in plastic flow is that of establishing simple relationships between stress and strain. Yet another problem is to be
able to establish a meaningful stress–strain curve up to the
values of large strain required to predict behaviour of material
subjected to large–deformation plastic flow. It is not possible to
derive that sort of information from the simple tensile test owing
to the onset of necking (the tensile plastic instability discussed in
Section 5.12 above), which occurs at extensions rarely exceeding
about 25 per cent. For that reason, true stress–true strain curves
generally have to be determined from compression or torsion
tests and even then present great difficulty, if precision is
needed. Probably the easiest compression test to visualize is the
simple compression of a short solid cylindrical rod. However,
there are difficulties with even that elementary geometry, due to
frictional end effects that can lead to non–uniform 'barrelling' of
the specimen. Most testing procedures are beset with such
problems, as discussed in more detail by Alexander, Brewer and
Rowe.

　　Suppose that it *is* possible to obtain a load–displacement
curve for such a cylindrical specimen. To determine from those
results the **basic true stress–true strain curve** requires some
knowledge of the parameter of **true strain,** which has been one
of the main topics of this chapter. For large deformations the
volume of the cylindrical specimen may be assumed, as before,
to remain constant, so that the product of the current cross–
sectional area A and the current height h remains constant. Thus
$Ah = A_0 h_0$, the initial volume, at all times. Hence, the true
stress $\bar{\sigma}$ is given by the equation $\bar{\sigma} = P/A$, P being the current
applied load, and since $A = A_0 h_0 / h$, $\bar{\sigma} = (P/A_0) \cdot (h/h_0)$.

　　The total true strain is:

$$\int_{h_0}^{h} \frac{dh}{h} = \ln \frac{h}{h_0}$$

which is negative, as required by convention for compression,
since $h < h_0$. The plastic true strain, measured after removal
of the load, is equal to the total strain less the elastic recovery,
namely:

$$\ln \frac{h}{h_0} + \frac{\sigma}{E}.$$

However, the <u>equivalent</u> plastic strain $\bar{\epsilon}_p$ must be positive in order to have any meaning, so it is made positive as:

$$\bar{\epsilon}_p = \ln \frac{h_0}{h} - \frac{\sigma}{E}.$$

In fact, the elastic strain is negligibly small in comparison with real plastic strains and is almost always neglected.

As discussed by Alexander, Brewer and Rowe, a realistic empirical relationship for fitting the stress–strain curves of many metals was suggested by Swift, as follows:

or simply:
$$\bar{\sigma} = A(B+\bar{\epsilon}_p)^n$$
$$\bar{\sigma} = A(B+\bar{\epsilon})^n. \tag{5.83}$$

The tangent modulus of that curve is:

$$E_p = d\bar{\sigma}/d\bar{\epsilon} = nA(B+\bar{\epsilon})^{n-1}. \tag{5.84}$$

Thus, the initial tangent modulus when the plastic strain $\bar{\epsilon}$ is zero is:

$$E_{po} = nAB^{n-1} \tag{5.85}$$

and:
$$B = [E_{po}/(nA)]^{n-1}. \tag{5.86}$$

If $B = 0$ in equation (5.83), the equation $\bar{\sigma} = A\bar{\epsilon}^n$ results, [equation (5.95), introduced later]. Equation (5.95) gives a reasonable fit to experimental data for annealed metals, whilst equation (5.83) fits data for strain–hardened metals. The problem of determining the best values of the three constants A, B and n of Swift's equation is difficult, but an estimate for B can be obtained by realizing that the stress–strain curve for the strain–hardened metal is effectively that for the annealed metal displaced to the left by a true strain $\epsilon = B$.

There are numerical methods involving rather complicated optimization routines which can be used. A simpler method which may appeal to engineers is to employ a simple search routine which seeks to minimize the sum of the squares of the differences between the given (measured) values and those calc–ulated by using different trial values of the 'constant' B in equation (5.83), as follows.

In Fig. 5.38 is illustrated the linear relationship between $\ln \bar{\sigma}$ and $\ln(B+\bar{\epsilon})$ of equation (5.83). If, as illustrated in the figure, there are m measured values of $\bar{\epsilon}$ and $\bar{\sigma}$ for any given value of B, then $y_m = \ln \bar{\sigma}$ can be plotted against $x_m = \ln(B+\bar{\epsilon})$, where x_m and y_m are the generic values of the abscissa and ordinate for the mth pair of measured values. The problem to be solved is simply that of finding the best fit to the measured values, of the straight line relationship $y=a+nx$, where $x=\ln(B+\bar{\epsilon})$, $y=\ln\bar{\sigma}$, $a=\ln A$, and n is the slope of the straight line which minimizes the sum of the squares of the differences (y_m-y).

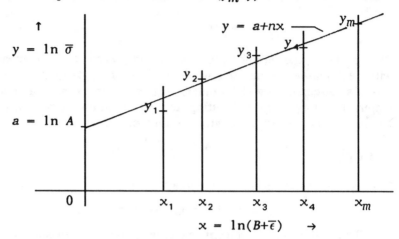

Fig. 5.38 A log–log plot of Swift's equation

Mathematically, it is necessary to find:

Now:

$$R_{min} = \left[\sum_1^m (y_m-y)^2 \right]_{min}$$

$$R = \sum_1^m \{y_m-(a+nx_m)\}^2$$

since it is convenient to make $x = x_m$, the measured value. To minimize R it is necessary to satisfy the conditions for determining a stationary value of R, viz:

$$\frac{\partial R}{\partial a} = \frac{\partial R}{\partial n} = 0$$

and to check whether they do in fact coincide with the desired minimum (not a maximum) value of R.

These partial differential coefficients are:

$$\frac{\partial R}{\partial a} = \sum_1^m 2\{y_m-(a+nx_m)\}(-1) = 0$$

$$\therefore \sum_1^m \{y_m-(a+nx_m)\} = 0$$

and:

$$\frac{\partial R}{\partial n} = \sum_1^m 2\{y_m-(a+nx_m)\}(-x_m) = 0$$

$$\therefore \sum_1^m x_m\{y_m-(a+nx_m)\} = 0.$$

These equations can be written in terms of the various

summations involved as follows:

$$\frac{\partial R}{\partial a} = \sum_1^m y_m - ma - n\sum_1^m x_m = 0 \qquad (5.87)$$

and:

$$\frac{\partial R}{\partial n} = \sum_1^m x_m y_m - a\sum_1^m x_m - n\sum_1^m x_m^2 = 0. \qquad (5.88)$$

These summations can most easily and quickly be found by using a computer, a convenient notation (finally in the FORTRAN language) being, for example:

$$\sum_1^m x_m y_m - a\sum_1^m x_m - n\sum_1^m x_m^2 =$$

$S_{xy} - aS_x - nS_{xx} = $ SUMXY–ALOG(A)*SUMX–XN*SUMXX = 0.

Using this notation, equations (5.87) and (5.88) may conveniently be expressed as:

$$S_y - ma - nS_x = 0 \qquad (5.87a)$$

and:

$$S_{xy} - aS_x - nS_{xx} = 0. \qquad (5.88a)$$

$\therefore \quad S_y/m - a - nS_x/m = S_{xy}/S_x - a - nS_{xx}/S_x = 0 \qquad (5.88b)$

and: $\quad S_y S_x - nS_x^2 = mS_{xy} - nmS_{xx} \qquad (5.88c)$

hence: $\quad n = (mS_{xy} - S_x S_y)/(mS_{xx} - S_x^2) \qquad (5.89)$

or, in FORTRAN, remembering that I,J,K,L,M,N are integers:

XN =

(XM*SUMXY–SUMX*SUMY)/(XM*SUMXX–SUMX**2),   (5.90)

and also remembering that: $a = \ln A = (S_y - nS_x)/m \qquad (5.91)$

or in FORTRAN: A = EXP((SUMY–XN*SUMX)/XM).   (5.92)

A computer program for solving this problem is given below as Program 5.3, with typical print–outs of the results of curve–fitting $m = 6$ pairs of values derived from actual compression tests on aluminium specimens, shown as Solutions 5.3(a) and 5.3(b). These test results have been taken from Johnson and Mellor, p.253. It may be observed that equations (5.90) and (5.92) appear as lines numbered 122: and 123: in the main program and as lines 170: 171: and 196: 197: in the SUBROUTINES SUM and SUM1.

Continuing with this discussion of empirical relationships, or constitutive equations, the total effective or equivalent true strain is:

$$d\bar{\epsilon} = d\bar{\epsilon}_e + d\bar{\epsilon}_p, \qquad (5.93)$$

the **total tangent modulus** is $d\bar{\sigma}/d\bar{\epsilon} = E_t$, $\qquad \therefore d\bar{\epsilon} = d\bar{\sigma}/E_t$,
the **plastic tangent modulus** is $d\bar{\sigma}/d\bar{\epsilon}_p = E_p$, $\therefore d\bar{\epsilon}_p = d\bar{\sigma}/E_p$,
the **elastic tangent modulus** is $d\bar{\sigma}/d\bar{\epsilon}_e = E$, $\therefore d\bar{\epsilon}_e = d\bar{\sigma}/E$

From equation (5.93), $1/E_t = 1/E + 1/E_p$ $\qquad (5.94)$

Taking typical values (e.g. for aluminium), $E_t = 67$ MPa, $E = 70$ GPa so, to a good approximation, $E_t = E_p$.

As already mentioned, another constitutive equation (often used to relate effective stress to effective strain *rate*), may also be used for relating effective stress to effective strain, namely:

$$\bar{\sigma} = A\bar{\epsilon}^n. \qquad (5.95)$$

This approximates well to the behaviour of annealed metals, except close to $\bar{\epsilon} = 0$, but for work–hardened alloys an intercept or initial value is preferable, such as:

$$\bar{\sigma} = \sigma_0 + A\bar{\epsilon}^n \qquad (5.96)$$

It should be noted that E_{p_0} is infinite for both these relationships. Equations (5.83), (5.95) and (5.96) are all included in the single generic equation:

$$\bar{\sigma} = \sigma_0 + A(B+\bar{\epsilon})^n \qquad (5.97)$$

Program 5.3 which follows is arranged to find the best solutions for equations (5.83), (5.95) and (5.96) and indicates which of these gives the best fit. Solutions 5.3(a) and (b) relate to half–hard and annealed aluminium repectively, with stresses in lbf/in^2.

PROGRAM 5.3 Curve–fitting Swift's equation and others covered by the generic equation $\bar{\sigma} = \sigma_0 + A(B+\bar{\epsilon})^n$

```
 1:*          PROGRAM SWIFT
 2:           DIMENSION E(20),S(20),X(20),Y(20),XY(20),XX(20),B(20),R(20),
 3:        1  SS(20),S0(20),R1(20)
 4:           WRITE (*,1)
 5: 1         FORMAT(1X' SWIFT FITS GIVEN S/E DATA POINTS TO VARIANTS OF',/1X,
 6:        1  ' EQUATION S = S0+A(B+E)**N  BY A LEAST SQUARE FIT OF M',/1X,
 7:        2  ' LOG-LOG VALUES FOR A NUMBER (J) OF ASSUMED VALUES OF B',/1X,
 8:        3  '   OR ALTERNATIVELY A NUMBER (K) OF ASSUMED VALUES OF S0 ',/1X,
 9:        4  ' WHICH ARE OPTIMIZED BY DIRECT SEARCH, CHOOSING SOLUTIONS',/1X,
10:        5  ' TO MINIMIZE R = SUMMATION OF (S(GIVEN)-S(CALCULATED))**2',/1X,
11:        6  ' TRY NOT TO USE ANY STRAIN VALUE LESS THAN E=0.01 ',/)
12:           WRITE(*,2)
13: 2         FORMAT( 1X' TYPE IN M ')
14:           READ(*,*) M
15:           DO 4 I = 1,M
16:           WRITE(*,3) I,I
17: 3         FORMAT( 1X' TYPE IN E(',I2,'),S(',I2,')')
18:           READ(*,*) E(I),S(I)
19: 4         CONTINUE
20:           N = 1
21:           B(1) = .001
22:           F = 2.
23:           IF (E(1).EQ.0.) E(1) =.1E-01
24:           IF (S(1).EQ.0.) S(1) =.1E-01
25:           WRITE (*,59)
26: 59        FORMAT(/2X,' FITTING EQUATION S=A(B+E)**N GIVES:- ',/)
27: 12        DO 6 J = 1,9
28:           CALL SUM(E,S,B,A ,EP0,XN,M,J)
29:           R(J) = 0.
30:           DO 5 I = 1, M
31:           SS(I) = A*(B(J)+E(I))**XN
32: 5         R(J) = R(J)+(S(I)-SS(I))**2
33: 6         B(J+1) = F*B(J)
34:           DO 7 J = 1,8
35:           IF (R(J+1)-R(J)) 7,8,8
36: 7         CONTINUE
37: 8         WRITE(*,9) R(J), B(J)
38: 9         FORMAT( 1X,'LATEST MINIMUM VALUE OF R IS ',E12.6,' FOR B = ',
39:        1  E12.6,)
40:           IF (ABS((R(J+1)-R(J))/R(J+1))-.5E-6)11,11,10
41: 10        IF(N.EQ.20)GO TO 19
42:           B(1) = B(J)/F
43:           F = F**.25
44:           N = N+1
45:           GO TO 12
46: 11        CALL SUM(E,S,B,A ,EP0,XN,M,J)
47:           WRITE(*,13)
48: 13        FORMAT(/ 2X,' (R(J+1)-R(J))/R(J+1) LESS THAN .5E-6 NEAR',/1X,
49:        1  ' MINIMUM R(J), SO THE BEST SOLUTION IS AS FOLLOWS ',/)
50:           WRITE(*,14)
51: 14        FORMAT( 6X,' N',10X,' A',10X,'EP0', 9X,'B',11X,'R',)
52:           WRITE(*,15) XN, A,EP0,B(J),R(J)
53: 15        FORMAT(6E12.6,I2)
54:           WRITE(*,16)
55: 16        FORMAT(/11X,'STRAIN',10X,'STRESS(GIVEN)',5X,'STRESS(CALCULATED)'
56:        1  ,)
57:           DO 18 I = 1,M
58:           SS(I) = A*(B(J)+E(I))**XN
59:           WRITE (*,17) E(I),S(I),SS(I)
60: 17        FORMAT(3E20.7)
61: 18        CONTINUE
62:           N = 0
63:           NEND = 0
64:           S0(1) = 0.
65:           SR = S(1)
66:           WRITE (*,60)
67: 60        FORMAT(/2X,' FITTING EQUATION S=S0+AE**N GIVES:- ',/)
```

```
68: 42        DO 36 K = 1,9
69:           CALL SUM1(E,S,SO,A,XN,M,K)
70:           IF(NEND.EQ.1) GO TO 23
71:           R1(K) = 0.
72:           DO 35 I = 1,M
73:           SS(I) = SO(K)+A*E(I)**XN
74: 35        R1(K) = R1(K)+(S(I)-SS(I))**2
75: 36        SO(K+1) = SO(K)+SR*.125
76:           DO 37 K = 1,8
77:           IF (R1(K+1)-R1(K)) 37,38,38
78: 37        CONTINUE
79: 38        WRITE(*,39) R1(K), SO(K)
80: 39        FORMAT(1X,' LATEST MINIMUM VALUE OF R IS ',E12.6,' FOR SO = ',
81:         1 E12.6,)
82:           IF (K.EQ.1) GO TO 22
83:           IF (ABS((R1(K+1)-R1(K))/R1(K+1))-.5E-6) 41,41,40
84: 40        IF (N.EQ.20) GO TO 19
85:           SO(1) = SO(K)-SR*.125
86:           SR = SR*.25
87:           N = N+1
88:           GO TO 42
89: 22        NEND = 1
90: 41        CALL SUM1(E,S,SO,A,XN,M,K)
91: 23        WRITE(*,43)
92: 43        FORMAT(/2X,'(R(K+1)-R(K)/R(K+1)LESS THAN .5E-6 NEAR',/1X,
93:         1 ' MINIMUM R(K), SO THE BEST SOLUTION IS AS FOLLOWS ',/)
94:           WRITE(*,44)
95: 44        FORMAT( 6X, 'N',10X,'SO',10X,'A',11X,'R',6X,/)
96:           WRITE(*,45) XN,SO(K),A,R1(K)
97: 45        FORMAT(4E12.6,I2)
98:           WRITE(*,46)
99: 46        FORMAT(/11X,'STRAIN',10X,'STRESS(GIVEN)'5X,'STRESS(CALCULATED)'
100:        1 ,)
101:          DO 48 I = 1,M
102:          SS(I) = SO(K)+A*E(I)**XN
103:          WRITE(*,47) E(I),S(I),SS(I)
104: 47       FORMAT(3E20.7)
105: 48       CONTINUE
106:          WRITE (*,61)
107: 61       FORMAT(/2X,' FITTING EQUATION S=AE**N  GIVES:- ',/)
108:          SUM X = 0.
109:          SUM Y = 0.
110:          SUMXY = 0.
111:          SUMXX = 0.
112:          DO 49 I = 1,M
113:          X(I) = ALOG (E(I))
114:          Y(I) = ALOG (S(I))
115:          XY(I) = X(I)*Y(I)
116:          XX(I) = X(I)**2
117:          SUMX = SUMX  + X(I)
118:          SUMY = SUMY  + Y(I)
119:          SUMXY = SUMXY + XY(I)
120: 49       SUMXX = SUMXX + XX(I)
121:          XM = M
122:          XN = (XM*SUMXY-SUMX*SUMY)/(XM*SUMXX-SUMX**2)
123:          A = EXP((SUMY-XN*SUMX)/XM)
124:          R2 = 0.
125:          DO 50 I = 1,M
126:          SS(I) = A*E(I)**XN
127: 50       R2 = R2+(S(I)-SS(I))**2
128:          WRITE (*,51) XN,A,R2
129: 51       FORMAT(1X,' N = ',E12.6,' A = ',E12.6,' R  = ',E12.6,/1X,
130:        1 ' WITH THE FOLLOWING SOLUTION:- ',/)
131:          WRITE(*,46)
132:          DO 52 I = 1,M
133:          SS(I) = A*E(I)**XN
134:          WRITE (*,47) E(I),S(I),SS(I)
135: 52       CONTINUE
```

```
136:            IF (R(J).LE.R1(K).AND.R(J).LE.R2) GO TO 53
137:            IF (R2.LE.R1(K)  .AND.R2.LE.R(J)) GO TO 57
138:            IF (R1(K).LE.R(J).AND.R1(K).LT.R2)GO TO 55
139: 53         WRITE(*,54)
140:            GO TO 21
141: 55         WRITE(*,56)
142:            GO TO 21
143: 57         WRITE(*,58)
144: 54         FORMAT(//9X,29HS = A(B+E)**N GIVES BEST FIT./)
145: 56         FORMAT(//9X,29HS = S0+AE**N  GIVES BEST FIT./)
146: 58         FORMAT(//9X,29HS = AE**N    GIVES BEST FIT./)
147:            GO TO 21
148: 19         WRITE(*,20)
149: 20         FORMAT(/1X,' NUMBER OF ITERATIONS = 20',/)
150: 21         STOP
151:            END
152:
153:            SUBROUTINE SUM (E,S,B, A,EPO,XN,M,J)
154:            DIMENSION E(20),S(20),X(20),Y(20),XY(20),XX(20),B(20),R(20),
155:          1 SS(20),S0(20),R1(20)
156:            SUMX  =  0.
157:            SUMY  =  0.
158:            SUMXY =  0.
159:            SUMXX =  0.
160:            DO 1 I = 1,M
161:            X(I) = ALOG(B(J)+E(I))
162:            Y(I) = ALOG(S(I))
163:            XY(I) = X(I)*Y(I)
164:            XX(I) = X(I)**2
165:            SUMX  = SUMX  +  X(I)
166:            SUMY  = SUMY  +  Y(I)
167:            SUMXY = SUMXY + XY(I)
168: 1          SUMXX= SUMXX + XX(I)
169:            XM = M
170:            XN = (XM*SUMXY-SUMX*SUMY)/(XM*SUMXX-SUMX**2)
171:            A  = EXP((SUMY-XN*SUMX)/XM)
172:            EPO = XN*A*(B(J))**(XN-1.)
173:            RETURN
174:            END
175:
176:            SUBROUTINE SUM1(E,S,S0,A,XN,M,K)
177:            DIMENSION E(20),S(20),X(20),Y(20),XY(20),XX(20),B(20),R(20),
178:          1 SS(20),S0(20),R1(20)
179:            SUMX  =  0.
180:            SUMY  =  0.
181:            SUMXY =  0.
182:            SUMXX =  0.
183:            DO 4 I = 1,M
184:            X(I) = ALOG(E(I))
185:            IF (S(I)-S0(K)) 1,1,2
186: 1          Y(I) = ALOG(ABS(S(I)-S0(K))+.1E-6)
187:            GO TO 3
188: 2          Y(I) = ALOG(S(I)-S0(K))
189: 3          XY(I) = X(I)*Y(I)
190:            XX(I) = X(I)**2
191:            SUMX  = SUMX  +  X(I)
192:            SUMY  = SUMY  +  Y(I)
193:            SUMXY = SUMXY + XY(I)
194: 4          SUMXX= SUMXX + XX(I)
195:            XM = M
196:            XN = (XM*SUMXY-SUMX*SUMY)/(XM*SUMXX-SUMX**2)
197:            A  = EXP((SUMY-XN*SUMX)/XM)
198:            RETURN
199:            END
```

SOLUTION 5.3(a) Half-Hard Aluminium

```
A>SWIFT -FOR HALF-HARD ALUMINIUM:- UNITS LBF. IN.
SWIFT FITS GIVEN S/E DATA POINTS TO VARIANTS OF
EQUATION S = S0+A(B+E)**N  BY A LEAST SQUARE FIT OF M
LOG-LOG VALUES FOR A NUMBER (J) OF ASSUMED VALUES OF B
  OR ALTERNATIVELY A NUMBER (K) OF ASSUMED VALUES OF S0
WHICH ARE OPTIMIZED BY DIRECT SEARCH, CHOOSING SOLUTIONS
TO MINIMIZE R = SUMMATION OF (S(GIVEN)-S(CALCULATED))**2
TRY NOT TO USE ANY STRAIN VALUE LESS THAN E=0.01

TYPE IN M
    6
TYPE IN E( 1),S( 1)
    .1     16800.
TYPE IN E( 2),S( 2)
    .2     18040.
TYPE IN E( 3),S( 3)
    .3     19000.
TYPE IN E( 4),S( 4)
    .4     19770.
TYPE IN E( 5),S( 5)
    .5     20420.
TYPE IN E( 6),S( 6)
    .6     21000.

  FITTING EQUATION S=A(B+E)**N GIVES:-

LATEST MINIMUM VALUE OF R IS   .125542E+04 FOR B =  .128000E+00
LATEST MINIMUM VALUE OF R IS   .100261E+03 FOR B =  .107635E+00
LATEST MINIMUM VALUE OF R IS   .100241E+03 FOR B =  .107635E+00
LATEST MINIMUM VALUE OF R IS   .988577E+02 FOR B =  .106475E+00
LATEST MINIMUM VALUE OF R IS   .984848E+02 FOR B =  .106764E+00
LATEST MINIMUM VALUE OF R IS   .984583E+02 FOR B =  .106764E+00
LATEST MINIMUM VALUE OF R IS   .985118E+02 FOR B =  .106710E+00
LATEST MINIMUM VALUE OF R IS   .985106E+02 FOR B =  .106696E+00
LATEST MINIMUM VALUE OF R IS   .985136E+02 FOR B =  .106692E+00
LATEST MINIMUM VALUE OF R IS   .985357E+02 FOR B =  .106691E+00
LATEST MINIMUM VALUE OF R IS   .985184E+02 FOR B =  .106690E+00
LATEST MINIMUM VALUE OF R IS   .985431E+02 FOR B =  .106690E+00
LATEST MINIMUM VALUE OF R IS   .985431E+02 FOR B =  .106690E+00

  (R(J+1)-R(J))/R(J+1) LESS THAN .5E-6 NEAR
MINIMUM R(J), SO THE BEST SOLUTION IS AS FOLLOWS

   N            A          EP0          B           R
.181523E+00 .223643E+05 .253481E+05 .106690E+00 .985431E+02

        STRAIN           STRESS(GIVEN)      STRESS(CALCULATED)
        .1000000E+00      .1680000E+05        .1679847E+05
        .2000000E+00      .1804000E+05        .1804593E+05
        .3000000E+00      .1900000E+05        .1899448E+05
        .4000000E+00      .1977000E+05        .1976783E+05
        .5000000E+00      .2042000E+05        .2042483E+05
        .6000000E+00      .2100000E+05        .2099842E+05

  FITTING EQUATION S=S0+AE**N GIVES:-

LATEST MINIMUM VALUE OF R IS   .181946E+04 FOR S0 =  .126000E+05
LATEST MINIMUM VALUE OF R IS   .723207E+03 FOR S0 =  .131250E+05
LATEST MINIMUM VALUE OF R IS   .723207E+03 FOR S0 =  .131250E+05
LATEST MINIMUM VALUE OF R IS   .711520E+03 FOR S0 =  .130922E+05
LATEST MINIMUM VALUE OF R IS   .710504E+03 FOR S0 =  .130758E+05
LATEST MINIMUM VALUE OF R IS   .710438E+03 FOR S0 =  .130799E+05
LATEST MINIMUM VALUE OF R IS   .710466E+03 FOR S0 =  .130778E+05

  (R(K+1)-R(K))/R(K+1) LESS THAN .5E-6 NEAR
MINIMUM R(K), SO THE BEST SOLUTION IS AS FOLLOWS

    N          S0          A           R

.422923E+00 .130778E+05 .984121E+04 .710466E+03

        STRAIN           STRESS(GIVEN)      STRESS(CALCULATED)
        .1000000E+00      .1680000E+05        .1679426E+05
        .2000000E+00      .1804000E+05        .1806023E+05
        .3000000E+00      .1900000E+05        .1899224E+05
        .4000000E+00      .1977000E+05        .1975743E+05
        .5000000E+00      .2042000E+05        .2041851E+05
        .6000000E+00      .2100000E+05        .2100693E+05

  FITTING EQUATION S=AE**N  GIVES:-

N =  .124760E+00 A =  .222217E+05 R  =  .772034E+05
WITH THE FOLLOWING SOLUTION:-

        STRAIN           STRESS(GIVEN)      STRESS(CALCULATED)
        .1000000E+00      .1680000E+05        .1667310E+05
        .2000000E+00      .1804000E+05        .1817913E+05
        .3000000E+00      .1900000E+05        .1912239E+05
        .4000000E+00      .1977000E+05        .1982119E+05
        .5000000E+00      .2042000E+05        .2038075E+05
        .6000000E+00      .2100000E+05        .2084965E+05


  S = A(B+E)**N GIVES BEST FIT.
```

SOLUTION 5.3(b) Annealed Aluminium

```
A>SWIFT -FOR ANNEALED ALUMINIUM:- UNITS LBF. IN.
SWIFT FITS GIVEN S/E DATA POINTS TO VARIANTS OF
EQUATION S = S0+A(B+E)**N  BY A LEAST SQUARE FIT OF M
LOG-LOG VALUES FOR A NUMBER (J) OF ASSUMED VALUES OF B
  OR ALTERNATIVELY A NUMBER (K) OF ASSUMED VALUES OF S0
WHICH ARE OPTIMIZED BY DIRECT SEARCH, CHOOSING SOLUTIONS
TO MINIMIZE R = SUMMATION OF (S(GIVEN)-S(CALCULATED))**2
TRY NOT TO USE ANY STRAIN VALUE LESS THAN E=0.01

TYPE IN M
        6
TYPE IN E( 1),S( 1)
        .01    6700.
TYPE IN E( 2),S( 2)
        .04    9750.
TYPE IN E( 3),S( 3)
        .1    12500.
TYPE IN E( 4),S( 4)
        .2    15000.
TYPE IN E( 5),S( 5)
        .3    16750.
TYPE IN E( 6),S( 6)
        .4    18100.

   FITTING EQUATION S=A(B+E)**N GIVES:-

LATEST MINIMUM VALUE OF R IS  .172905E+05 FOR B =  .100000E-02
LATEST MINIMUM VALUE OF R IS  .717625E+04 FOR B =  .500000E-03
LATEST MINIMUM VALUE OF R IS  .598878E+04 FOR B =  .420448E-03
LATEST MINIMUM VALUE OF R IS  .574339E+04 FOR B =  .402623E-03
LATEST MINIMUM VALUE OF R IS  .568310E+04 FOR B =  .398286E-03
LATEST MINIMUM VALUE OF R IS  .567211E+04 FOR B =  .397209E-03
LATEST MINIMUM VALUE OF R IS  .566165E+04 FOR B =  .396940E-03
LATEST MINIMUM VALUE OF R IS  .566352E+04 FOR B =  .396873E-03
LATEST MINIMUM VALUE OF R IS  .566126E+04 FOR B =  .396860E-03
LATEST MINIMUM VALUE OF R IS  .566353E+04 FOR B =  .396856E-03

   (R(J+1)-R(J))/R(J+1) LESS THAN .5E-6 NEAR
MINIMUM R(J), SO THE BEST SOLUTION IS AS FOLLOWS

     N          A          EP0          B          R
.271783E+00 .232502E+05 .189488E+07 .396856E-03 .566353E+04

            STRAIN          STRESS(GIVEN)      STRESS(CALCULATED)
         .1000000E-01       .6700000E+04        .6721309E+04
         .4000000E-01       .9750000E+04        .9719750E+04
         .1000000E+00       .1250000E+05        .1244832E+05
         .2000000E+00       .1500000E+05        .1502076E+05
         .3000000E+00       .1675000E+05        .1676766E+05
         .4000000E+00       .1810000E+05        .1812967E+05

   FITTING EQUATION S=S0+AE**N GIVES:-

LATEST MINIMUM VALUE OF R IS  .187498E+04 FOR S0 =  .000000E+00

   (R(K+1)-R(K)/R(K+1) LESS THAN .5E-6 NEAR
MINIMUM R(K), SO THE BEST SOLUTION IS AS FOLLOWS

     N          S0          A          R
.269168E+00 .000000E+00 .231700E+05 .187498E+04

            STRAIN          STRESS(GIVEN)      STRESS(CALCULATED)
         .1000000E-01       .6700000E+04        .6707942E+04
         .4000000E-01       .9750000E+04        .9741924E+04
         .1000000E+00       .1250000E+05        .1246688E+05
         .2000000E+00       .1500000E+05        .1502399E+05
         .3000000E+00       .1675000E+05        .1675651E+05
         .4000000E+00       .1810000E+05        .1810561E+05

   FITTING EQUATION S=AE**N GIVES:-

N =  .269168E+00 A =  .231700E+05 R  =  .187498E+04
WITH THE FOLLOWING SOLUTION:-

            STRAIN          STRESS(GIVEN)      STRESS(CALCULATED)
         .1000000E-01       .6700000E+04        .6707942E+04
         .4000000E-01       .9750000E+04        .9741924E+04
         .1000000E+00       .1250000E+05        .1246688E+05
         .2000000E+00       .1500000E+05        .1502399E+05
         .3000000E+00       .1675000E+05        .1675651E+05
         .4000000E+00       .1810000E+05        .1810561E+05


     S = AE**N     GIVES BEST FIT.
```

Finally, it should be pointed out here that the above discussion and simple equations apply only to the mathematical modelling of the **work–hardening** characteristics of the material of interest. In real examples, such as in finite element analysis, it is preferable to use polynomial fitting to the $\bar{\sigma}$ vs. $\bar{\epsilon}$ curve. Certain ranges of both the **strain rate** and the **homologous temperature** of working profoundly affect the flow stress $\bar{\sigma}$ and cannot be neglected.

A discussion of the various standard methods of testing materials as well as of the whole problem of establishing meaningful **constitutive equations, deformation and fracture mechanism maps** and general **mapping information systems** was discussed at the beginning of this chapter. For the present purpose, namely that of obtaining realistic true stress–true strain curves incorporating strain–hardening which can be used in some of the analyses of deformation processes, equations (5.83) and (5.95) above are quite suitable.

It is hoped that the simple computer Program 5.3 written in the FORTRAN code will be found useful, for fitting these equations to measured stress–strain data. To obtain the initial data, tests should be undertaken with two values of platen width for each height h, as described by Rowe, to provide a correction for the inevitable friction effect.

Chapter 6

FINITE ELEMENT METHOD

6.1 INTRODUCTION

The finite element method (FEM) is a widely accepted numerical technique for the solution of a wide variety of problems found in engineering. It approximates the governing differential equations for a given system with a set of algebraic equations relating a finite number of variables to specific points called nodes. This method can be used for solving structural frameworks utilizing discrete elements. The solutions obtained in this case for joint displacements and member forces are identical to solutions obtained using structural analysis.

The advantages of using the finite element method are numerous. The method is easily applied to any irregular-shaped body made up of one or more materials and having a mixed set of boundary conditions. However, the method is cumbersome without the aid of digital computers. Fortunately, the finite element method is particularly well suited to computer applications and numerous general purpose finite element programs are currently available.

Thus, problems having complex geometries and boundary conditions which cannot usually be solved using classical 'strength of materials' approaches can be solved using the finite element method. This chapter provides a brief overview of the method as applied to the solution of 'strength of material' and 'structural analysis' problems for continua.

6.2 GENERAL THEORY OF THE FINITE ELEMENT METHOD

Finite element analysis can be divided into three specific methods:

1. The direct stiffness or displacement method where the nodal displacements are the basic unknowns in the set of algebraic equations.

2. The force method in which the internal nodal forces are unknown.

3. The mixed method where the equation may be

expressed in terms of nodal displacement and internal forces.

Throughout this chapter, the displacement method of finite element analysis is used.
The continuum is first idealized into a number of discrete elements inter-connected at suitable nodal points. The displacements and the forces on the continuum are usually represented at these nodal points, the basic unknowns for the problem being the nodal displacements. The other parameters are usually obtained from these displacements. A displacement mode with as many undetermined coefficients (generalized displacements) as there are nodal point displacements is chosen for the type of element being used.

Consider the two-dimensional case in which the coordinates are given by x and y; then

$$\rho(x,y) = (u,v)^T \tag{6.1}$$

where ρ is the displacement at any point and u,v are the respective displacement vectors in the x and y directions. Hence,

$$\rho(x,y) = f(x,y)\alpha \tag{6.2}$$

where α is the generalized displacement, and $f(x,y)$ the chosen displacement function. Substituting the displacements and the coordinates at the nodal points gives:

$$\rho = F\,\alpha \tag{6.3}$$

where

$$\rho = (u_1,\ v_1,...u_n,\ v_n)^T \tag{6.4}$$

n being the number of nodal points per element and F being an $(n \times n)$ matrix. (note: bold letter denotes a matrix).

Inverting the previous equation gives:

$$\alpha = F^{-1}\rho \tag{6.5}$$

A differential operator, A, which transforms the generalized displacements into strains is defined. Therefore:

$$\epsilon = AF^{-1}\rho \qquad \text{i.e.} \quad \epsilon = B\rho \tag{6.6}$$

where B is the strain displacement matrix. The external forces, P, are assumed to be acting at the nodal points and using the Principle of Virtual Work, we have

$$\delta\rho^T P = \int \delta\epsilon^T \sigma \, dv \qquad (6.7)$$

Substituting equation (6.6) and cancelling the non-zero displacement vector, gives:

$$\rho = \int B^T \sigma \, dv \qquad (6.8)$$

The stresses and strains are related to each other by

$$\sigma = D\epsilon \qquad (6.9)$$

where D, an elasticity matrix, contains material properties. For details refer to *Strength of Materials* vol I chapter 5. Now equation (6.8) becomes

$$P = K\rho \qquad (6.10)$$

where

$$K = \int B^T DB dv \qquad (6.11)$$

and is known as the element stiffness matrix.

For the general two-dimensional case the stresses and strains can be expressed as:

$$\sigma^T = \left\{ \sigma_x \ \sigma_y \ \sqrt{2}\sigma_{xy} \right\} \qquad (6.12)$$

$$\epsilon^T = \left\{ \epsilon_x \ \epsilon_y \ 1/\sqrt{2} \ \gamma_{xy} \right\} \qquad (6.13)$$

If the displacements are assumed to be small, then

$$\epsilon_x = \frac{\partial u}{\partial x} \qquad (6.14)$$

$$\epsilon_{xy} = 1/2 \ \gamma_{xy} \qquad (6.15)$$

$$\epsilon_{xy} = \frac{1}{2} \left[\frac{\partial u}{\partial x} + \frac{\partial v}{\partial y} \right] \tag{6.16}$$

Using equation (6.6) gives:

$$B = AF^{-1} \tag{6.17}$$

The stress-strain matrix for an isotropic elastic material in plane strain has been shown in *Strength of Materials* Vol 1 to be:

$$D = \frac{E}{1 - \nu^2} \begin{bmatrix} 1 & \nu & 0 \\ \nu & 1 & 0 \\ 0 & 0 & (1-\nu)/2 \end{bmatrix} \tag{6.18}$$

The overall stiffness matrix, K, is obtained by the appropriate assembly of all the element stiffness matrices, thus

$$K \delta = Q \tag{6.19}$$

where Q = the nodal forces and δ = the nodal displacements. This matrix, relating all the nodal displacements to the various nodal loads, is generally singular as it includes the rigid body motions and rotations.

6.3 DEVELOPMENT OF A ROD ELEMENT

In the following section a practical problem dealing with the analysis of a rod element will be discussed. The rod of length l, cross-sectional area A, and modulus of elasticity E, will have an axial force F as shown in the Fig. 6.1

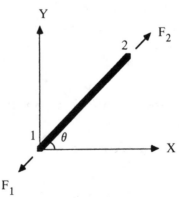

Fig. 6.1 Illustrating a rod element

The rod element will have nodes 1 and 2 present at each end. The displacement of these nodes will occur along the axial direction creating a change in length according to the relation shown below.

$$\delta_2 - \delta_1 = \delta = FL/AE \qquad (6.20)$$

The stiffness (K) of the member will be defined as the force present divided by the change in length of the member:

$$K = \frac{F}{\delta} \qquad (6.21)$$

Combining equations (6.20) and (6.21) and solving for K yields

$$K = \frac{AE}{L} \qquad (6.22)$$

Applying equation (6.22) to the two-force member shown in Fig. 6.1, gives the force at each node in terms of the displacement, δ_1, δ_2, and the stiffness K of the member

$$F_1 = K\delta_1 - K\delta_2 \qquad (6.23)$$

$$F_2 = K\delta_2 - K\delta_1 \qquad (6.24)$$

Similarly, the above two equations can be written in matrix form as shown below.

$$\left\{ \begin{array}{c} F_1 \\ F_2 \end{array} \right\} = K \left[\begin{array}{cc} 1 & -1 \\ -1 & 1 \end{array} \right] \left\{ \begin{array}{c} \delta_2 \\ \delta_1 \end{array} \right\} \qquad (6.25)$$

To summarize the matrix format above

$$Q = K\, q \qquad (6.26)$$

where, K is the element stiffness matrix and the load vector Q and the displacement vector q are given as follows,

$$Q = \left\{ \begin{array}{c} F_1 \\ F_2 \end{array} \right\} \tag{6.27}$$

$$q = \left\{ \begin{array}{c} \delta_1 \\ \delta_2 \end{array} \right\} \tag{6.28}$$

However, the two-force member shown in the Fig. 6.1 has an arbitrary position and therefore requires a new coordinate system to be established with the origin at node 1 and x directed along the member's length. For a general orientation of the member in space, it can be shown that:

$$\delta_2 = \delta_{2x}\cos\theta + \delta_{2y}\sin\theta \tag{6.29}$$

$$\delta_1 = \delta_{1x}\cos\theta + \delta_{1y}\sin\theta \tag{6.30}$$

Similarly, as shown before, equations (6.29) and (6.30) can be written in matrix form as

$$\left\{ \begin{array}{c} \delta_1 \\ \delta_2 \end{array} \right\} = [T] \left[\begin{array}{c} \delta_{1x} \\ \delta_{1y} \\ \delta_{2x} \\ \delta_{2y} \end{array} \right] \tag{6.31}$$

where the transformation matrix T is shown below:

$$T = \left[\begin{array}{cccc} \cos\theta & \sin\theta & 0 & 0 \\ 0 & 0 & \cos\theta & \sin\theta \end{array} \right] \tag{6.32}$$

Using equation (6.22) for K along with equation (6.31) and the transformation matrix, and performing some basic matrix operations gives

$$K = \frac{AE}{L} \left[\begin{array}{cccc} \cos^2\theta & \sin\theta & -\cos^2\theta & -\cos\theta\sin\theta \\ \cos\theta\sin\theta & \sin^2\theta & -\cos\theta\sin\theta & -\sin^2\theta \\ -\cos^2\theta & -\cos\theta\sin\theta & \cos^2\theta & \cos\theta\sin\theta \\ -\cos\theta\sin\theta & -\sin^2\theta & \cos\theta\sin\theta & \sin^2\theta \end{array} \right] \tag{6.33}$$

Equation (6.33) is used in the formulation of the global stiffness matrix which is used to calculate the displacement of the respective nodes.

To calculate the forces along the member's x and y directions, the matrix for forces should be solved, as follows:

$$F_1 = \frac{AE}{L} [(\delta_{1x} - \delta_{2x})\cos\theta + (\delta_{1y} - \delta_{2y})\sin\theta)] \qquad (6.34)$$

$$F_2 = \frac{AE}{L} [(\delta_{2x} - \delta_{1x})\cos\theta + (\delta_{2y} - \delta_{1y})\sin\theta] \qquad (6.35)$$

The solution of the above equations gives the value of the force. It is a general practice to assume tension as a positive value and compression as a negative value.

Ex: 1
Calculate the unknown nodal displacements and axial forces in each member. Show that the calculated axial forces produce a system that is in equilibrium. There is no temperature change.

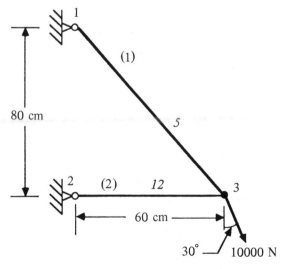

Member	A	L	cos θ	sin θ
1	5	100	0.6	-0.8
2	12	60	1.0	0.0

	Member	
	1	2
AE $\cos^2\theta/L$	$3.6(10^5)$	$4(10^6)$
AE $\cos\theta \sin\theta/L$	$-4.8(10^5)$	0
AE $\sin^2\theta/L$	$6.4(10^5)$	0

$$K^1 = \begin{bmatrix} 3.6 & -4.8 & -3.6 & 4.8 \\ -4.8 & 6.4 & 4.8 & -6.4 \\ -3.6 & 4.8 & 3.6 & -4.8 \\ 4.8 & -6.4 & -4.8 & 6.4 \end{bmatrix} 10^5 \begin{matrix} 1 \\ 2 \\ 5 \\ 6 \end{matrix}$$

$$\begin{matrix} 1 & 2 & 5 & 6 \end{matrix}$$

$$K^2 = \begin{bmatrix} 4 & 0 & -4 & 0 \\ 0 & 0 & 0 & 0 \\ -4 & 0 & 4 & 0 \\ 0 & 0 & 0 & 0 \end{bmatrix} 10^6 \begin{matrix} 3 \\ 4 \\ 5 \\ 6 \end{matrix}$$

$$\begin{matrix} 3 & 4 & 5 & 6 \end{matrix}$$

$$K = 10^5 \begin{bmatrix} 3.6 & -4.8 & 0 & 0 & -3.6 & 4.8 \\ -4.8 & 6.4 & 0 & 0 & 4.8 & -6.4 \\ 0 & 0 & 40 & 0 & -40.0 & 0 \\ 0 & 0 & 0 & 0 & 0 & 0 \\ -3.6 & 4.8 & -40.0 & 0 & 43.6 & -4.8 \\ 4.8 & -6.4 & 0 & 0 & -4.8 & 6.4 \end{bmatrix}$$

$$K U - P = 0$$

$$10^5 \begin{bmatrix} 43.6 & -4.8 \\ -4.8 & 64.0 \end{bmatrix} \begin{Bmatrix} U_5 \\ U_6 \end{Bmatrix} = \begin{Bmatrix} 5000.00 \\ -8660.3 \end{Bmatrix}$$

$$U_5 = -0.000374$$
$$U_6 = -0.013816$$

Member 1

$$i = 1 \quad j = 3$$

$$S_1 = AE/L[(U_5 - U_1)\cos\theta + (U_6 - U_2)\sin\theta]$$

$$= 5(20 \times 10^6)/100[(-0.000374)0.6 + (-0.01382)(-0.8)]$$

$$= 10944 \ N$$

$$S^2 = AE/L[(U_5 - U_2)\cos\theta + (U_6 - U_2)\sin\theta]$$

$$= -1496 \ N$$

6.4 DEVELOPMENT OF A BEAM ELEMENT

While the rod element previously studied can certainly be used to advantage in structural analysis, it is also a severely limiting element since it applies only to axial forces. For this reason, consider a beam element with flexural stiffness EI over a length L, where E is the Young's Modulus and I the moment of inertia.

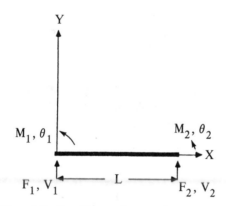

Fig. 6.2 Configurations of a beam element

The beam element will support shear and bending forces. Its configuration is shown in Fig. 6.2. As indicated, there are displacements, V, corresponding to the applied shear forces, F, as well as rotation, θ, relating to the applied moments, M, at the nodal points. The relationship can be expressed as

$$Q = K \ q \qquad (6.36)$$

or for this specific case,

$$\begin{Bmatrix} F_1 \\ M_1 \\ F_2 \\ M_2 \end{Bmatrix} = [K] \begin{Bmatrix} V_1 \\ \theta_1 \\ V_2 \\ \theta_2 \end{Bmatrix} \qquad (6.37)$$

where K represents a 4 × 4 matrix. It is now possible to determine the entries, K_{ij}, in the stiffness matrix using a strength of materials approach. This is accomplished by equating one of the displacements to one and assuming the rest to be equal to zero. Thus by setting $V_1 = 1$ and specifying the remaining deflections to be equal to zero gives:

$$\begin{Bmatrix} F_1 \\ M_1 \\ F_2 \\ M_2 \end{Bmatrix} = \begin{bmatrix} K_{11} & K_{12} & K_{13} & K_{14} \\ K_{21} & K_{22} & K_{23} & K_{24} \\ K_{31} & K_{32} & K_{33} & K_{34} \\ K_{41} & K_{42} & K_{43} & K_{44} \end{bmatrix} \begin{Bmatrix} 1 \\ 0 \\ 0 \\ 0 \end{Bmatrix} \qquad (6.38)$$

on multiplying:

$$\begin{Bmatrix} F_1 \\ M_1 \\ F_2 \\ M_2 \end{Bmatrix} = \begin{Bmatrix} K_{11} \\ K_{21} \\ K_{31} \\ K_{41} \end{Bmatrix} \qquad (6.39)$$

Thus the forces required for the imposed displacement are equal to the entries in the stiffness matrix. From 'strength of materials' we know that for the displacement V_1 the reactions will be:

$$F_1 = \frac{12EI}{L^3} V_1 \qquad\qquad M_1 = - \frac{6EI}{L^2} V_1 \qquad (6.40)$$

$$F_2 = - \frac{12EI}{L^3} V_1 \qquad\qquad M_2 = - \frac{6EI}{L^2} V_1 \qquad (6.41)$$

and since $V=1$, substituting equations (6.40) and (6.41) into the equation (6.39) gives the first column of the stiffness matrix.

By setting $\theta_1=1$ and all other deflection equal to zero it is possible to obtain the second column in a similar manner.

$$\begin{Bmatrix} F_1 \\ M_1 \\ F_2 \\ M_2 \end{Bmatrix} = \begin{bmatrix} K_{11} & K_{12} & K_{13} & K_{14} \\ K_{21} & K_{22} & K_{23} & K_{24} \\ K_{31} & K_{32} & K_{33} & K_{34} \\ K_{41} & K_{42} & K_{43} & K_{44} \end{bmatrix} \begin{Bmatrix} 0 \\ 1 \\ 0 \\ 0 \end{Bmatrix} \qquad (6.42)$$

and

$$\begin{Bmatrix} F_1 \\ M_1 \\ F_2 \\ M_2 \end{Bmatrix} = \begin{Bmatrix} K_{11} \\ K_{21} \\ K_{31} \\ K_{41} \end{Bmatrix} \qquad (6.43)$$

Again from 'strength of materials', for the rotation θ_1,

$$F_1 = - \frac{6EI}{L^2} \theta_1 \qquad\qquad M_1 = \frac{4EI}{L} \theta_1 \qquad (6.44)$$

$$F_2 = \frac{6EI}{L^2} \theta_1 \qquad\qquad M_2 = \frac{2EI}{L} \theta_1 \qquad (6.45)$$

Continuing in a similar manner for the displacement and rotation at node 2 yields the remaining items to be entered in the matrix finally giving:

$$K = \frac{EI}{L^3} \begin{bmatrix} 12 & -6L & 2L^2 & 6L \\ -6L & 4L^2 & -12 & 6L \\ -12 & 6L & 12 & 6L \\ -6L & 2L^2 & 6L & 4L^2 \end{bmatrix} \qquad (6.46)$$

6.5 THREE NODE TRIANGULAR ELEMENT

A rod element and a beam element have so far been considered. These two basic elements work well for applications involving discrete structures; however, when applying the finite element method, a two or three dimensional element is required (note: the discussion is restricted to 2D elements only).

There are several types of 2D elements which can be used for this purpose. The most common of these are the 3-node triangular and 4-node quadrilateral elements. For greater accuracy a 6-node triangular or 8-node quadrilateral element may be used. The extra nodes are placed at the mid-points of the sides. More importantly, the interpolation functions are no longer assumed linear, but of second order. For this reason, these elements are sometimes referred to as quadratic elements or in general, iso-parametric elements.

However, in the interest of simplicity, only the 3-node triangular element will be considered in this text.
    Assuming a linear interpolation function results in the following

$$\Phi(x,y) = \alpha_1 + \alpha_2 x + \alpha_3 y \tag{6.47}$$

Applying this to displacements u and v in the x and y directions yields

$$u(x,y) = \alpha_1 + \alpha_2 x + \alpha_3 y \tag{6.48}$$
$$v(x,y) = \beta_1 + \beta_2 x + \beta_3 y \tag{6.49}$$

where the orientation is as shown in Fig. 6.3

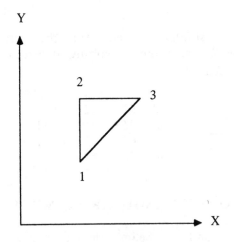

Fig. 6.3 Triangular element

Substituting nodal coordinates and displacements

$$\begin{bmatrix} u_1 \\ u_2 \\ u_3 \end{bmatrix} = \begin{bmatrix} 1 & x_1 & y_1 \\ 1 & x_2 & y_2 \\ 1 & x_3 & y_3 \end{bmatrix} \tag{6.50}$$

Symbolically,

$$q_1 = A\,\alpha \tag{6.51}$$

Similar expressions can be obtained for displacements v in the y-direction as follows:

$$\begin{bmatrix} v_1 \\ v_2 \\ v_3 \end{bmatrix} = \{ q_2 \} = [A] \{ \beta \} \tag{6.52}$$

Solving for α and β:

$$\alpha = A^{-1} q_1 \tag{6.53}$$

$$\beta = A^{-1} q_2 \tag{6.54}$$

where

$$A^{-1} = \frac{1}{\det} \begin{bmatrix} a_1 & a_2 & a_3 \\ b_1 & b_2 & b_3 \\ c_1 & c_2 & c_3 \end{bmatrix} \tag{6.55}$$

and

$$
\begin{array}{lll}
a_1 = x_2 y_3 - x_3 y_2 & a_2 = x_3 y_1 - x_1 y_3 & a_3 = x_1 y_2 - x_2 y_1 \\
b_1 = y_2 - y_3 & b_2 = y_3 - y_1 & b_3 = y_1 - y_2 \\
c_1 = x_3 - x_2 & c_2 = x_1 - x_3 & c_3 = x_2 - x_1
\end{array}
$$

and the determinant

$$\begin{bmatrix} 1 & x_1 & y_1 \\ 1 & x_2 & y_2 \\ 1 & x_2 & y_2 \end{bmatrix} = 2(\text{Area of the triangle}) = 2A$$

Substitution of equations (6.53) and (6.54) into equation (6.48) and (6.49) yields

$$u = \begin{bmatrix} 1 & x & y \end{bmatrix} \begin{bmatrix} A \end{bmatrix}^{-1} \{ q_1 \} \tag{6.56}$$

$$v = \begin{bmatrix} 1 & x & y \end{bmatrix} \begin{bmatrix} A \end{bmatrix}^{-1} \{ q_2 n \} \qquad (6.57)$$

which gives:

$$\begin{bmatrix} u \\ v \end{bmatrix} = \begin{bmatrix} N_1 & 0 & N_2 & 0 & N_3 & 0 \\ 0 & N_1 & 0 & N_2 & 0 & N_3 \end{bmatrix} \begin{Bmatrix} u_1 \\ v_1 \\ u_2 \\ v_2 \\ u_3 \\ v_3 \end{Bmatrix} \qquad (6.58)$$

where

$$N_1 = \frac{1}{2A} \begin{bmatrix} a_1 + b_1 x + c_1 y \end{bmatrix}$$

$$N_2 = \frac{1}{2A} \begin{bmatrix} a_2 + b_2 x + c_3 y \end{bmatrix} \qquad (6.59)$$

$$N_3 = \frac{1}{2A} \begin{bmatrix} a_3 + b_3 x + c_3 y \end{bmatrix}$$

N_1, N_2 and N_3 are called *interpolation functions* and a, b and c are as previously defined.

6.5.1 Strain and Strain Energy

Strain is derived from the gradient of the displacements. Hence,

$$\begin{Bmatrix} \epsilon_x \\ \epsilon_y \\ \gamma_{xy} \end{Bmatrix} = \begin{bmatrix} \partial/\partial x & 0 \\ 0 & \partial/\partial y \\ \partial/\partial x & \partial/\partial x \end{bmatrix} \begin{Bmatrix} u \\ v \end{Bmatrix} = \begin{bmatrix} \partial/\partial x & 0 \\ 0 & \partial/\partial y \\ \partial/\partial y & \partial/\partial x \end{bmatrix} N q$$

$$= \frac{1}{Det} \begin{bmatrix} y_2-y_3 & 0 & y_3-y_1 & 0 & y_1-y_2 & 0 \\ 0 & x_3-x_2 & 0 & x_1-x_3 & 0 & x_2-x_1 \\ x_3-x_2 & y_2-y_3 & x_1-x_3 & y_3-y_1 & x_2-x_1 & y_1-y_2 \end{bmatrix} \begin{Bmatrix} u_1 \\ v_1 \\ u_2 \\ v_2 \\ u_3 \\ v_3 \end{Bmatrix}$$

$$(6.60)$$

i.e. $\epsilon = B \, q$

where

$$B = \frac{1}{\text{Det}} \begin{bmatrix} b_1 & 0 & b_2 & 0 & b_3 & 0 \\ 0 & c_1 & 0 & c_2 & 0 & c_3 \\ c_1 & b_1 & c_2 & b_2 & c_3 & b_3 \end{bmatrix}$$

The strain energy in a continuum can be expressed as:

$$U = \frac{1}{2} \int \epsilon^T D \, \epsilon \, dv$$

$$= \frac{1}{2} \int q^T B^T D B \, q \, dv \qquad (6.61)$$

where D is the constitutive matrix or the stress-strain matrix. (For details refer to *Strength of Materials* Vol 1).
        For plane stress, the stress-strain matrix can be expressed as:

$$D = \frac{E}{1-\nu^2} \begin{bmatrix} 1 & \nu & 0 \\ \nu & 1 & 0 \\ 0 & 0 & (1-\nu)/2 \end{bmatrix} \qquad (6.62)$$

For plane strain, it is given by:

$$D = \frac{E}{(1+\nu)(1-2\nu)} \begin{bmatrix} 1-\nu & \nu & 0 \\ \nu & 1-\nu & 0 \\ 0 & 0 & (1-2\nu)/2 \end{bmatrix} \qquad (6.63)$$

6.5.2 Derivation of Applied Load Energy Functions

There are three types of element loading to be considered.

1. Concentrated forces applied to nodes:

        The potential energy function, V, for a force F having the components F_x, F_y is defined by:

$$V = -F_x u - F_y v \qquad (6.64)$$

For forces at node 1 with the components F_{1x} and F_{2x}, the potential energy is given by:

$$V_1 = -F_{1x} u_1 - F_{1y} v_1 \qquad (6.65)$$

By extending this development to all force components acting at all nodes of the element,

$$V_{NF} = -\, q^T\, Q_{NF} \qquad (6.66)$$

where $\quad Q^T_{NF} = -[\; F_{1x}\ F_{1y}\ F_{2x}\ F_{2y}\ F_{3x}\ F_{3y}\;]$

2. Forces distributed along the edge of the element

The force produced by the stress traction vector (the force per unit area at a particular point) acting in the direction defined by the unit vector, n on the area dA along one side of the element is given by:

$$dF = T\ dA\ n$$

The differential potential energy function, dV_T can be written for this differential force as:

$$dV_T = -(T_x dA)u - (T_y\ dA)v$$

where u,v are the displacement components for the arbitrary point on the element boundary

$$V_T = -\int_s [\; u\ \ v\;] \left\{ \begin{array}{c} T_x \\ T_y \end{array} \right\} h\ ds$$

$$= -\, h\int_s q\ N^T \left\{ \begin{array}{c} T_x \\ T_y \end{array} \right\} ds$$

$$= -\, h\ q^T \int_s N^T \left\{ \begin{array}{c} T_x \\ T_y \end{array} \right\} ds \qquad (6.67)$$

3. Body Forces

Defining B_x as the body force in the x-direction and B_y as the body force in the y-direction, the differential potential function for these forces can be written as;

$$dV_{BF} = - (B_x \, d\Omega \,)u - (B_y \, d\Omega \,)v$$

$$= - \int_A [\, u \, v \,] \left\{ \begin{matrix} B_x \\ B_y \end{matrix} \right\} h \, dA$$

$$V_{BF} = - h \, q^T \int_A N^T \left\{ \begin{matrix} B_x \\ B_y \end{matrix} \right\} dA \qquad (6.68)$$

6.5.3 *Summation of Energy Terms*

The total energy in a single element system is expressed as:

$$\Pi = U + V_{NF} + V_T + V_{BF}$$

$$= \frac{1}{2} \int_\Omega q^T \, B^T \, C \, B \, q \, d\Omega - q^T \left\{ \begin{matrix} F_{1x} \\ F_{1y} \\ F_{2x} \\ F_{2y} \\ F_{3x} \\ F_{3y} \end{matrix} \right\}$$

$$- h \, q^T \int_S N^T \left\{ \begin{matrix} T_x \\ T_y \end{matrix} \right\} ds - h \, q^T \int N^T \left\{ \begin{matrix} B_x \\ B_y \end{matrix} \right\} dA$$

$$(6.69)$$

6.5.4 *Application of the Energy Minimization Principle*

The total energy in the element is minimized relative to the nodal displacements. This can be accomplished by differentiating the scalar expression with respect to each and every one of the nodal displacements and setting each resulting equation equal to zero.

$$B^T \, C \, B \, d\Omega \left\{ \begin{array}{c} u_1 \\ v_1 \\ u_2 \\ v_2 \\ u_3 \\ v_3 \end{array} \right\} = \left\{ \begin{array}{c} F_{1x} \\ F_{1y} \\ F_{2x} \\ F_{2y} \\ F_{3x} \\ F_{3y} \end{array} \right\} + h \int N^T \left\{ \begin{array}{c} T_x \\ T_y \end{array} \right\} ds$$

$$+ h \int N^T \left\{ \begin{array}{c} B_x \\ B_y \end{array} \right\} dA \qquad (6.70)$$

i.e. $K \, q \;\; = Q_{NF} + Q_T + Q_{BF} = Q$ (6.71)

where K - element stiffness matrix
q - element nodal displacement velocity
Q - total element load vector

q can be calculated from the equation $K \, q = Q$ by using the Gaussian elimination method and satisfying the boundary conditions.

Then element strain, $\epsilon = B \, q$ (6.72)

and element stress, $\sigma = D \, \epsilon$ (6.73)

6.6 MATERIAL NON-LINEARITY(PLASTICITY)

The development of the finite element method and the associated derivation of element stiffness matrices have so far being restricted to materials with linear elastic stress-strain relationship. Diagramatic stress-strain curves for mild steel and aluminium in tension are shown in Fig. 2.1 and 2.2 of *Strength of Materials* Vol 1. Beyond the elastic limit the material undergoes a permanent plastic deformation.

The gradient of the stress-strain curve within the limit of proportionality is known as the Young's Modulus. Outside this limit the gradient is referred to as the plastic modulus. The plastic modulus may change as a function of the strain. In addition to this form of non-linearity, the basic stress-strain relationship is also non-linear and is governed by the yield criteria and associated flow rules.

Plastic flow takes place when the 'equivalent' or 'effective' flow stress exceeds the initial yield stress of the material (refer to *Strength of Materials* Vol 1 Pg 175 for more details).

For a complex 3D state of stress the effective stress must be a function of the invariant of the stress tensor. It has been shown that for most practical purposes the yield stress is unaffected by the invariants J_1 and J_3. The two most important criteria are those of Tresca and von Mises (Maxwell).

von Mises' yield criteria is stated as

$$J_2 = \frac{S_{ij}\ S_{ij}}{2} = k^2 \tag{6.74}$$

where k is the yield shear stress of the given material, and S_{ij} is the stress tensor.

The associated elastic plastic stress-strain incremental relationship is known as Prandtl-Reuss equation. From *Strength of Materials* Vol 1 the total strain increment $d\epsilon_{ij}$ is assumed to be the sum of an increment of elastic strain and an increment of plastic strain. These can be written as

$$d\epsilon_{ij}{}^e = \frac{ds_{ij}}{2G} + \frac{\delta_{ij}d\sigma_m}{3K}$$

$$d\epsilon_{ij}{}^e = \frac{1}{E}\left[(1 + \nu)d\sigma_{ij} - 3\nu\,\delta_{ij}d\sigma_m\right]$$

$$d\epsilon_{ij}{}^P = s_{ij}\ d\lambda \tag{6.75}$$

$$d\epsilon_{ij} = \frac{1}{E}\left[(1 + \nu)d\sigma_{ij} - 3\nu\,\delta_{ij}\ d\sigma_m\right]$$

$$+ \frac{9}{4\sigma^2\ E_p}s_{ij}\ s_{kl}\ d\sigma_{kl}$$

In matrix notation the above equations can be written as

$$d\epsilon = d\epsilon^e + d\epsilon^P = (\ F^e + F^P\ )\ d\epsilon \tag{6.76}$$

$$= F^t\ d\sigma$$

where F^t is the tangential flexibility matrix.

F^e is the elastic flexibility matrix and is given by equation (5.106) of *Strength of Materials* Vol 1.

The plastic flexibility matrix can be written as

$$F^p = \frac{9}{4\sigma^2 E_p} \, S \, S^T \, e_6 \qquad (6.77)$$

where $S^T = [\, S_{11} \;\; S_{22} \;\; S_{33} \;\; S_{12} \;\; S_{23} \;\; S_{31}]$ are the deviatoric stresses

E_p = plastic modulus $d\sigma/d\epsilon_p$

σ = effective stress

The full matrix is given in *Strength of Materials* Vol 1 equation (5.109). The elastic-plastic stiffness matrix can be written as

$$D_{ep} = (F^t)^{-1}$$

D_{ep} is used in place of D_e when deriving the element stiffness matix.

6.7 FIELD EQUATIONS - TORSION OF NON-CIRCULAR SECTIONS

A slightly different equation has to be solved for the solution of 2D field equation problems uch as torsion, heat transfer and fluid flow. The general 2D field equation has the following form:

$$D_x \frac{\partial^2 \varphi}{\partial x^2} + D_y \frac{\partial^2 \varphi}{\partial y^2} - P\varphi + Q = 0 \qquad (6.78)$$

However, for the torsion problem using Prandtl's theory the governing differential equation reduces to:

$$\frac{1}{G} \frac{\partial^2 \varphi}{\partial x^2} + \frac{1}{G} \frac{\partial^2 \varphi}{\partial y^2} + 2\theta = 0 \qquad (6.79)$$

where G is the shear modulus of the material and θ is the angle of twist. Equation (6.79) is obtained from equation (6.78) by noting that $D_x = D_y = 1/G$, $P=0$ and $Q=2\theta$. The variable φ is a stress function, and the shear stresses within the shaft are related to the derivatives of φ with respect to x and y as follows:

$$\tau_{zx} = \frac{\partial \varphi}{\partial y} \quad \text{and} \quad \tau_{zy} = -\frac{\partial \varphi}{\partial x}$$

where $\varphi = 0$ on the boundary.

Prandtl's formulation does not have the applied torque M_t in the governing equation. Instead M_t is calculated using

$$M_t = 2\int_A \varphi dA$$

once $\varphi(x,y)$ is known.

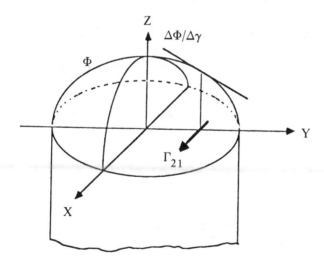

Fig. 6.4 The φ surface and the related shear stress components

The stress function represents a surface covering the cross-section of the shaft as shown in the Fig. 6.4. The torque is proportional to the volume under the surface while the shear stresses are related to the surface gradients in the x and y coordinates.

Equation (6.79) is usually written as

$$\frac{\partial^2 \varphi}{\partial x^2} + \frac{\partial^2 \varphi}{\partial y^2} + 2G\theta = 0 \qquad (6.80)$$

when the shaft is composed of a single material.

6.7.1 Derivation of the Triangular Element Matrix

Following the derivation of the B matrix for the elastic case, the value of φ at any point within the element can be written as:

$$\varphi = N \{ \varphi \} \qquad (6.81)$$

where $N = [N_i \ N_j \ N_k]$

and $\{\varphi\}^t = \{ \varphi_i \ \varphi_j \ \varphi_k \}$

and i,j,k are the three nodes.

The gradient of N yields the B matrix. Thus:

$$B = \begin{bmatrix} \frac{\partial N_i}{\partial x} & \frac{\partial N_j}{\partial x} & \frac{\partial N_k}{\partial x} \\ \frac{\partial N_i}{\partial y} & \frac{\partial N_j}{\partial y} & \frac{\partial N_k}{\partial y} \end{bmatrix}$$

$$= \frac{1}{2A} \begin{bmatrix} b_i & b_i & b_k \\ c_i & c_j & c_k \end{bmatrix}$$

The D matrix for this case can be shown to be the identity matrix (because $D_x = D_y = 1$). Hence:

$$D = \begin{bmatrix} 1 & 0 \\ 0 & 1 \end{bmatrix}$$

The element stiffness matrix for this case becomes:

$$K = \int B^T \ D \ B \ dA$$

$$= B^T \ B \ \int dA$$

$$= B^T \ B \ A \tag{6.82}$$

Expanding the matrix yields

$$K_D = \frac{1}{4A} \begin{bmatrix} b_i^2 & b_i b_j & b_i b_k \\ b_i b_j & b_j^2 & b_j b_k \\ b_i b_k & b_j b_k & b_k^2 \end{bmatrix} + \frac{1}{4A} \begin{bmatrix} c_i^2 & c_i c_j & c_i c_k \\ c_i c_j & c_j^2 & c_j c_k \\ c_i c_k & c_j c_k & c_k^2 \end{bmatrix} \tag{6.83}$$

The load vector can be determined in a similar manner as:

$$f = \frac{2g\theta A}{3} \begin{bmatrix} 1 \\ 1 \\ 1 \end{bmatrix} \tag{6.84}$$

6.7.2 *Derivation of the Rectangular Element Matrix*

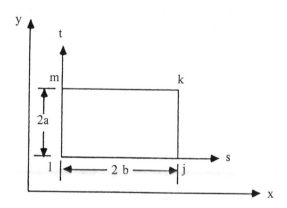

Fig. 6.5 Rectangular element

The rectangular element as shown in the Fig. 6.5 has a length $2b$ and a height $2a$. The local coordinate sytem is given in terms of s and t and its origin is at the lower left-hand corner and coincides with node i. The value of φ at any point within the element can be written as

$$\varphi = N_i \Phi_i + N_j \Phi_j + N_k \Phi_k + N_m \Phi_m \tag{6.85}$$

where

$$N_i = \left[1 - \frac{s}{2b} \right] \left[1 - \frac{s}{2a} \right]$$

$$N_j = \frac{s}{2b} \left[1 - \frac{t}{2a} \right]$$

$$N_k = \frac{st}{4ab}$$

$$N_m = \frac{t}{2a} \left[1 - \frac{s}{2b} \right]$$

These shape functions have properties similar to those of
a triangular element.   Each shape function varies linearly
along the edges between its node and the two adjacent nodes.
The values of $[K]$ and $\{f\}$ are evaluated by using specific
integral in each case.   A simple example is shown below,

$$\{f^e\} = \int_A Q \, [N]^T \, dA$$

$$= \int_0^{2b}\int_0^{2a} Q \begin{Bmatrix} N_i \\ N_j \\ N_k \\ N_m \end{Bmatrix} dt \; ds \qquad (6.86)$$

which after integeration yields

$$\{f^e\} = \frac{QA}{4} \begin{Bmatrix} 1 \\ 1 \\ 1 \\ 1 \end{Bmatrix} \qquad (6.87)$$

The integral associated with $[K_G]$ is

$$[K_G] = \int_A G \, [N]^T [N] \, dA$$

$$= \int_A G \begin{bmatrix} N_i{}^2 & N_i N_j & N_i N_k & N_i N_m \\ N_i N_j & N_j{}^2 & N_j N_k & N_j N_m \\ N_i N_k & N_j N_k & N_k{}^2 & N_k N_m \\ N_i N_m & N_j N_m & N_k N_m & N_m{}^2 \end{bmatrix} dA$$

(6.88)

The complete set of coefficients is given by:

$$[K_G] = \frac{GA}{36} \begin{bmatrix} 4 & 2 & 1 & 2 \\ 2 & 4 & 2 & 1 \\ 1 & 2 & 4 & 2 \\ 2 & 1 & 2 & 4 \end{bmatrix}$$

(6.89)

The gradient matrix $[B]$ is given by

$$[B] = \begin{bmatrix} \dfrac{\partial N_i}{\partial x} & \dfrac{\partial N_j}{\partial x} & \dfrac{\partial N_k}{\partial x} & \dfrac{\partial N_m}{\partial x} \\ \dfrac{\partial N_i}{\partial y} & \dfrac{\partial N_j}{\partial y} & \dfrac{\partial N_k}{\partial y} & \dfrac{\partial N_m}{\partial y} \end{bmatrix}$$

(6.90)

The complete result for $[K_D]$ is

$$[K_D] = \frac{D_x a}{6b} \begin{bmatrix} 2 & -2 & -1 & 1 \\ -2 & 2 & 1 & -1 \\ -1 & 1 & 2 & -2 \\ 1 & -1 & -2 & 2 \end{bmatrix} + \frac{D_y b}{6a} \begin{bmatrix} 2 & 1 & -1 & -2 \\ 1 & 2 & -2 & -1 \\ -1 & -2 & 2 & 1 \\ -2 & -1 & 1 & 2 \end{bmatrix}$$

(6.91)

6.7.3 *Torsion of a square shaft*

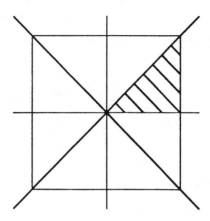

Fig. 6.6a Torsion of a square shaft

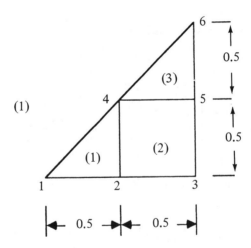

Fig. 6.6b Element subdivision of the shaft

The square shaft (Fig. 6.6a) is used to illustrate the evaluation and the assembly of the element matrices into a set of linear equations. This shaft has four axes of symmetry; therefore, it is sufficient to consider only one-eighth of the cross-section. This is divided into three elements as shown in Fig. 6.6b.

Three elements are not sufficient to obtain an answer, but they are enough to illustrate the calculations.

The element node numbers are written in a convenient tabular form as follows:

e	i	j	k	m
1	1	2	4	
2	2	3	5	4
3	4	5	6	

Elements 1 and 3 have the same orientation and the same dimensions; therefore, their matrices are identical.

The matrices for the triangular element are given by equation (6.83) and (6.84) while for the rectangular element they are given by equation (6.87) and (6.91). The matrices become

$$[K_D{}^e] = \begin{bmatrix} b_i{}^2 & b_i b_j & b_i b_k \\ b_i b_j & b_j{}^2 & b_j b_k \\ b_i b_k & b_j b_k & b_k{}^2 \end{bmatrix} + \frac{1}{4A} \begin{bmatrix} c_i{}^2 & c_i c_j & c_i c_k \\ c_i c_j & c_j{}^2 & c_j c_k \\ c_i c_k & c_j c_k & c_k{}^2 \end{bmatrix} \tag{6.92}$$

where

$$\{f^e\} = \frac{2g\theta A}{3} \begin{Bmatrix} 1 \\ 1 \\ 1 \end{Bmatrix} \tag{6.93}$$

for the triangular element and

$$[K] = \frac{1}{6} \begin{bmatrix} 4 & -1 & -2 & -1 \\ -1 & 4 & -1 & -2 \\ -2 & -1 & 4 & -1 \\ -1 & -2 & -1 & 4 \end{bmatrix} \tag{6.94}$$

and

$$\{f^e\} = \frac{2g\theta A}{4} \begin{Bmatrix} 1 \\ 1 \\ 1 \\ 1 \end{Bmatrix} \tag{6.95}$$

for the rectangular element. For elements 1 and 3 , the element area is 1/8 and $4A = 1/2$. The b and c for element 1 are,

$$
\begin{aligned}
b_1 &= Y_2 - Y_4 = -0.5, & c_1 &= X_4 - X_2 = 0 \\
b_2 &= Y_4 - Y_1 = 0.5, & c_2 &= X_1 - X_4 = -0.5 \\
b_4 &= Y_1 - Y_2 = 0, & c_4 &= X_2 - X_1 = 0.5
\end{aligned}
$$

Substituting these values into equation (6.92) gives

$$
[K^1] = \frac{1}{2}\begin{bmatrix} 1 & -1 & 0 \\ -1 & 1 & 0 \\ 0 & 0 & 0 \end{bmatrix} + \frac{1}{2}\begin{bmatrix} 0 & 0 & 0 \\ 0 & 1 & -1 \\ 0 & -1 & 1 \end{bmatrix} \tag{6.96}
$$

Adding the two matrices yields

$$
[K^1] = \frac{1}{2}\begin{bmatrix} 1 & -1 & 0 \\ -1 & 2 & -1 \\ 0 & -1 & 1 \end{bmatrix} = [K^3] \tag{6.97}
$$

Using the given values of g and θ we get

$$
2g\theta = 2(10)(10^6)(0.01)(\pi/180)
$$

$$
= 3490
$$

Substitution of 3490 and $A^1 = 1/8$ into equation (6.93) gives

$$
\{f^1\}^T = [\ 145 \quad 145 \quad 145\ ] = \{f^3\}^T \tag{6.98}
$$

The stiffness matrix for element 2 is obtained from equation 6.94. The force vector is given as

$$
\{f^2\}^T = [\ 218 \quad 218 \quad 218 \quad 218 \quad 218\ ] \tag{6.99}
$$

since we have $A = 1/16$ and $2g\theta = 3490$

The element matrices are given below. The node numbers represents the rows and columns of $[K]$ and $\{F\}$.

$$[K^1] = \frac{1}{2} \begin{bmatrix} \overset{1}{1} & \overset{2}{-1} & \overset{4}{0} \\ -1 & 2 & -1 \\ 0 & -1 & 1 \end{bmatrix} \qquad \{f^1\} = \begin{Bmatrix} 145 \\ 145 \\ 145 \end{Bmatrix} \begin{matrix} 1 \\ 2 \\ 4 \end{matrix}$$

$$(6.100)$$

$$[K^2] = \frac{1}{6} \begin{bmatrix} \overset{2}{4} & \overset{3}{-1} & \overset{5}{-2} & \overset{4}{-1} \\ -1 & 4 & -1 & -2 \\ -2 & -1 & 4 & -1 \\ -1 & -2 & -1 & 4 \end{bmatrix} \qquad \{f^2\} = \begin{Bmatrix} 218 \\ 218 \\ 218 \\ 218 \end{Bmatrix} \begin{matrix} 2 \\ 3 \\ 5 \\ 4 \end{matrix}$$

$$(6.101)$$

$$[K^1] = \frac{1}{2} \begin{bmatrix} \overset{4}{1} & \overset{5}{-1} & \overset{6}{0} \\ -1 & 2 & -1 \\ 0 & -1 & 1 \end{bmatrix} \qquad \{f^3\} = \begin{Bmatrix} 145 \\ 145 \\ 145 \end{Bmatrix} \begin{matrix} 4 \\ 5 \\ 6 \end{matrix}$$

$$(6.102)$$

Adding the element contributions using direct stiffness procedure and simplifying gives the following system of equations:

$$\begin{bmatrix} 3 & -3 & 0 & 0 & 0 & 0 \\ -3 & 10 & -1 & -4 & -2 & 0 \\ 0 & -1 & 4 & -2 & -1 & 0 \\ 0 & -4 & -2 & 10 & -4 & 0 \\ 0 & -2 & -1 & -4 & 10 & -3 \\ 0 & 0 & 0 & 0 & -3 & 3 \end{bmatrix} \begin{Bmatrix} \Phi_1 \\ \Phi_2 \\ \Phi_3 \\ \Phi_4 \\ \Phi_5 \\ \Phi_6 \end{Bmatrix} = \begin{Bmatrix} 875 \\ 2180 \\ 1310 \\ 3055 \\ 2180 \\ 875 \end{Bmatrix} \qquad (6.103)$$

Since the nodal values Φ_3, Φ_5, and Φ_6 are on the external boundary, they are equal to zero. Hence, the corresponding three equations 3,5 and 6 are eliminated. However, columns 3,5 and 6 must be incorporated into $\{F\}$. Since $\Phi_3 = \Phi_5 = \Phi_6 = 0$, they contribute nothing to $\{F\}$ and the modified system of equations is given by:

$$\begin{bmatrix} 3 & -3 & 0 \\ -3 & 10 & -4 \\ 0 & -4 & 10 \end{bmatrix} \begin{Bmatrix} \Phi_1 \\ \Phi_2 \\ \Phi_4 \end{Bmatrix} = \begin{Bmatrix} 875 \\ 2180 \\ 3055 \end{Bmatrix}$$

Solution yields

$$\Phi_1 = 1085, \qquad \Phi_2 = 795, \quad \text{and} \quad \Phi_4 = 625$$

The Φ surface for these nodal values is shown in Fig. 6.7.

Shear stress components

The gradients of the nodal parameters, Φ, are important because the shear stress components are related to these gradients by:

$$\tau_{zx} = \frac{\partial\varphi}{\partial y} \qquad \text{and} \qquad \tau_{zy} = -\frac{\partial\varphi}{\partial x} \qquad (6.104)$$

The gradient vector for the triangular element is given as:

$$\{gv\} = \frac{1}{2A} \begin{bmatrix} b_i & b_j & b_k \\ c_i & c_j & c_k \end{bmatrix} \begin{Bmatrix} \Phi_1 \\ \Phi_2 \\ \Phi_3 \end{Bmatrix} \qquad (6.105)$$

The gradient vector for the rectangular element is given as:

$$\{gv\} = \frac{1}{4ab} \begin{bmatrix} -(2a-t) & (2a-t) & t & -t \\ -(2b-s) & -s & s & (2b-s) \end{bmatrix} \begin{Bmatrix} \Phi_1 \\ \Phi_2 \\ \Phi_3 \\ \Phi_4 \end{Bmatrix}$$

$$(6.106)$$

The values for the area as well as the b and c coefficients are the same for both triangular and rectangular elements. Using these values and the values of Φ_1, Φ_2, and Φ_4 gives

$$\{gv'\} = \frac{8}{4} \begin{bmatrix} -1 & 1 & 0 \\ 0 & -1 & 1 \end{bmatrix} \begin{Bmatrix} 1085 \\ 795 \\ 625 \end{Bmatrix} = \begin{Bmatrix} 2720 \\ 4640 \end{Bmatrix}$$

and

$$\tau_{zx} = \frac{\partial \varphi}{\partial y} = -1360 \ \text{N/cm}^2$$

$$\tau_{zy} = -\frac{\partial \varphi}{\partial x} = 2320 \ \text{N/cm}^2$$

for element one.
The stress components for element three are calculated in a similar manner giving

$$\tau_{zx} = 0 \qquad \text{and} \qquad \tau_{zy} = 5000 \ \text{N/cm}^2$$

The local coordinates of node two are $s = 2b$ and $t = 0$; also, $2a = 2b = 0.5$. For element two, the gradient vector is given as

$$\{gv\} = \frac{8}{4} \begin{bmatrix} -1 & 1 & 0 & 0 \\ 0 & -1 & 1 & 0 \end{bmatrix} \begin{Bmatrix} \Phi_2 \\ \Phi_3 \\ \Phi_5 \\ \Phi_4 \end{Bmatrix} = \begin{Bmatrix} -636 \\ 0 \end{Bmatrix}$$

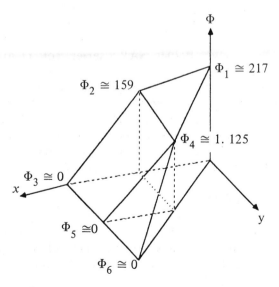

Fig. 6.7 The φ surface for the given problem

or

$$\{gv^2\} = \begin{bmatrix} -4 & 4 & 0 & 0 \\ 0 & -4 & 4 & 0 \end{bmatrix} \begin{Bmatrix} 795 \\ 0 \\ 0 \\ 625 \end{Bmatrix} = \begin{Bmatrix} -636 \\ 0 \end{Bmatrix}$$

and

$$\tau_{zx} = 0 \quad \text{and} \quad \tau_{zy} = 6360 \text{ N/cm}^2$$

6.8 COMMERCIALLY AVAILABLE FEA SOFTWARE

For larger structural and strength of materials problems involving large number of degrees of freedom, it is best to use a commercially available FEM programs such as NASTRAN. These programs offer a variety of element shapes for 2D, 3D and axisymmetric problems with complex boundary conditions. Some of the problems handle non-linearity arising from large displacements (geometric non-linearity) or non-linear material behaviour such as plasticity or non-linear elasticity (material non-linearity).

Microcomputer based finite element programs are becoming increasingly popular both in industry as well as in universities. In this section a typical mainframe/mini finite element software and a PC based finite element software are described. In early 1970s SAP series of finite element programs were pioneered by Professors K.J. Bathe and E.L. Wilson at the University of California, Berkley. Similar work has been carried out by Professor O.C. Zienkiewicz in the U.K., Professor J.H. Argyris in Germany and others. Some of the other well known computer codes in use include: ANSYS, NASTRAN, MARC, ABAQUS. The discussion in the book is limited to MSC/NASTRAN and MSC/pal 2.

6.8.1 *MSC/NASTRAN*

MSC/NASTRAN is developed by The MacNeal-Schwendler Corporation. In its early development, NASTRAN was primarly used in the aerospace industry and became an industrial standard.

NASTRAN supports some 20 standard element types plus user defined element options for non-standard analysis. The direct Matrix Abstraction Program(DMAP) allows users to solve problems using their own customized analysis routines within NASTRAN. This feature is intended for advanced finite element users. Non-linear analysis, aeroelasticity, and heat transfer analysis are supproted by NASTRAN in addition to statics and dynamics. Interactive graphics capabilities are supported by NASTRAN along with a variety of pre- and postprocessing utilities. NASTRAN is primarily operated in batch mode and runs on most mainframe and minicomputers and engineering workstations.

NASTRAN has powerful capabilities for almost any engineering applications. Again, there are less complex and expensive finite element codes for standard linear static and dynamic analysis. The MacNeal-Schwendler Corporation also markets a finite element program for microcomputers called MSC/pal 2, which is discussed in the next section.

6.8.2. *MSC/pal 2*

The MacNeal-Schwendler Corporation supports a microcomputer statics and dynamics program called MSC/pal 2. This is not a version of NASTRAN but is a separate code written specifically for IBM PC and compatible microcomputers and also for Apple Macintosh. An option is provided for translating MSC/pal 2 files to NASTRAN format. MSC/pal 2 uses an 'in-core' solver and thus is limited to problems of roughly 500 to 1000 DOF, depending on the type of analysis. In-core solvers are much faster than 'out-core' solvers, and MSC/pal 2 is a good example of this fact.

MSC/pal 2 supports 2D and 3D truss and beam elements, a curved beam element, quadrilateral and triangular plate elements, along with shear panel, discrete mass, spring, and damper elements, and a generalized stiffness element. The program performs static analysis and dynamic, modal and transient response analysis. Solid elements and response spectrum analysis are not supported by MSC/pal 2.

MSC/pal 2 has strong interactive graphics capabilities, including deflected shapes superimposed on the original shapes and the stress contour plots. MSC/pal 2 is designed as a companion to larger finite element codes, allowing quick analysis of small- to medium-sized structural problems.

6.9 PC BASED COMPUTER PROGRAM FOR 2D ELASTICITY PROBLEMS

The finite element method has very little use if a computer is not available. Two graduate students at Ohio University, Tod Bosel and Wadu Jayasuriya developed the computer program 'TSAP' given in this section.   The program is written in Microsoft Fortran and runs on popular IBM or compatible personal computers.   The program is based on the formulation developed in this chapter for the 3-node triangular element using either plane stress or plane strain conditions.   The intended use is for solving small educational problems rather than large commercial problems.

This program does not have a mesh generator program. However, a small separate program creates input data files interactively. The flow chart of the program is shown in Fig. 6.8.   The program is broken down into several subroutines. The function of each subroutine is listed in Fig. 6.8.

Flowchart

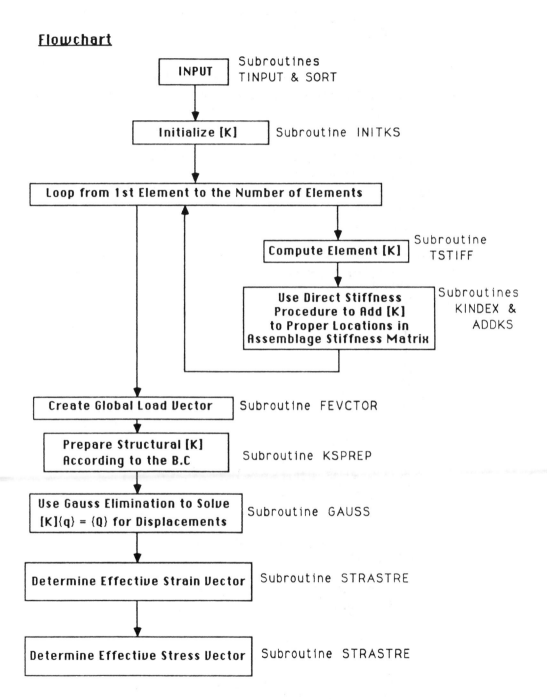

Fig 6.8 Flow chart for TSAP

```
c+++  input: creates input data files (data,restr, loads, misc)
c+++  needed for TSAP program
c
c
c+++  written by W. M. L. Jayasuriya and Tod Bosel
c+++  written in Microsoft Fortran version 4.01
c
c
      character*8 file
      character*12 file1, file2, file3, file4
      integer restr(100,2)
      real coord(300,3), prop(100,6),data(100,12), misc(10),
     *loads(200,3)
c
c
      print *,'enter name to store file under'
      print *,'cannot exceed 8 characters in length'
      read(5,111) file
111   format(a8)
c
      print *,' '
      file1 = file//'.1'
      File2 = file//'.2'
      File3 = file//'.3'
      File4 = file//'.4'
c
c
      open(1, file=file1, status='new')
      open(2, file=file2, status='new')
      open(3, file=file3, status='new')
      open(4, file=file4, status='new')
c
c
      print *,'enter the total number of elements'
      read(5,*) misc(1)
      print *,' '
      print *,'enter the total number of nodes'
      read(5,*) misc(2)
      print *,' '
      print *,'enter the thickness'
      read(5,*) misc(4)
      print *,' '
      print *,'enter 1 for plane stress or 2 for plane strain'
      read(5,*) misc(3)
      print *,' '
```

```
c
c
      do 10 I = 1, misc(2)
        print *,'enter node #, x-coordinate, y-coordinate'
        read(5,*) coord(i,1), coord(i,2), coord(i,3)
        print *,' '
  10  continue
c
      do 20 I = 1, misc(1)
        print *,'enter element #, node #1, node #2, node #3'
        read(5,*) prop(i,1), prop(i,2),prop(i,3), prop(i,4)
        print *,' '
        print *,'enter element modulus and poisson ratio'
        read(5,*) prop(i,5), prop(i,6)
        print *,' '
  20  continue
c
c
      do 40 I = 1, misc(1)
        do 30 j = 1, misc(2)
          if(prop(i,2).Eq.Coord(j,1)) then
              data(i,2) = coord(j,2)
              data(i,3) = coord(j,3)
          elseif(prop(i,3).Eq.Coord(j,1)) then
              data(i,4) = coord(j,2)
              data(i,5) = coord(j,3)
          elseif(prop(i,4).Eq.Coord(j,1)) then
              data(i,6) = coord(j,2)
              data(i,7) = coord(j,3)
          end if
  30    continue
c
        data(i,1) = prop(i,1)
        data(i,8) = prop(i,2)
        data(i,9) = prop(i,3)
        data(i,10) = prop(i,4)
        data(i,11) = prop(i,5)
        data(i,12) = prop(i,6)
  40  continue
c
      print *,'enter the total number of restrained nodes'
      read(5,*) misc(5)
      print *,' '
      print *,'enter node number and degree of restraint'
      print *,'where degree of restraint is defined as:'
```

```
      print *,'1-for x and y restrained'
      print *,'2-for x restrained'
      print *,'3-for y restrained'
      print *,' '
c
c
      do 50 I = 1, misc(5)
        print *,'enter node number and degree of restraint'
        read(5,*) restr(i,1), restr(i,2)
        print *,' '
 50   continue
c
c
      print *,'enter total number of concentrated loads'
      read(6,*) misc(6)
      print *,' '
      do 60 I = 1, misc(6)
        print *,'enter node number, 1-x dir or 2-y dir, magnitude'
        read(5,*) loads(i,1), loads(i,2), loads(i,3)
        print *,' '
 60   continue
c
c+++  call subroutine sort to sort data
c
      call sort(data, misc)
      do 70 I = 1, misc(1)
 70      write(1,*) (data(i,j), j = 1, 12)
      do 80 I = 1, misc(5)
 80      write(2,*) (restr(i,j), j = 1, 2)
      do 90 I = 1, misc(6)
 90      write(3,*) (loads(i,j), j = 1, 3)
      do 100 I = 1, 6
100      write(4,*) misc(i)
c
c
      rewind 1
      rewind 2
      rewind 3
      rewind 4
c
      stop
      end
c
c
```

```
      subroutine sort(data, misc)
c
c+++  sorts data in ascending order
c
      real misc(10), data(100,12)
c
      n = misc(1)
      jj = 12
      n1 = n - 1
      do 20 I = 1, n1
      k = I + 1
c
c
      do 20 ii = k, n
        if(data(i,1).Lt.Data(ii,1)) go to 20
          temp = data(i,1)
          data(i,1) = data(ii,1)
          data(ii,1) = temp
          do 10 j = 2, jj
            temp = data(i,j)
            data(i,j) = data(ii,j)
            data(ii,j) = temp
 10        continue
 20   continue
      return
      end
c
c
c+++  TSAP: structural analysis program for constant
c+++  thickness triangular elements
c
c+++  written by W. M. L. Jayasuriya and Tod Bosel
c+++  written in Microsoft Fortran version 4.01
C
      character*8 file
      character*12 file1, file2, file3, file4, file10
      integer idest(6), restr(100,2)
      real data(100,12), k(6,6), ks(200,200), loads(200,3),
     *misc(10), fv(200), displ(200)
c
c+++  obtain input files
c
      print *,'enter name of input file'
      read(5,111) file
 111  format(a8)
```

```
       print *,' '
       file1 = file//'.1'
       File2 = file//'.2'
       File3 = file//'.3'
       File4 = file//'.4'
       File10 = file//'.Ans'
       write(6,222) file10
222    format(' ','output can be found in the file: ',' ',a12)
c
c
       open(1, file=file1, status='old')
       open(2, file=file2, status='old')
       open(3, file=file3, status='old')
       open(4, file=file4, status='old')
       open(7, status='scratch')
       open(8, status='scratch')
       open(9, status='scratch')
       open(10, file=file10)
c
c
       do 10 I = 1, 100
10     read(1,*,end=20) (data(i,j), j = 1,12)
20     do 30 I = 1, 100
30      read(2,*,end=40) (restr(i,j), j = 1, 2)
40     do 50 I = 1, 100
50      read(3,*,end=60) (loads(i,j), j = 1, 3)
60     do 70 I = 1, 10
70      read(4,*,end=80) misc(i)
c
c+++  start analysis
c
80     call initks(misc, ks)
       do 90 I = 1, misc(1)
          call tstiff(data, misc, i, k)
          call kindex(i, data, idest)
          call addks(k, idest, ks)
90     continue
       rewind 7
       rewind 8
       rewind 9
       call fvector(loads, misc, fv)
       call ksprep(ks, restr, misc, fv)
       call gause(ks, fv, displ, misc)
       call strastre(displ, misc, file)
       stop
```

```
      end
c
c
c
      subroutine initks(misc, ks)
c
c+++  initializes the complete stiffness matrix (ks) to zero
c
      real ks(200,200), misc(10)
c
      do 10 I = 1, misc(2)*2
      do 10 j = 1, misc(2)*2
 10      ks(i,j) = 0
c
      return
      end
c
c
c
      subroutine tstiff(data, misc, i, k)
c
c+++  creates constitutive (d) and stiffness (k) matrices
c+++  for a triangular element
c
      real data(100,12), a, b(3,6), d(3,3), mult, initk(3,6),
     *k(6,6), misc(10)
c
c+++  create d matrix
c
      do 10 j = 1, 3
      do 10 I = 1, 3
 10      d(j,I) = 0.0
      If(misc(3).Eq.1.0) Then
      mult = data(i,11)/(1-data(i,12)*data(i,12))
        d(1,1) = mult
        d(2,2) = mult
        d(3,3) = mult*(1-data(i,12))/2
        d(1,2) = mult*data(i,12)
        d(2,1) = d(1,2)
      else
        mult = data(i,11)/ ((1+data(i,12))*(1-2*data(i,12)))
        d(1,1) = mult*(1-data(i,12))
        d(2,2) = d(1,1)
        d(3,3) = mult*(1-2*data(i,12))/2
        d(1,2) = mult*data(i,12)
```

```
            d(2,1) = d(1,2)
        endif
        do 20 j = 1, 3
 20       write(9,*) (d(j,kk), kk = 1, 3)
c
c+++  create k matrix
c
        a = (data(i,4)*data(i,7) - data(i,6)*data(i,5) -
            data(i,2)*data(i,7) +  data(i,2)*data(i,5) +
            data(i,3)*data(i,6) -  data(i,3)*data(i,4))/2
        do 30 j = 1, 3
        do 30 l = 1, 6
 30       b(j,l) = 0.0
        B(1,1) = (data(i,5) - data(i,7))/(2*a)
        b(1,3) = (data(i,7) - data(i,3))/(2*a)
        b(1,5) = (data(i,3) - data(i,5))/(2*a)
        b(2,2) = (data(i,6) - data(i,4))/(2*a)
        b(2,4) = (data(i,2) - data(i,6))/(2*a)
        b(2,6) = (data(i,4) - data(i,2))/(2*a)
        b(3,1) = (data(i,6) - data(i,4))/(2*a)
        b(3,2) = (data(i,5) - data(i,7))/(2*a)
        b(3,3) = (data(i,2) - data(i,6))/(2*a)
        b(3,4) = (data(i,7) - data(i,3))/(2*a)
        b(3,5) = (data(i,4) - data(i,2))/(2*a)
        b(3,6) = (data(i,3) - data(i,5))/(2*a)
        do 40 j = 1, 3
 40       write(7,*) (b(j,l), l = 1, 6)
        do 60 j = 1, 3
        do 60 l = 1, 6
          initk(j,l) = 0.0
          Do 50 m = 1, 3
 50         initk(j,l) = initk(j,l) + d(j,m)*b(m,l)
 60     continue
        do 80 j = 1, 6
        do 80 l = 1, 6
          k(j,l) = 0.0
          Do 70 m = 1, 3
 70         k(j,l) = k(j,l) + b(m,j)*initk(m,l)
          k(j,l) = k(j,l)*misc(4)*a
 80     continue
        return
        end
c
c
c
```

```
      subroutine kindex(i, data, idest)
c
c+++  creates element index or destination vector (idest)
c
      integer idest(6)
      real data(100,12)
c
      idest(1) = (data(i,8)-1)*2+1
      idest(2) = idest(1)+1
      idest(3) = (data(i,9)-2)*2+3
      idest(4) = idest(3)+1
      idest(5) = (data(i,10)-3)*2+5
      idest(6) = idest(5)+1
      do 10 j = 1, 6
 10      write(8,111) idest(j)
111   format(i5)
      return
      end
c
c
c
      subroutine addks(k, idest, ks)
c
c+++  creates structural stiffness matrix (ks)
c
      integer idest(6)
      real k(6,6), ks(200,200)
c
      do 10 j = 1, 6
        l = idest(j)
        do 10 m = 1, 6
          n = idest(m)
          ks(l,n) = ks(l,n) + k(j,m)
 10   continue
      return
      end
c
c
c
      subroutine fvector(loads, misc, fv)
c
c+++  creates structural force vector (fv)
c
      real loads(200,3), fv(200), misc(10)
c
```

```
      do 10 I = 1, misc(2)*2
10       fv(i) = 0
      do 20 I = 1, misc(6)
         if(loads(i,2).Eq.1) Then
            fv((loads(i,1)-1)*2+1) = loads(i,3)
         else
            fv((loads(i,1)-1)*2+2) =  loads(i,3)
         end if
20    continue
      return
      end
c
c
c
      subroutine ksprep(ks, restr, misc, fv)
c
c+++  prepares structural stiffness matrix (ks) and the
c+++  force vector (fv) for solution
c
      integer restr(100,2)
      real ks(200,200), misc(10), fv(200)
c
      do 10 I = 1, misc(5)
         m = ((restr(i,1)-1)*2+1)
         n = m + 1
         if(restr(i,2).Eq.1) Then
            ks(m,m) = ks(m,m)*1.E20
            fv(m) = fv(m)*1.E20
            ks(n,n) = ks(n,n)*1.E20
            fv(n) = fv(n)*1.E20
         else if(restr(i,2).Eq.2) Then
            ks(m,m) = ks(m,m)*1.E20
            fv(m) = fv(m)*1.E20
         else
            ks(n,n) = ks(n,n)*1.E20
            fv(n) = fv(n)*1.E20
         end if
10    continue
      return
      end
c
c
c
      subroutine gause(a, b, x, misc)
c
```

```
c+++  solves a system of linear equations
c
      real a(200,200), b(200), x(200), ratio, misc(10)
c
      n = misc(2)*2
      do 30 I = 1, n-1
        do 30 j = i+1, n
          if(a(j,i)) 10,30,10
10            ratio = a(j,i) / a(i,i)
              do 20 k = i, n
              a(j,k) = a(i,k)*ratio - a(j,k)
20            continue
            b(j) = b(i)*ratio - b(j)
30    continue
      do 50 I = 0, n-1
      k = n - i
        do 40 j = k+1, n
40        b(k) = b(k) - x(j)*a(k,j)
        x(k) = b(k) / a(k,k)
50    continue
      return
      end
c
c
c
      subroutine strastre(displ, misc, file)
c
c+++  determines element strain and stress vectors and
c+++  writes output to a file
c
      character*8 file
      integer idest(6)
      real displ(200), misc(10), b(3,6), d(3,3), strain(3),stress(3)
c
      do 10 I = 1, misc(2)*2
10    if(abs(displ(i)).Lt.1E-15) displ(i) = 0.
      Write(10,111) file
111   format(1x,'output for file: ',' ',a8//)
      write(10,222)
222   format('global displacements:'/)
      write(10,333)
333   format(' node      x-displ            y-displ')
      I = 1
      do 20 j = 1, misc(2)*2, 2
        write(10,444) i, displ(j), displ(j+1)
```

```
       I = I + 1
444      format(i5, 2x, e15.6, 2X, e15.6)
 20  Continue
     do 80 I = 1, misc(1)
       do 30 j = 1, 6
 30       read(8,555) idest(j)
555      format(i5)
       do 40 j = 1, 3
 40       read(7,*) (b(j,k), k = 1, 6)
       do 50 j = 1, 3
 50       read(9,*) (d(j,k), k = 1, 3)
c
c+++  strain vector
c
       do 60 j = 1, 3
       strain(j) = 0.
       Do 60 k = 1, 6
         kk = idest(k)
         strain(j) = strain(j) + b(j,k)*displ(kk)
 60      continue
c
c+++  stress vector
c
       do 70 j = 1, 3
       stress(j) = 0.
       Do 70 k = 1, 3
         stress(j) = stress(j) + d(j,k)*strain(k)
 70      continue
       write(10,666) i
666      format(/'element: ', i5/)
       write(10,777)
777      format('   strain-x        strain-y         strain-xy')
       write(10,888) strain(1), strain(2), strain(3)
888      format(e15.6, 2X, e15.6, 2X, e15.6)
       Write(10,999)
999      format(/'  stress-x         stress-y        stress-xy')
       write(10,123) stress(1), stress(2), stress(3)
123      format(f15.3, 2X, f15.3, 2X, f15.3/)
 80  Continue
     return
     end
```

6.9.1 *Example Problems for TSAP*

A solution to a very simple problem consisting of 2 elements and 4 nodes with two different loading conditions are given here. The results obtained from TSAP are compared with the theoretical solution and are found to be in good agreement.

<u>Ex.1:</u>

$h = 1.0$ inch (25 mm)
$E = 30 \times 10^6$ psi (200 GPa)
$v = 0.5$

The loading conditions are as shown in the following figure.

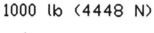

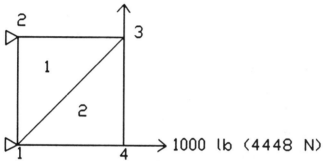

OUTPUT FOR FILE: EXAMPLE1

GLOBAL DISPLACEMENTS:

NODE	X-DISPL	Y-DISPL
1	.000000E+00	.000000E+00
2	.000000E+00	.000000E+00
3	.774194E-04	.158065E-03
4	-.161290E-04	.135484E-03

ELEMENT: 1

STRAIN-X	STRAIN-Y	STRAIN-XY
-.161290E-05	.000000E+00	.135484E-04

STRESS-X	STRESS-Y	STRESS-XY
-64.516	-32.258	135.484

ELEMENT: 2

STRAIN-X   STRAIN-Y   STRAIN-XY
.774194E-05   -.225806E-05   .645161E-05

STRESS-X   STRESS-Y   STRESS-XY
264.516   64.516   64.516

Ex 2:

h = 0.2 inch (5 mm)
E = 30 x 10^6 psi (200 GPa)
v = 0.3

The loading conditions are as shown in the following figure.

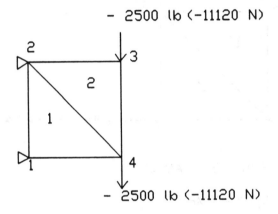

OUTPUT FOR FILE: EXAMPLE2

GLOBAL DISPLACEMENTS:

NODE	X-DISPL	Y-DISPL
1	.000000E+00	.000000E+00
2	-.110790E-02	-.672783E-02
3	.000000E+00	.000000E+00
4	.153677E-02	-.722817E-02

ELEMENT: 1

STRAIN-X	STRAIN-Y	STRAIN-XY
-.276976E-03	.000000E+00	-.168196E-02

STRESS-X	STRESS-Y	STRESS-XY
-11185.560	-4793.80	-19407.190

ELEMENT: 2

STRAIN-X	STRAIN-Y	STRAIN-XY
.384192E-03	-.250171E-03	-.484707E-03

STRESS-X	STRESS-Y	STRESS-XY
11185.560	-3453.602	-5592.777

References

Alexander, J.M., *Strength of Materials − Vol.1: Fundamentals*,
Ellis Horwood, Chichester, 1981.

Alexander, J.M., Brewer, R.C., Rowe, G.W.R., *Manufacturing
Technology − Vol.1: Engineering Materials, Vol.2:
Engineering Processes*, Ellis Horwood, Chichester, 1986.

Argon, A.S. (Ed.), *Constitutive Equations in Plasticity*,
M.I.T. Press, 1975, p.117.

Atkins, A.G., Mai, Y.W., *Elastic and Plastic Fracture*,
Ellis Horwood, Chichester, 1985.

Avner,S.H., *Introduction to Physical Metallurgy*,
McGraw−Hill, New York, 1987.

Bailey, A.R., *A Textbook of Metallurgy*,
Macmillan, London, 1964.

Baker, J.F., Horne, M.R., Heyman, J., *The Steel Skeleton*,
Cambridge University Press, 1956.

Beer, F.P., Johnson, E.R.Jr., *Mechanics of Materials*,
McGraw−Hill, New York, 1981.

Bickley, W.G., Thompson, R.S.H.G., *Matrices−Their Meaning
and Manipulation*, Hodder & Stoughton, London, 1964.

Borg, S.F., *Matrix−tensor Methods in Continuum Mechanics*,
Van Nostrand, 1963.

Braithwaite, E.R. (Ed.), *Lubrication and Lubricants*,
Elsevier, Amsterdam, 1967.

Bridgman, P.W., *Studies in Large Plastic Flow and Fracture*,
McGraw−Hill, New York, 1952.

Briggs, C.W., *The Metallurgy of Steel Castings*,
McGraw−Hill, New York, 1946.

Brinell, J.A., "Methods of Testing Steel",
Cong. Int. Methodes d'Essai, Paris 1900,**2**.

Caddell, R.M., *Deformation and Fracture of Solids*,
Prentice−Hall, New Jersey, 1980.

Case, J, Chilver, A.H., *Strength of Materials and Structures*,
Arnold, 1971

Chou, P.C., Pagano, N.J., *Elasticity*, Van Nostrand, 1967.

Collins, J.A., *Failure of Materials in Mechanical Design*,
John Wiley, New York, 1981.

Conte, S.D., *Elementary Numerical Analysis*, McGraw–Hill, 1965.

Crandall, S.H., Dahl, N.C., *An Introduction to the Mechanics
of Solids*, McGraw–Hill, 1959.

Crane, F.A.A., Charles, J.A., *Selection and Use of Engineering
Materials*, Butterworths, 1984.

Den Hartog, J.P., *Advanced Strength of Materials*,
McGraw–Hill, 1952

Dieter, G.E., *Mechanical Metallurgy*,
McGraw–Hill, New York, 1961.

Faupel, J.H., *Engineering Design*, John Wiley, 1964.

Fenner, R.T., *Computing for Engineers*, Macmillan, 1974.

Fenner, R.T., *Finite Element Methods for Engineers*,
Macmillan, 1975.

Ford, H., Alexander, J.M., *Advanced Mechanics of Materials*,
Ellis Horwood, Chichester, 1977 (first published by
Longman's Green, 1963)

Frederick, D., Chang, T.S., *Vectors and Cartesian Tensors*,
Allyn and Bacon, 1965.

Gerald, C.F., *Applied Numerical Analysis*,
Addison–Wesley, 1970.

Golden, G.T., *Fortran–IV – Programming and Computing*,
Prentice–Hall, 1975.

Gopinath, S., *Automation of the Data Analysis System Used in
Process Modelling Applications*, M.S. Thesis, Ohio
University, Athens, Ohio 45701, 1986.

Gorjunov, I.I. *Castings with Increased Dimensional Accuracy*,
Mashgiz, Moscow, 1958.

Goult, R.J., *Applied Linear Algebra*, Ellis Horwood, 1978.

Graham, A., *Matrix Theory and Applications for Engineers and
Mathematicians*, Ellis Horwood, 1979.

Halbart, G. *Eléments d'une Théorie Mathématique de
la Fonderie*, H. Vaillant, Liège, 1945.

Hartley, G. Jr., *Fundamentals of the Finite Element Method*,
Macmillan Publishing Company, New York, 1986.

Hertzberg, R.W., *Deformation and Fracture Mechanics of
Engineering Materials*, Wiley, New York, 1976.

Hill, R., *The Mathematical Theory of Plasticity*,
Clarendon Press, Oxford, 1950.

Hinton, E., Owen, D.R.J., *An Introduction to Finite Element
Computations*, Pineridge Press, 1979.

Hinton, E., Owen, D.R.J., *Finite Element Programming*,
Academic Press, 1977.

Hoffman, O., Sachs, G., *Introduction to the Theory of Plasticity for Engineers*, McGraw–Hill, 1953.

Holister, G.S., *Experimental Stress Analysis*, Cambridge University Press, 1967.

Hosford, W.F., Caddell, R.M., *Metal Forming: Mechanics and Metallurgy*, Prentice–Hall, New Jersey, 1983.

Hunter, S.C., *Mechanics of Continuous Media*, Ellis Horwood, 1976.

Irons, B., Ahmad, S., *Techniques of Finite Elements*, Ellis Horwood, 1980.

James, E., O'Brien, F., Whitehead, P., *A Fortran Programming Course*, Prentice–Hall, 1970.

Jenkins, I. *Controlled Atmospheres for the Heat–Treatment of Metals*, Chapman and Hall, London, 1946.

Johnson, W., *Impact Strength of Materials*, Edward Arnold, 1972.

Johnson, W., Mellor, P.B., *Plasticity for Mechanical Engineers*, Van Nostrand, London, 1962 (Second Edition:– *'Engineering Plasticity'* 1973).

Johnson, W., Sowerby, R., Haddow, J.B., *Plane Strain Slip Line Fields: Theory and Bibliography*, Edward Arnold, London, 1970.

Jones, W.D. *Fundamental Principles of Powder Metallurgy*, Edward Arnold, London, 1961.

Journal of Machine Design, February and June 1951.

Kelly, A., *Strong Solids*, Clarendon Press, Oxford,1966.

Knott, J.F., *Fundamentals of Fracture Mechanics*, Butterworths, 1973.

Long, R.L., *Mechanics of Solids and Fluids*, Prentice–Hall, 1961.

Mase, G.E., *Continuum Mechanics*, McGraw–Hill, 1970.

McCracken, D.D., *A Guide to Fortran IV Programming*, John Wiley, 1965.

Morley, A., *Strength of Materials*, Longmans, 1908.

Nadai, A., *Plasticity*, McGraw–Hill, New York, 1931.

Organick, E.I., *A Fortran IV Primer*, Addison–Wesley, 1966.

Owen, D.R.J., Hinton, E., *A Simple Guide to Finite Elements*, Pineridge Press, 1980.

Pearson, C.E. *The Extrusion of Metals*, Chapman and Hall, London, (2nd Ed.), 1960.

Peterson, M.B., Winer, W.O. (Eds.), *Wear Control Handbook*, A.S.M.E., New York, 1980.

Petersen, R.E., *Stress Concentration Factors*, John Wiley, New York, 1973.

Phillips, A., *Introduction to Plasticity*, Ronald Press, 1966.

Pipes, L.A., Hovanessian, S.A., *Matrix-computer Methods in Engineering*, John Wiley, 1969.

Polushkin, E.P. *Defects and Failures of Metals*, Elsevier, Amsterdam, 1956.

Popov, E.P., *Mechanics of Materials*, Prentice-Hall, 1976.

Prager, W. *Introduction to Plasticity*, Addison-Wesley, Reading, Mass., 1959.

Prager, W., Hodge, P.G., *The Theory of Perfectly Plastic Solids*, John Wiley, New York, 1951.

Richards, C.E., Lynch, A.C. (Eds.) *Soft Magnetic Materials for Telecommunications*, Pergamon, London, 1953.

Richards, T.H., *Energy Methods in Stress Analysis*, Ellis Horwood, 1977.

Roark, R.J., *Formulas for Stress and Strain*, McGraw-Hill, 1965.

Roffe, S.T. and Barson, J.M., *Fracture and Fatigue Control in Structures*, Prentice Hall, New Jersey, 1977.

Rollason, E.C., *Metallurgy for Engineers*, Edward Arnold, London (4th Ed.), 1973.

Rowe, G.W., *Elements of Metalworking Theory*, Edward Arnold, London, 1979.

Rowe, G.W., Sturgess, C.E.N., Hartley, P., Pillinger, I., *Finite Element Plasticity and Metalforming Analysis*, Cambridge University Press, 1987.

Ruddle, R.W. *The Solidification of Castings*, Inst. of Metals Monograph No.7, May 1957.

Ryder, G.H., *Strength of Materials*, Macmillan, 1969.

Sachs, G. *Fundamentals of the Working of Metals*, Pergamon, London, 1954.

Salvadori, M.G., Baron, M.L., *Numerical Methods in Engineering*, Prentice-Hall, 1961.

Scheidegger, A.E. *The Physics of Flow Through Porous Media*, University of Toronto Press, (2nd Ed.) 1962.

Scheuer, E. "Mechanical Properties of Castings and Relation to Nominal Strength of Alloys", in *Non-Ferrous Foundry Metallurgy*. A.J. Murphy (Ed.), Pergamon, London, 1954.

Schey, J.A. (Ed.), *Metal Deformation Processes: Friction and Lubrication*, Marcel Dekker, New York, 1970.

Schultz, D.G., Melsa, J.L., *State Functions and Linear Control Systems*, McGraw-Hill, 1967, 155-195.

Séférian, D. *Metallurgie de la Soudure*, Dunrod, Paris, 1960.

Segerlind, L.J., *Applied Finite Element Analysis*, 2nd. Ed. John Wiley, New York, 1984.

Shaw, M.C. *Metal Cutting Principles*,
Massachusetts Institute of Technology, (3rd Ed.), 1954.

Shigley, J.E. and Mitchell, L.D. *Mechanical Engineering Design*, 4th. Ed. McGraw-Hill, New York, 1983.

Sines, G., *Elasticity and Strength*, Allyn and Bacon, 1969.

Stephens, R.C., *Strength of Materials*, Edward Arnold, 1970.

Swann, P.R., Ford, F.P., Westwood, A.R.C., *Mechanisms of Environment-Sensitive Cracking of Materials*, Metals Society, 1977.

Tabor, D, *The Hardness of Metals*, Oxford University Press, 1951.

Tetelman, A.S. and McEvily, J., *Fracture of Structural Materials*, John Wiley, New York, 1967.

Timoshenko, S.P., Gere, J.M., *Theory of Elastic Stability*, McGraw-Hill Kogakusha, 1961.

Timoshenko, S.P., Goodier, J.N.,*Theory of Elasticity*, McGraw-Hill Kogakusha, 1970.

Timoshenko, S.P., Woinowsky-Kreiger, S., *Theory of Plates and Shells*, McGraw-Hill Kogakusha, 1959.

Tipper, C.F. *The Brittle Fracture Story*, Cambridge University Press, 1962

Van Vlack, L.H., *Elements of Materials Science*, Addison-Wesley, Reading, Mass:, 1959.

Wainwright, S.A., Biggs, W.D., Currey, J.D., Gosline, J.M., *Mechanical Design in Organisms*, Arnold, London, 1976.

Wang, C.T., *Applied Elasticity*, McGraw-Hill, 1953.

Warnock, F.V., Benham, P.P., *Mechanics of Solids and Strength of Materials*, Pitman, 1965.

Williams, J.G., *Stress Analysis of Polymers*, Ellis Horwood, 1980.

Woldman, N.E. *Metal Process Engineering*, Reinhold, New York, 1948.

Young, B.W., *Essential Solid Mechanics*, Macmillan, 1976.

Zienkiewicz, O.C., *The Finite Element Method*, McGraw-Hill, New York, 1977.

Zlatin, N., Merchant, M.E. *The Distribution of Hardness in Chips and Machined Surfaces*, The Cincinnati Milling Machine Company, Cincinnati, Ohio, 1947.